MILITARY Living's

MW01701846

U.S. Forces Travel Guide U.S.A. & Caribbean Areas

This book has everything the military family needs to know to... "Travel on Less Per Day...The Military Way."

by
William "Roy" Crawford, Ph.D.
and
Lela Ann Crawford

Cover Design: Sarah Jane Valdez
Layout Artist: Sarah Jane Valdez
Editorial Staff: Bryce D. Thompson
Pamela Greer, Rose C. McLain

Marketing Manager: William R. Crawford, Jr.
Office Staff: Anna Belle Causey
Helen Henderson, Irene Kearney

Military Living Publications
P. O. Box 2347
Falls Church, Virginia 22042
(703) 237-0203

NOTICE

The information in this book has been compiled and edited either from the activity/facility listed, their superior headquarters or from other sources which may or may not be so noted. All listed facilities, their composition, contact information, rates and availability could change. This book should be used as a guide to the listed facilities with the above understanding. Please forward any corrections or additions to **Military Living Publications, P.O. Box 2347, Falls Church, VA 22042.**

This directory is published by **Military Marketing Services, Inc.**, a private firm in no way connected with the U.S. Federal or other governments. The book is copyrighted by William Roy and Lela Ann Crawford. Opinions expressed by the publisher and writers herein are their own and are not to be considered an official expression by any government agency or official.

The information and statements contained in this directory have been compiled from sources believed by the publisher to be reliable and to represent the best current opinion on the subject. No warranty, guarantee or representation is made by **Military Marketing Services, Inc.**, as to the absolute correctness or sufficiency of any representation contained in this or other publications and we can assume no responsibility.

First Printing--January 1987
Second Printing--July 1987
Third Printing--August 1988

MILITARY MARKETING SERVICES, INC.
(MILITARY LIVING PUBLICATIONS)

Library of Congress Cataloging in Publication Data
Crawford, William Roy, 1932-
U.S. forces travel guide.
Includes index.
1. Soldiers--United States--Recreation--Directorie. 2. Military dependents--United States--Recreation--Directories. 3. Recreation areas--United States--Directories. 4. Recreation areas-Caribbean Area--Directories. 5. United States-Armed Forces--Barracks and quarters--Directories. I. Crawford, Ann Caddell. II. Title. III. Title: US forces travel guide.
UC403.C744 1987 917'.04538'088355 86-23664
ISBN 0-914862-11-1

HOW TO USE THIS DIRECTORY

NAME OF INSTALLATION (AL01R2) — **Comm:** 000-000-0000
Street/P O Box (if required) — **ATVN:** 000 0000
City/APO/FPO, State, ZIP Code — FTS: 000-0000

Location Identifier: Example **(AL01R2).** The first two digits (letters) are Country/State abbreviations used in **Military Living's** books (Appendix A). The next two digits (numbers) are random numbers (00-99) assigned to a specific location. The fifth digit (letter) is an R indicating region and the sixth digit (number) (0-9) is the region location. The location identifiers for each of the more than 300 listings in this book are keyed to regional maps which were not printed in this Military Living title. For your convenience we have included **Military Living Maps** at the beginning of each state/possession/country listing which shows the location of each individual listing in this book. Some minor installations, which have very limited support facilities for the military traveler, are shown on the maps but are not detailed in this title. Detailed driving instructions are contained under location in each listing.

Comm: This is the commercial telephone service for the installation's main or information/operator assistance number. Within the US Area Code System, the first 3-digit group is the area code. The next 3-digit group is the area telephone exchange/switch number. The last 4-digit group is the information or operator assistance number.

ATVN: This is the Department of Defense, world-wide, Automatic Voice Network (ATVN). The number given is for information/operator assistance.

FTS: This is the Federal Telephone System. The number given is for information/operator assistance.

LOCATION: This is the specific driving instruction to the installation from local major cities, interstate highways and routes. More than one routing may be provided. NMC-nearest major city: Distance in miles and **DIRECTION FROM THE INSTALLATION TO THE NMC** is given.

GENERAL INFORMATION: Major activities, organizations, military units and special features concerning the installation, if any, are provided in this section.

TEMPORARY MILITARY LODGING: The billeting office or check-in point mailing address (combined with the installation name, city, state and ZIP code), physical location, complete Comm/ATVN/FTS telephone contact/reservation numbers are provided. Hours of operation are noted. Category of persons accommodated are specified and special contact numbers for **DV/VIP** (Distinguished Visitor/Very Important Person, both Officer and Enlisted grades) are provided. **Detailed information on TML at over 480 installations world-wide is contained in Military Living's all time best selling title, Temporary Military Lodging Around The World. See coupons in this book for ordering information.**

LOGISTICAL SUPPORT: The scope of support facilities available at the installation is specified and the key telephone exchange and number of the facility are provided. Space limitations preclude the listing of all support numbers; however, the telephone numbers for facilities not listed are available through the information operator number in each listing.

HEALTH & WELFARE: The installation or nearest medical emergency and appointment service numbers are listed. Also, the installation or servicing medical inpatient numbers are given. Chaplain support and chapel service contact numbers are specified.

RECREATION: The contact telephone numbers for most rest, recreation, athletic facilities and opportunities are given for each installation. Specific location, addresses, contact telephone numbers, facilities available and scope of activities are specified for camping and recreation areas on the installation, controlled by the installation or near the installation. **Complete camping and recreation area details are contained in Military Living's Military RV, Camping and Rec Areas Around the World. See coupons in this book for ordering information.**

SPACE-A: The location, hours of operation, Comm and ATVN contact telephone numbers and flight location opportunities are specified for departure locations on each installation, or Space-A opportunities at nearby installations or other departure locations. The numbers provided are the Space-A information numbers at specified departure locations. **Full Space-A information such as arrival locations, equipment flown, routes, schedules, departure/arrival terminal facilities, eligibility, rules, travel documentation and more are all contained in Military Living's best selling book, Military Space-A Air Opportunities Around The World. See coupons in this book for ordering information.**

ATTRACTIONS: This section provides helpful information to the traveler on key attractions on the installation or in the nearby area.

Please review the Appendices for abbreviations used in this book and for other helpful information. **Please address your questions regarding this title or other Military Living books to: Military Living Publications, PO Box 2347, Falls Church, VA 22042, ATTN: Ann or Roy Crawford. You may telephone us at 703-237-0203.** No collect calls accepted.

-NOTES-

CONTENTS

CONTINENTAL UNITED STATES (CONUS)

FLORIDA (CONT)

GEORGIA

IDAHO

ILLINOIS

INDIANA

IOWA

NONE

KANSAS

KENTUCKY

LOUISIANA

LOUISIANA (CONT)

MAINE

MARYLAND

MASSACHUSETTS

MICHIGAN

MINNESOTA

MISSISSIPPI

MISSOURI

MONTANA

NEBRASKA

NEVADA

NEW HAMPSHIRE

NEW JERSEY

NEW MEXICO

NEW YORK

NORTH CAROLINA

NORTH DAKOTA

OHIO

OKLAHOMA

OREGON

PENNSYLVANIA

RHODE ISLAND

SOUTH CAROLINA

SOUTH DAKOTA

TENNESSEE

TEXAS

TEXAS (CONT)

UTAH

VERMONT

None

VIRGINIA

WASHINGTON

WASHINGTON (CONT)

WEST VIRGINIA

None

WISCONSIN

WYOMING

OUTSIDE CONTINENTAL UNITED STATES (OCONUS)

ALASKA

GUAM

HAWAII

MIDWAY ISLAND

PUERTO RICO

WAKE ISLAND

FOREIGN COUNTRIES

ANTIGUA

BERMUDA

CANADA

CUBA

DENMARK (GREENLAND)

PANAMA

APPENDICES

INFO ON MILITARY LIVING

MILITARY LIVING PUBLICATIONS
(MILITARY MARKETING SERVICES, INC.)

Military Living was founded by Ann Crawford in 1969. The President of the parent company, Military Marketing Services, Inc. is William "Roy" Crawford, Sr., Ph.D. The company publishes **Military Living,** a monthly magazine with a circulation of more than 30,000 which is distributed on more than twenty military installations in the Washington, DC, area.

Our travel newsletter, Military Living's **R&R Report,** has been published since 1971 and is available worldwide by subscription.

In addition to five (5) books currently available, Military Living published **U.S. Forces Travel Guide Europe** which is out of print but is under revision at this time. The company is also researching additional titles to improve the "quality of life" for military personnel and their families. See the Central Order Coupons on pages 259 and 261.

HOW TO RECOGNIZE MILITARY LIVING'S BOOKS

All of Military Living's books carry the famous ML logo. Military Living is known as "The Morale Booster Publisher." The slogan, "Travel on Less Per Day, the Military Way," is copyrighted by Military Living.

Our company offices are located at 137 N. Washington Street, Suite 201, Falls Church, Virginia 22046. The mailing address is P. O. Box 2347, Falls Church, Virginia 22042. Our phone number is (703) 237-0203.

MILITARY Living's™

FAMILY OF PUBLICATIONS

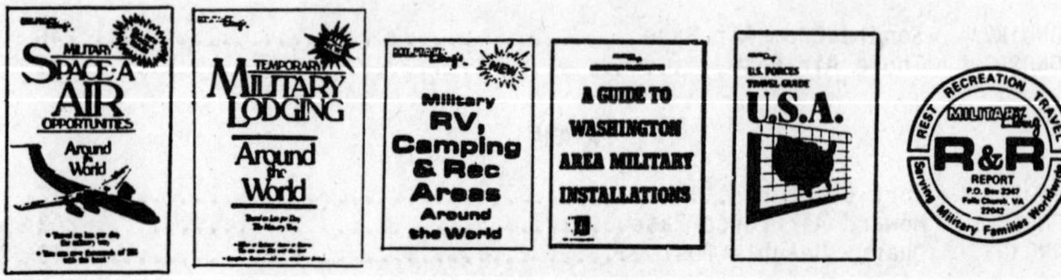

See Central Order Coupons & Description of

Publications on Pages 259-262

CONTINENTAL UNITED STATES

ALABAMA

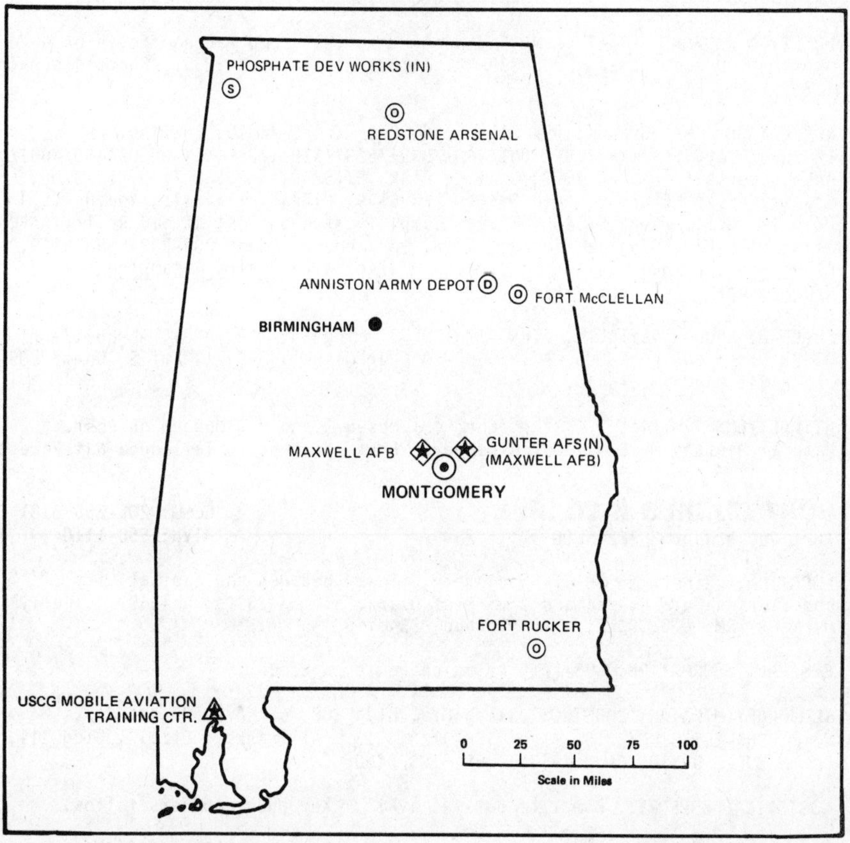

FORT McCLELLAN (AL01R2)

Fort McClellan, AL 36205-5000

Comm: 205-238-4611
ATVN: 865-4611

LOCATION: Nine mi N of I-20. Take AL-21 N to Fort. Also loc 25 mi SE of I-59. Take US-431 to Fort. RM: p-4, G/11. NMC: Anniston, 3 mi SE.

GENERAL INFORMATION: United States Army Chemical and Military Police Centers and Fort McClellan. In NE AL with easy access to Birmingham and Atlanta, GA via I-20.

TEMPORARY MILITARY LODGING: Biltg Ofc, Bldg 163, 20th St, ATTN: Chief Housing Div, Comm: 205-238-4338, ATVN: 865-4338, 0730-2300 hrs dly. All ranks. DV/VIP call Comm: 205-238-3800, ATVN: 865-3800.

Fort McClellan, Cont'd

LOGISTICAL SUPPORT: Complete spt fac avail. Key phone numbers follow.

PX-820-2190	Caf-820-2000	Chap-238-5351
Chld C-238-4857	Commiss-238-3130	Conv S-820-9530
EN C-238-5294	Fam Svc-238-4525	Gas S-820-9310
Legal-238-5435/3414	Loc-238-3281	Med-238-2315
NCO C-238-5294	Off C-238-5391	Pkg S-820-9280
Pol-238-5555	Snack B-820-2000	Theat-238-3861

HEALTH & WELFARE: Noble Army Community Hospital, comp range of clinics & depts. 100 bed. Emerg, 205-238-2315/4830, Appoint, EX-3835. Chapels serving all faiths, EX-5311.

RECREATION: Rec Svc Ofc, EX-3119/5106, Rec Ctr, EX-4819, Bowling, EX-3035/5149, Swimming, EX-3091/3401/3391/5297/3102, Gym, EX-3860/4160/4681/4656, Sports Arena, EX-4802, Skating Rink, EX-5249, Skeet & Trap, EX-3700, Crafts Center, EX-5642, Auto Crafts, EX-5146, Library, EX-3175, Youth Act, EX-4217, Golf, EX-3905/5492, Travel Camp, located on Post at old Reilly AAF, ATTN: ATZN-PAR, Bldg 421 for check-in, year round, Comm: 205-238-5663/5649, ATVN: 865-5663/5649, 9 camping spaces w/elec hookup only, 4 camping spaces w/o hookups. Full activities.

SPACE-A: None available. Try 117 TRW, Birmingham, MAP (ANG), AL 35217-0198, Comm: 205-599-9200, ATVN: 694-2210 or Maxwell AFB, 110 mi S, Comm: 205-293-6454, ATVN: 875-6454.

ATTRACTIONS: Military Police, Chemical Museum, and WAC Museum on Post. Natural History Museum located nearby. Fort adjacent to Talladega Nat Forest.

FORT RUCKER (AL02R2)

Fort Rucker, AL 36362-5000

Comm: 205-255-6181
ATVN: 558-1110

LOCATION: Ninety mi SE of Montgomery, midway between the capital city and Florida Gulf Coast, and 7 mi S of Ozark, off US-231 on AL-249. Clearly marked. RM: p-5, Q/12. NMC: Dothan, 26 mi SE.

GENERAL INFORMATION: US Army Aviation Center & School.

TEMPORARY MILITARY LODGING: Biltg Ofc, Bldg 308, 6th Ave, 24 hrs dly, Comm: 205-255-3782/3213, ATVN: 558-3782/3213. All ranks. DV/VIP, Bldg 114, Comm: 205-255-3100/3400, ATVN: 558-3100/3400.

LOGISTICAL SUPPORT: Complete spt fac avail. Key phone numbers follow.

PX-598-6384	Chap-255-2989	Chld C-255-4813
Commiss-255-3610	Fam Svc-255-3817	Gas S-598-6384
Legal-255-5491	Loc-255-4580	Med-255-4186
NCO C-255-5191	Off C-255-5187	Pkg S-598-2566
Pol-255-4175	Theat-255-2484	

HEALTH & WELFARE: Lyster Army Community Hospital, comp clinics and depts, 75 bed. Emerg svc, 205-255-2800/4186. Chapels serving all faiths, EX-3544/3614.

RECREATION: Bowling, EX-4304, Golf, EX-5845, Spec Svc, EX-5201, Auto Crafts, EX-4143, Gym, EX-2296, Libraries, EX-5014/3812, Swimming, EX-3998/5782, NCO, EX-3489, Off, EX-5178. On post camp area at Lake Tholocco Engineer Beach, Outdoor Rec, EX-4305, 18 camper spaces w/hookups, complete activities and fac on lake. Also, operates Fort Rucker Florida Rec Area on Choctawhatchee Bay near Niceville, FL, EX-5816, 20 camper spaces w/hookups, 20 trailers furn,

Fort Rucker, Cont'd

complete fac and activities available.

SPACE-A: Army Airfield, Comm: 205-255-6181, ATVN: 558-1110, flts via exec aircraft to CONUS locations.

ATTRACTIONS: National Peanut Festival, 3rd wk Oct in Dothan. Monument to Boll Weevil, Enterprise, 7 mi W. Army Aviation Museum on Post.

GUNTER AIR FORCE STATION (AL04R2)

Gunter Air Force Station, AL 36114-5000

Comm: 205-279-1110
ATVN: 446-1110

LOCATION: Take I-65 to Northern bypass, 6 miles to exit on AL-231, continue west 1 mi to AFS. Coming from the opposite direction, from I-85, follow signs and take Eastern Bypass north one mile to AL-231. Then west one mile to AFS. RM: p-5, X/7. NMC: Montgomery, 2 mi SW.

GENERAL INFORMATION: Standard Information Systems Center, Log Mgt Ctr, and educational activities. Near the capital of AL.

TEMPORARY MILITARY LODGING: Biltg Ofc, Bldg 1503. Comm: 205-279-3660, ATVN: 446-3360, 24 hrs dly. Extensive facilities. DV/VIP, call Maxwell AFB, Comm: 205-293-2095, ATVN: 875-2095.

LOGISTICAL SUPPORT: Complete spt fac avail. Key phone numbers follow.

BX-279-9776	**Chap-279-4131**	**Chld C-293-6943**
Commiss-279-4942	**Fam Svc-279-3222**	**Gas S-279-9774**
Loc-293-5027	**Med-279-4211**	**NCO C-277-2856**
Off C-264-1731	**Pkg S-264-7472**	**Pol-279-4250**

HEALTH & WELFARE: USAF Regional Hospital, Maxwell AFB. Chapels serving all faiths.

RECREATION: Arts & Crafts, EX-3118, Auto Hobby, EX-3119, Wood Hobby at Maxwell, Gym, EX-3175, Cadet Pool, EX-3178, Off C Pool, EX-4986, Picnic Area, EX-4888, Bowling Center, EX-3186, Library, EX-3179, Rec Ctr, EX-4802, Youth Act, EX-4802. See Maxwell AFB listing for off-Base recreation facilities.

SPACE-A: See Maxwell AFB listing for Space-A in this area.

ATTRACTIONS: Alabama state capital, Coliseum, Montgomery Theater where "Dixie" was set to music, and the new Alabama State Farmers Market- "largest in the southeastern U.S."

MAXWELL AIR FORCE BASE (AL03R2)

Maxwell Air Force Base, AL 36112-5000

Comm: 205-293-1110
ATVN: 875-1110

LOCATION: Take I-85 S to I-65, exit on Day St which leads to main gate of Base. RM: p-5, Y/4. NMC: Montgomery, 1.5 mi SE.

GENERAL INFORMATION: Hq Air University, professional education center for USAF and AFRES units stationed here.

TEMPORARY MILITARY LODGING: Biltg Ofc, 3800 ABW/SVH, Bldg 157, West Dr, Comm: 205-293-2430, ATVN: 875-2430, 24 hrs dly. Extensive facilities. DV/VIP, Comm: 205-293-2095, ATVN: 875-2095.

LOGISTICAL SUPPORT: Complete spt fac avail. Key phone numbers follow.

Maxwell AFB, Cont'd

BX-834-5946	Caf-263-6044	Chap-293-2109
Chld C-293-6667	Commiss-293-6209	Conv S-265-7892
Fam Svc-293-5002	Gas S-265-7773	Legal-293-2786
Loc-293-5027	Med-293-2333	NCO C-262-8364
Off C-264-6423	Pkg S-265-7472	Pol-293-7222
Snack B-265-2089/2079	Theat-293-7411	

HEALTH & WELFARE: USAF Regional Hospital, Bldg 50, complete range of medical clinics and depts, 90 bed. Emerg, 293-2333. Chapels serving all faiths, EX-2109.

RECREATION: Aero Club, EX-7342, Library, EX-5947, Rec Svc, EX-7320/6369, Arts & Crafts, EX-6426, Gym and Pools, EX-5954, Bowling Lanes, EX-5049, Golf Course, EX-2209, Rec Ctr, EX-7370, Youth Act, EX-6292, Riding Club, EX-7365. Fam-Camp, on Base, EX-6904, year round, 24 camper spaces w/full hook-ups. Also operates Lake Martin Rec Area, off Base near Dadeville, Bldg 834, EX-5496, year round, 42 camper spaces w/full hookups and 14 trailers, complete fac and full act. Also, operates Lake Pippin Maxwell/Gunter Rec Area on Choctawhatchee Bay near Niceville, year round, 30 trailers, A/C, full furnished, EX-5496, complete facilities and full activities.

SPACE-A: Pax Term and Lounge in Bldg 844, 0700-1700 hrs dly, Comm: 205-293-6454, ATVN: 875-6454. DV/VIP lounge, EX-6961. Flights scheduled 24 hrs in advance, most via exec aircraft.

ATTRACTIONS: State capital, Shakespeare Theater, historic houses and buildings in Montgomery, including First White House of the Confederacy.

MOBILE COAST GUARD AVIATION TRAINING CENTER (AL08R2)

Mobile, AL 36608-5000

Comm: 205-694-6110
ATVN: 436-3635
FTS: 537-6110

LOCATION: Take Airport Blvd exit, west from I-65, 8 mi from Base, turn right on Schllinger Road, turn left on Tanner Williams Road, 1 mi. Clearly marked. RM: p-5, Y/11. NMC: Mobile, 1 mi NW.

GENERAL INFORMATION: The Aviation Training Center for the US Coast Guard.

TEMPORARY MILITARY LODGING: Spc Svc, USCG Avn Tng Ctr, dty hrs, Comm: 205-694-6130. Lodging at Dauphin Island Rec Complex and extremely limited BEQ/BOQ at CG Base.

LOGISTICAL SUPPORT: Limited spt fac avail. Key phone numbers follow.

CGX-694-6398	Conv S-694-6377	EM C-694-6358
Gas S-694-6496	Pkg S-694-6372	

HEALTH & WELFARE: Outpatient clinic only, EX-6400. Inpatient, Pensacola Naval Hosp, FL, 50 mi E, Comm: 904-452-2353.

RECREATION: Some rec on Base, Racquet Courts, Auto Hobby Shop, Picnic Area. Operates the Dauphin Island Family Rec Facility, on Dauphin Island, 40 mi S of Mobile. Take AL-163 S. Comm: 205-694-6130, ATVN: 436-6130, FTS: 537-6130, 13 furnished cottages, 6 camper spaces w/hookups, comp fac and act.

SPACE-A: Limited. Try CGATC Base Ops, EX-6161. Ask for Ops.

ATTRACTIONS: Mobile is one of the largest comm ports in the country. Historic houses and beautiful flowers in the Mobile Bay area.

REDSTONE ARSENAL (AL06R2)

Redstone Arsenal, AL 35898-5020

Comm: 205-876-2151
ATVN: 746-0011

LOCATION: Off US-231 S on Martin Rd for main gate with visitor control. For uniformed pers, Gate 8 is on Drake Ave, take US-72 onto Jordan Lane, then W on Drake. Drake becomes Goss Rd on the Arsenal. RM: p-5, E/15. NMC: Huntsville, adjacent N and E sides.

GENERAL INFORMATION: Hq US Army Missile Command, US Army Ordnance Missile & Munitions Center & School and National Aeronautic and Space Administration.

TEMPORARY MILITARY LODGING: Biltg Ofc, ATTN: DRSMI-KNB, Bldg 3467, Vincent Dr, Comm: 205-876-5713, ATVN: 746-5713, 0700-1630 hrs dly, other SDO, Bldg 5250, EX-8351. All ranks. DV/VIP call Comm: 205-876-7135, ATVN: 746-7135.

LOGISTICAL SUPPORT: Complete spt fac avail. Key phone numbers follow.

PX-883-6100	**Caf-881-9973/6210**	**Chap-876-5751**
Chld C-837-6464	**Commiss-876-8429**	**Conv S-881-7862**
EM C-881-6525	**Fam Svc-876-2859**	**Gas S-881-7588**
Legal-876-2296	**Loc-876-3331**	**Med-876-6110/5854**
NCO-876-3331	**NCO C-837-0750**	**Off C-830-2582**
Pkg S-883-7761/6525	**Pol-876-2222**	**Snack B-881-1591**
SDO-876-8351	**Theat-876-6524**	

HEALTH & WELFARE: Fox Army Community Hospital, Bldg 4100, Comm: 205-876-5854, Emerg and walk-in clinic, complete range of clinics and depts, 40 bed. Chaplain, EX-2337/3433.

RECREATION: Rec Svc, EX-2501, Craft Shops, EX-7924/5841, Golf Course, EX-6888, Rec Ctr, EX-4531, Gym, EX-6701, Swimming-NCO, EX-6758, Off, EX-6713, Troop, EX-6605, Bowling Lanes, EX-6634, Martial Arts, EX-2501, Sports Off, EX-2255, Youth Act, EX-5437. Arsenal campground on Post (Tenn River), Year round, Outdoor Recreation, Bldg 5129, EX-4868, 24 camper spaces w/hookups.

SPACE-A: Limited. Space-A via exec type aircraft, MEDEVAC, C-9, C-130, Arsenal Army Airfield, EX-2186.

ATTRACTIONS: Alabama Space & Rocket Center, Hwy 20, Tenn River, and Southern BBQ is wonderful.

MILITARY RETIRED HOMES

Air Force Village
4917 Ravenswood Drive
San Antonio, TX 78227
(512) 673-2721

Army Distaff Hall
6200 Oregon Ave, NW
Washington, DC 20015
(202) 541-0400

US Naval Home
01800 East Beach Blvd
Gulfport, MS 39501
(601) 896-3110

U.S. Soldiers' & Airmen's Home
3700 N Capitol St
Washington, DC 20317
(202) 722-3183

ARIZONA

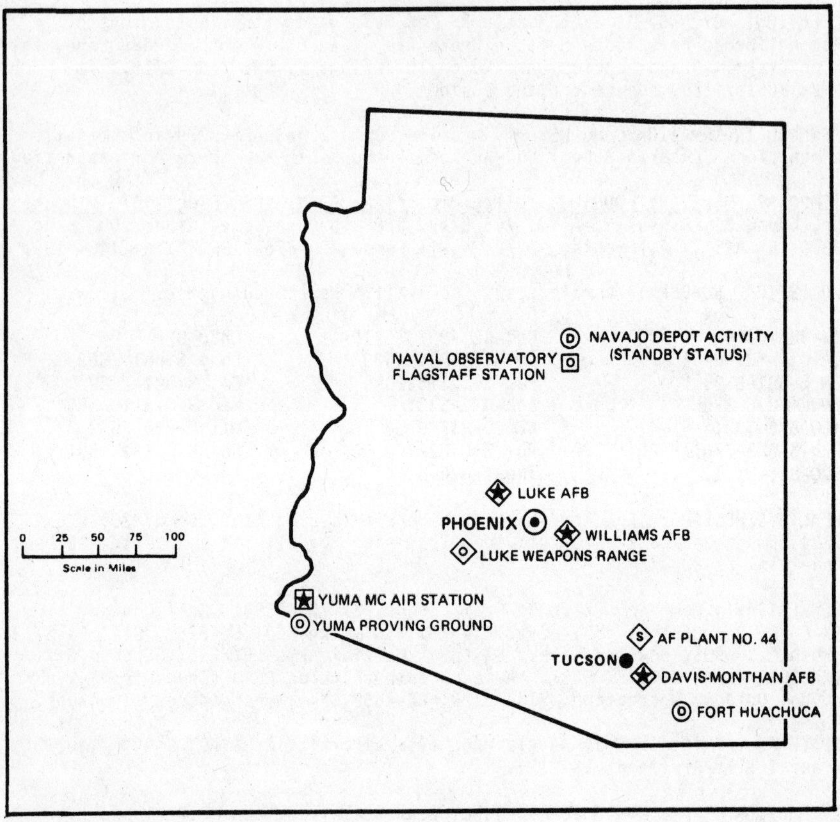

DAVIS-MONTHAN AIR FORCE BASE (AZ01R4)

Comm: 602-748-3900
ATVN: 361-1110

Davis-Monthan Air Force Base, AZ 85707-5000

LOCATION: Exit I-10 N Palo Verde, right on Irvington, left on Alvernon to Golf Links then to main gate. RM: p-7. P/10. NMC: Tucson, 3 mi NW.

GENERAL INFORMATION: Tactical Air Command Base, TAC Air Ops, Missile Training, and Aerospace Maintenance and Regeneration Center (AMARC).

TEMPORARY MILITARY LODGING: Biltg Ofc, Bldg 2350, 10th St. ATTN: 836SVS/SVH Box 15013, 0700-1600 hrs dly, Comm: 602-748-3309, ATVN: 361-3309. All ranks. DV/VIP call Comm: 602-748-3600, ATVN: 361-3600.

LOGISTICAL SUPPORT: Complete spt fac avail. Key phone numbers follow.

BX-748-7887	Caf-790-6150	Chap-748-5411
Chld C-748-3336	Commiss-790-4341	Conv S-748-8076
Fam Svc-748-3368	Gas S-748-8212	Legal-748-5242
Loc-748-3347	Med-748-3878	NCO C-748-3100
Off C-748-3301	Pkg S-747-1365	Pol-748-3517
Theat-748-1157		

Davis-Monthan AFB, Cont'd

HEALTH & WELFARE: USAF Hospital, complete clinics and depts, 65 bed, Emerg, EX-3878. Chapels serving all faiths, EX-5411.

RECREATION: Library, EX-4381, Rec Ctr/Tickets, EX-3717/3700, Swimming, EX-3579, Gym, EX-3714, Aero Club, EX-3603, Golf, EX-3734, Bowling, EX-3461, Ceramics Shop, EX-4385/4028, Saddle Club, 885-9049, Wood Shop, EX-3578, Picnic Area, EX-3846, Auto Shop, EX-3614. Fam-Camp on Base, Comm: 602-748-3717, year round, 27 camper spaces w/comp hookups.

SPACE-A: Pax Lounge and Ops in Bldg 4820, 24 hrs, Comm: 602-748-3641, ATVN: 361-3641. Most flights unscheduled.

ATTRACTION: Old Tucson, movie sets, Reid Park, zoo, Arizona-Sonora Desert, and Pima Air Museum. Also, twice weekly tours at AMARC.

FORT HUACHUCA (AZ02R4)

Fort Huachuca, AZ 85613-6000

Comm: 602-538-2151
ATVN: 879-2151
FTS: 769-2151

LOCATION: From I-10 take AZ-90 S to Sierra Vista and main gate of Fort. RM: p-8, O/10. NMC: Tucson, 75 mi NW.

GENERAL INFORMATION: Hq US Army Information Systems Command, Intelligence Center & School and US Army Electronic Proving Ground.

TEMPORARY MILITARY LODGING: Biltg Ofc, ATTN: ASH-HS-BH, Bldg 43083, Service Rd, 24 hrs, Comm: 602-538-5950, ATVN: 879-5950. Extensive fac. All ranks. DV/VIP call Comm: 602-538-6742, ATVN: 879-6742.

LOGISTICAL SUPPORT: Complete spt fac avail. Key phone numbers follow.

PX-458-7210	**Caf-458-7982**	**Chap-538-5827**
Chld C-538-5209	**Commiss-538-3360**	**Conv S-458-8389**
EM C-458-2113	**Fam Svc-538-3234**	**Fire-538-2111**
Gas S-458-9735	**Legal-538-2263**	**Loc-538-3636**
Med-538-5121	**NCO-538-2922**	**NCO C-538-3802**
Off C-538-2195	**Pkg S-458-5863**	**Pol-538-2181/2182/3000**
Snack B-538-5759	**Theat-538-2950**	

HEALTH & WELFARE: Raymond W. Bliss Army Community Hospital, comp clinics & depts, 110 bed, Bldg 45001, Emerg, 538-2111. Chapels serving all faiths, Center 538-5559.

RECREATION: Bowling, 538-2849, Golf, 538-7160, Gym, 538-2948, Library, 538-3041, Museum, 538-5736, Picnic & Campsite Resv, 538-3317, Riding, 538-5220, Swimming, 538-3858/3853, Arts and Craft Shops, 538-2015, Auto Craft, 538-5024, Youth Act, 538-3027/3212, Outdoor Rec, 538-3317/3249, Aviation Club, 538-3633. Outdoor Rec Center on Fort, Comm: 602-538-3317, ATVN: 879-3317, FTS: 769-3317, year round, 20 camper spaces w/hookups, complete fac and all act.

SPACE-A: Limited flts from Libby Army Airfield on Fort via exec aircraft, 538-2373.

ATTRACTIONS: Fort museum. Original cantonment a National Historical Landmark. Tombstone, 25 min drive on the Charleston Rd.

LUKE AIR FORCE BASE (AZ03R4)

Luke Air Force Base, AZ 85309-5000

Comm: 602-856-7411
ATVN: 853-1110

Luke AFB, Cont'd

LOCATION: From Phoenix, W on I-10 to Litchfield Rd, N on Litchfield Rd approx 7 mi. Also, from Phoenix, on I-17 to Glendale Ave, W on Glendale Ave to intersection of Glendale Ave and Litchfield Rd, approx 10 mi. RM: p-8, K/6. NMC: Phoenix, 20 mi E.

GENERAL INFORMATION: Tactical Air Command Base. Largest tactical fighter training base in free world. Near Glendale, Sun City, and Phoenix.

TEMPORARY MILITARY LODGING: Biltg Ofc, ATTN: 832 CSG/SV, Bldg 666, K St, 24 hrs dly, Comm: 602-856-6336, ATVN: 853-6336. All ranks. DV/VIP, Comm: 602-856-6101, ATVN: 853-6101.

LOGISTICAL SUPPORT: Complete spt fac avail. Key phone numbers follow.

BX-935-2671	**Chap-856-6211**	**Chld C-856-6339**
Commiss-935-9710	**Conv S-935-2414**	**EM C-856-7136**
Fam Svc-856-6415	**Gas S-935-4953**	**Legal-856-6901**
Loc-856-6405	**Med-856-7506**	**NCO-856-6022**
NCO C-856-7136	**Off C-856-6446**	**Pkg S-856-6447**
Pol-856-6322	**Snack B-935-4029/9209**	**Theat-856-6461**

HEALTH & WELFARE: USAF Hospital with full clinics and depts, 105 bed, Emerg Svc, EX-7506, Appoint, EX-7513/7571. Chapels serving all faiths, EX-6211/6212.

RECREATION: Arts & Crafts, EX-6502, Auto Hobby, EX-6107, Photo Shop, EX-6502, Wood Shop, EX-6566, Bowling Ctr, EX-6946, Gym, EX-6953, Rec Ctr, EX-6630/6152, Tickets, EX-6000, Swimming, Rec Ctr or Club Pools, Library, EX-7191, Youth Ctr, EX-6225. Operates Luke Rec Ctr at Fort Tuthill, 150 mi N on I-17, 4 mi S of Flagstaff, Comm: 602-774-8893/6000, ATVN: 853-8893/6000. 26 camper spaces w/hookups, 16 mobile homes, 9 wood huts, comp fac and full act. Also operates Gila Bend Fam-Camp at Aux Field, Comm: 602-683-5275, ATVN: 853-5275. 15 spaces w/hookups.

SPACE-A: Pax Lounge and Ops in Bldg 439, Comm: 602-853-7131, ATVN: 853-7131, flts to mid-west and West Coast.

ATTRACTIONS: Colorful Scottsdale nearby. AZ state capital, fairgrounds, and coliseum in Phoenix. ASU in Tempe. Lots of sunshine.

WILLIAMS AIR FORCE BASE (AZ09R4)

Williams Air Force Base, AZ 85240-5000 — **Comm: 602-988-2611** **ATVN: 474-1110**

LOCATION: From I-10 exit go east on Chandler Blvd, 17 mi to AFB. RM: p-8 K/7. NMC: Phoenix, 30 mi NW.

GENERAL INFORMATION: Air Training Command Base, largest undergraduate Pilot Training Base.

TEMPORARY MILITARY LODGING: Biltg Ofc, ATTN: 82-ABG/SVH, Bldg 321, 7th St at Off C Circle. 24 hrs dly. Comm: 602-988-5546, ATVN: 474-5546. All ranks. DV/VIP call Comm: 602-988-5212, ATVN: 474-5212.

LOGISTICAL SUPPORT: Complete spt fac avail. Key phone numbers follow.

BX-988-1321	**Caf-988-1970**	**Chap-988-6761**
Chld C-988-5578	**Commiss-988-5585**	**Conv S-988-1900**
EM C-988-6635	**Fam Svc-988-5284**	**Gas S-988-1885**
Legal-988-6866	**Loc-988-5230**	**Med-988-5452**
NCO C-988-1501	**Off C-988-1581**	**Pkg S-988-1501**
Pol-988-6693	**Theat-988-6589**	

Williams AFB, Cont'd

HEALTH & WELFARE: USAF Hospital, complete clinics and depts, 25 bed. Emerg, EX-5452, Appoint, EX-6814. Chapels serving all faiths, EX-6761/6762 or 6811.

RECREATION: Svc Club, EX-6635, Rec Ctr, EX-6635, Teen Club, EX-6745, Outdoor Rec, EX-5592, Bowling Ctr, EX-5316, Golf, EX-5493, Gym, EX-6672, Hobby Shops, EX-6632/5281, Youth Ctr, EX-6745, Library, EX-5279. Operates Rec Annex (Waterdog), Comm: 602-988-5592, ATVN: 474-5592, 8 camper spaces w/hookups, 18 log cabins w/kit, comp fac and activities.

SPACE-A: Base Ops, Bldg 19, EX-5297. Limited Space-A opportunities. Call before you go to the Base.

ATTRACTIONS: In the greater Phoenix area, a unique combination of the old and new west. Clear, warm weather year round.

YUMA MARINE CORPS AIR STATION (AZ04R4)

Comm: 602-726-2011
ATVN: 951-2011

Yuma Marine Corps Air Station, AZ 85369-5000

LOCATION: From I-8 take Ave 3E S, for 1 mi to Base on right, adjacent to Yuma IAP. RM: p-8, M/1. NMC: Yuma, 3 mi NW.

GENERAL INFORMATION: The major Marine Corps Combat Air Training Base. Near perfect flying conditions and range facilities are available.

TEMPORARY MILITARY LODGING: Biltg Ofc, STA S-4, Biltg fund, Bldg 1020 Thomas Ave. 24 hrs dly. Comm: 602-726-4970, ATVN: 951-3094. All ranks. DV/VIP call Comm: 602-726-2253, ATVN: 951-2253.

LOGISTICAL SUPPORT: Complete spt fac avail. Key phone numbers follow.

PX-726-2294/2364	**Caf-726-2369**	**Chap-726-2371**
Chld C-726-2350	**Commiss-726-2245**	**Conv S-726-2770**
EM C-726-2458	**Fam Svc-726-3421**	**Gas S-726-2110**
Legal-726-2468	**Loc-726-2009**	**Med-726-2772**
Off C-726-2712	**Pkg S-726-2134**	**Pol-726-2205/2361**
SNCO C-726-2171/2406	**Theat-726-2358**	

HEALTH & WELFARE: Large medical and dental clinics, Emer, EX-2111, Appoint, EX-2353/2194. Inpatient, San Diego Naval Hosp, Comm: 619-233-2417. Chapels serving all faiths, EX-2371.

RECREATION: Spec Svc, EX-2278/2279, Athletic/Picnic Fac, EX-2278/9, Camp Gear Issue, EX-2848, Gym, EX-2727, Library, EX-2785, Auto Hobby, EX-2395, Bowling, EX-2221, Swimming, EM, EX-2477, Off, EX-3474. Operates Lake Martinez Rec Area, Spec Svc, Bldg 633, MCAS, Comm: 602-726-2278, ATVN: 951-2279, on the Colorado River, 38 mi N of MCAS off US-95. 17 camper spaces w/hookups, 8 cottages, 3 camp sites, complete fac and activities year round.

SPACE-A: Term Lounge and Ops, Bldg 151, 0700-1530 hrs dly, Comm: 602-726-2729, ATVN: 951-2729, flts to other MCASs.

ATTRACTIONS: Desert type climate, greyhound racing, baseball training.

YUMA ARMY PROVING GROUND (AZ05R4)

Comm: 602-328-2151
ATVN: 899-2151

Yuma Army Proving Ground, AZ 85365-5000

Yuma Army Prov Gd, Cont'd

LOCATION: NE of I-8 turn right on US-95. SW of I-10 turn left on US-95. US-95 is N/S route which bisects YPG. RM: p-8, K/1. NMC: Yuma, 27 mi SW.

GENERAL INFORMATION: Army Materiel Test Fac with instrumented ranges.

TEMPORARY MILITARY LODGING: Biltg Ofc, Bldg 1003, 5th St & Barranca Rd, 0700-1800 hrs dly, Comm: 602-328-2174, ATVN: 899-2174. All ranks. DV/VIP call Comm: 328-2020, ATVN: 899-2020.

LOGISTICAL SUPPORT: Complete spt fac avail. Key phone numbers follow.

PX-328-2788
Commiss-328-2214
Gas S-328-2868
Med-328-2646
Pkg S-328-2969
Theat-328-2391
Chap-328-2750
EM C-328-2097
Legal-328-2608
NCO C-328-2097
Pol-328-2641
Chld C-328-2576
Fam Svc-328-2852
Loc-328-2741
Off C-328-2097
Snack B-328-2680

HEALTH & WELFARE: Dispensary, outpatient only, Emer, EX-2646, Appoint, EX-2502. Inpatient San Diego Naval Hosp, Comm: 619-233-2417. Chapel serving all faiths, EX-2750.

RECREATION: Spec Svc, EX-2238, Bowling, EX-2308, Swimming, Gym, Outdoor Rec, and Courts, EX-2238. Boating on the Colorado River. See Yuma, MCAS listing for info on Lake Martinez Rec Area nearby.

SPACE-A: Available at Yuma, MCAS. See that listing for details.

ATTRACTIONS: Isolated post, adjacent to KOFA Nat'l Wildlife Refuge.

SMALL INSTALLATIONS

Many small military installations have exchanges with limited inventories. They may also have emergency medical, some recreational and other support facilities. The following is a list in alphabetical order by states of these installations.

King Salmon Airport, USAF
705 ACWS, Bldg 603
APO Seattle 98723-5000

Galena Airport, USAF
5072 ABS, Bldg 202
APO Seattle 98723-5000

Clear Air Force Station
Bldg 201, Clear MEWS
Fairbanks AK 99704-5000

Juneau Coast Guard Station
345 Egan Drive
Juneau AK 99801-5000

Ketchikan
Coast Guard Base
Ketchikan AK 99901-5000

Fort Chaffee
Fort Smith, Bldg 144
Fort Chaffee AR 72923-5000

Gila Bend Air Force Aux Field
Bldg 31
Gila Bend AZ 85337-5000

Camp Roberts
Bldg T-6036
Paso Robles CA 93446-5000

Presidio of Monterey
Bldg T-267
Monterey CA 93940-5000

Sunnyvale Air Force Station
Bldg 1018
Sunnyvale CA 94088-5000

Skaggs Island
Naval Security Group Activity
Sonoma CA 95476-5000

Centerville Beach
Naval Facility
Ferndale CA 95536-5000

Cont'd page 45

ARKANSAS

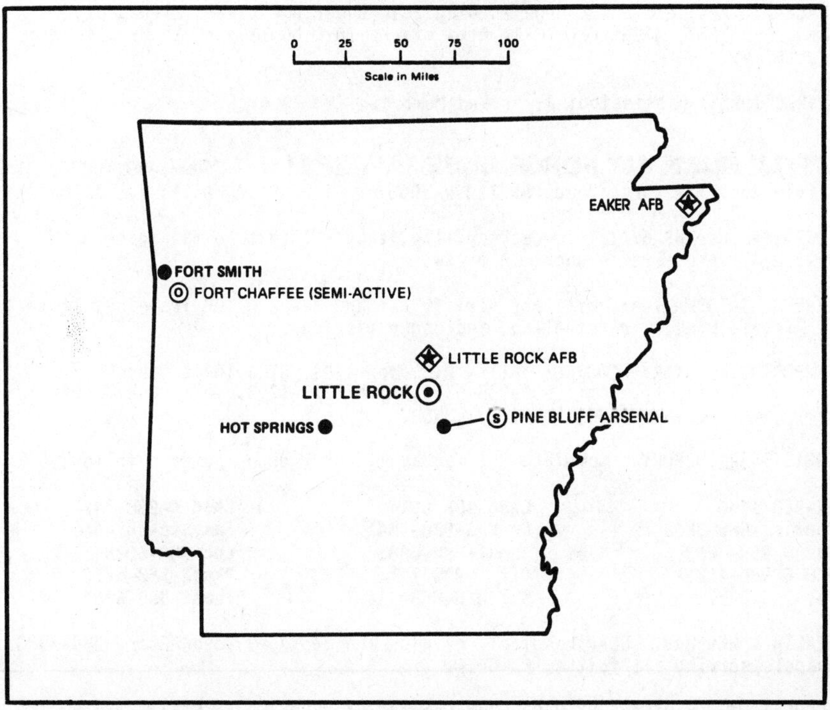

EAKER AIR FORCE BASE (AR01R2)

Eaker Air Force Base, AR 72317-5000

Comm: 501-762-7000
ATVN: 637-1110

LOCATION: From I-55 exit to US-61 N through Blytheville to Gosnell to main gate of AFB. RM: p-9, D/12. NMC: Memphis, TN, 70 mi S.

GENERAL INFORMATION: Strategic Air Command Base. 97th Bomb Wing located here.

TEMPORARY MILITARY LODGING: Biltg Ofc, Bldg 702, Memorial Dr, 24 hrs dly. Comm: 501-762-7461/7463, ATVN: 637-7461/7463. DV/VIP-same as Biltg Ofc.

LOGISTICAL SUPPORT: Complete spt fac avail. Key phone numbers follow.

BX-532-8534	**Chap-762-7365**	**Chld C-762-7898**
Commiss-762-7129	**Conv S-532-8484**	**Fam Svc-762-7522**
Gas S-532-8484	**Legal-762-7133**	**Loc-762-7260**
Med-762-7219	**NCO-762-7232**	**NCO C-762-7110**
Off C-762-7111	**Pkg S-762-7583**	**Pol-762-7208**
Snack B-532-8967	**Theat-762-7412**	

HEALTH & WELFARE: USAF Hospital, clinics and depts, Emer, EX-7219, Appoint, EX-7426. Chapels serving all faiths, EX-7365.

RECREATION: Call EX-7237 for the following Rec Svc: Gym, Swimming, Picnic Area, Library, Hobby Shops, and Youth Center. Bowling, EX-7109, Golf,

Eaker AFB, Cont'd

EX-7108. Boating on Mississippi River nearby. No Rec Area.

SPACE-A: Pax Term and Lounge in Bldg 201, under ATC tower, 24 hrs dly, Comm: 762-7272, ATVN: 637-7272, flts via KC-10 & KC-135 A/C to US & foreign countries.

ATTRACTIONS: Mississippi River and Memphis, TN, 70 mi S.

LITTLE ROCK AIR FORCE BASE (AR02R2)

Little Rock Air Force Base, AR 72099-5000

Comm: 501-988-3131
ATVN: 731-1110

LOCATION: Use US-67/167 to Jacksonville, take AFB exit to main gate. RM: p-9, G/7. NMC: Little Rock, 18 mi SW.

GENERAL INFORMATION: Military Airlift Command Base, C-130 Training, Titan II Missile Wing, air refueling, and other missions.

TEMPORARY MILITARY LODGING: Biltg Ofc, Box 1192, Bldg 1024, Cannon Circle, 24 hrs dly, Comm: 501-988-6753, ATVN: 731-6753. All ranks. DV/VIP call Comm: 501-988-3588, ATVN: 731-3588.

LOGISTICAL SUPPORT: Complete spt fac avail. Key phone numbers follow.

BX-988-1150	Chap-988-6014	Chld C-988-7422
Commiss-988-3202	Conv S-988-4841	Fam Svc-988-6451
Gas S-988-3263	Legal-988-6881	Loc-988-6025
NCO C-988-4121	Off C-988-1111	Pkg S-988-6763
Pol-988-3221	Snack B-988-1148	Theat-988-6461

HEALTH & WELFARE: USAF Hospital, clinics and depts, 35 bed, Emer, 988-7333. Chapels serving all faiths, EX-6014.

RECREATION: Bowling, EX-3793, Rec Center, EX-3220. Aero Club, Library, Golf, Swimming, Youth Center, Hobby Shops, Gym and Rec Supplies, call main number for EX's. Fam-Camp on Base, yr round, 10 camper spaces w/hookups, near lake, full fac and act, Comm: 501-988-6164, ATVN: 731-6164. Also operates DeGray Lake Rec Area, 93 mi S, Comm: 501-988-3216.

SPACE-A: Pax Term and Lounge, Bldg 120, 0730-1630 hrs dly, Comm: 501-988-3684, ATVN: 731-3684, mostly KC-10 & KC-135 flts CONUS & foreign countries.

ATTRACTIONS: Little Rock, Arkansas River, and Lake Conway.

PINE BLUFF ARSENAL (AR03R2)

Pine Bluff Arsenal, AR 71602-9500

Comm: 501-541-3000
ATVN: 966-3000
FTS: 740-5892

LOCATION: Off US-65 NW of Pine Bluff. Take AR-256, cross AR-365 into main gate of Arsenal. Or S on US-65 from Little Rock, 35 mi, follow signs. RM: p-9, H/7. NMC: Pine Bluff, 8 mi SE.

GENERAL INFORMATION: An Engineer, Production, Storage, and Test Facility for chemical munitions and related equipment.

TEMPORARY MILITARY LODGING: Biltg Ofc, ATTN: SMCPB-HO, Bldg 10-020, 0730-1600 hrs dly, Comm: 501-541-3008, ATVN: 966-3008. No DV/VIP fac.

LOGISTICAL SUPPORT: Limited spt fac avail. Key phone numbers follow.

Pine Bluff Arsenal, Cont'd

PX-535-1707	Legal-541-3131	Loc-541-3000
Med-541-3409	PAO-541-3421	Pol-541-3505/3506

HEALTH & WELFARE: Army Medical Clinic, EX-3409. Inpatient, USAF Hospital, Little Rock AFB, Comm: 501-988-3131. No chapels on the Arsenal, all faiths served in Pine Bluff.

RECREATION: Rec Bldg, Golf (9 hole course), Rod and Gun Club, Courts and Fields. The on Post rec area is at Lake Tulley, limited fac.

SPACE-A: Little Rock AFB, Bldg 120, 1730-1630 hrs dly, Comm: 501-988-3684, ATVN: 731-3684, flts to CONUS and overseas.

ATTRACTIONS: Pine Bluff and the Arkansas River. (Arkansas is the Natural State.)

CALIFORNIA

ALAMEDA COAST GUARD SUPPORT CENTER (CA28R4)

Alameda Coast Guard Island, CA 94501-5000

Comm: 415-437-3151
FTS: 536-3151

Alameda CG Spt Ctr, Cont'd

LOCATION: Take US-17 S to 23rd St exit, right and straight to causeway and Island. Signs posted. RM: p-11, ND/22. NMC: Oakland, 1 mi NE.

GENERAL INFORMATION: Major CG Support Base in San Francisco Bay Area.

TEMPORARY MILITARY LODGING: Biltg Ofc, Bldg 21, McCullough Dr, 0730-1530 hrs dly, other hrs OOD, Bldg 3, Comm: 415-437-3180, FTS: 536-3180. Limited facilities. DV/VIP, EX-3151.

LOGISTICAL SUPPORT: Most spt fac avail. No commissary. Key phone numbers follow.

CGX-437-3167	**Chap-437-3067**	**EM C-437-3304**
Gas S-437-3166	**Legal-437-3330**	**Loc-437-3628**
Med-437-3581	**NCO/CPO C-437-3303**	**Off C-437-3628**
Pkg S-437-3167	**Pol-437-3151**	

HEALTH & WELFARE: Outpatient clinics only, Comm: 437-3581. Inpatient, Oakland Naval Hosp, Comm: 466-0112, ATVN: 836-0111. Chapels serving all faiths, EX-3067.

RECREATION: Gym, Swimming, Auto & Woodworking Hobby Shops, Tennis and Racquetball Courts. Lake Tahoe Rec Fac, Spec Svc, year round, Comm: 415-437-3577, FTS: 536-3577, two A-frame cottages comp furn, full military or complete facilities and all activities.

SPACE-A: Available at Oakland IAP, Comm: 415-635-8482, 24 hrs, flts to CONUS and overseas.

ATTRACTIONS: San Francisco and the exciting San Francisco Bay Area nearby.

ALAMEDA NAVAL AIR STATION (CA33R4)

Alameda Naval Air Station, CA 94501-5000

Comm: 415-869-0111
ATVN: 686-0111

LOCATION: From Nimitz Hwy, CA-17 S, take the Broadway/Alameda exit. CA-17 N, take Broadway exit (17 N does not say Alameda). Directions to NAS clearly marked. RM: p-11, ND/22. NMC: Oakland, 2 mi NE.

GENERAL INFORMATION: Naval Air Rework Facility, Fleet Support, and Reserve units. Near the San Francisco-Oakland Bay Bridge on the Oakland side. 6 home-ported ships, 53 tenant commands, largest Nav Air Reserve in US.

TEMPORARY MILITARY LODGING: Biltg Ofc, Comm: 415-869-2032, ATVN: 686-2036 Bldg 101, D St, 0800-1600 hrs dly. All ranks. Navy Lodge, Bldg 531 on C St. Call Comm: 415-869-3206, ATVN: 686-3206. DV/VIP. Call Comm: 869-2384, ATVN: 686-2384.

LOGISTICAL SUPPORT: Complete spt fac avail. Key phone numbers follow.

NX-522-5988	**Chap-869-4386**	**Chld C-869-2435**
Commiss-869-2580	**Conv S-869-6310**	**EM C-869-2931**
Fam Svc-869-4111	**Gas S-869-6280**	**Legal-869-4516**
Loc-869-0111	**Med-869-4444/Emerg only**	**NCO-869-4197**
NCO/CPO C-869-2101	**Off C-869-3241**	**Pkg S-869-3000**
Pol-869-3053	**Theat-869-3040**	**Top Four Club-869-4441**

HEALTH & WELFARE: Dispensary only, EX-3166. Inpatient, Oakland Naval Hosp, Comm: 415-639-2418, ATVN: 855-0111. Chapels serving all faiths, EX-4386.

RECREATION: Boat Marina, EX-2840, Bowling Lanes, EX-2110, Golf Course,

Alameda NAS, Cont'd

EX-2946, Rec Gear Issue, EX-2407, Gym, EX-2407, Hobby Shop, EX-2962, Auto Hobby, EX-2789, Skeet Range, EX-2407/2365, Swimming, EX-2350, Spec Svc Off, EX-3536, Youth Ctr, EX-4042-2389, Library, EX-3504.

SPACE-A: Pax Term and Lounge, Bldg 77, Comm: 415-869-2727, ATVN: 686-2727, extensive flts to CONUS NAS and other locations.

ATTRACTIONS: The best seafood on the West Coast is at Tadich Grill, 240 California St. Closed Sun. Tell Chris, the Manager, Military Living sent you, 415-391-2373. No reservations and cash only.

ARMED FORCES RESERVE CENTER (CA39R4)

Comm: 213-493-8000
ATVN: 972-2000

California National Guard
Los Alamitos, CA 90720-5001

LOCATION: Off I-605 E of Long Beach. Clearly marked, follow signs. RM: p-15, K/12. NMC: Los Angeles, 5 mi N.

GENERAL INFORMATION: Armed Forces Reserve Center, supporting reservists from all branches of Armed Forces.

TEMPORARY MILITARY LODGING: Biltg Ofc, Hq AFRC, duty hrs, Comm: 213-493-8125, ATVN: 972-2125.

LOGISTICAL SUPPORT: Limited spt fac avail. Exchange, 430-1077, and military clothing sales store.

HEALTH & WELFARE: Medical spt at Long Beach Naval Hospital, Comm: 213-420-5201, ATVN: 873-9201. Chapel spt also available at Naval Hosp.

RECREATION: None at the AFRC, beaches nearby. See other Bases in the Los Angeles and Orange County area.

SPACE-A: None. See Los Angeles IAP, Comm: 213-643-0714, ATVN: 833-0714, flts to CONUS and overseas.

ATTRACTIONS: Los Angeles, Hollywood and Orange County.

BARSTOW MARINE CORPS LOGISTICS BASE (CA13R4)

Comm: 619-577-6211
ATVN: 282-0111

Barstow Marine Corps Logistics Base, CA 92311-5000

LOCATION: On I-40, 1.5 mi E of Barstow. Take I-15 NE from San Bernardino, or W from Las Vegas, NV. Signs mark direction to MCLB. RM: p-13, SI/17. NMC: San Bernardino, 75 mi SW.

GENERAL INFORMATION: The West Coast Marine Corps Logistics Base.

TEMPORARY MILITARY LODGING: Biltg Ofc, Bldg 362, Cape Gloucester Ave, 0700-1530 hrs dly, Comm: 619-577-6706, ATVN: 282-6706. All ranks. DV/VIP call Comm: 577-6555, ATVN: 282-6555.

LOGISTICAL SUPPORT: Complete spt fac avail. Key phone numbers follow.

MX-256-0412	**Caf-577-6428**	**Chap-577-6846**
Chld C-577-6287	**Commiss-577-6403**	**Conv S-256-8974**
EM C-577-6495	**Fam Svc-577-6533**	**Gas S-256-9411**
Legal-577-6771	**Loc (Civ)-577-6485**	**Loc (Mil)-577-6663**

Barstow MC Log Base, Cont'd

Med-577-6591	NCO C-577-6532	Off C-577-6432
Pkg S-577-6410	Pol-577-6666	Snack B-256-8976
Theat-577-6527		

HEALTH & WELFARE: Medical clinics. Outpatient, Emerg, 577-6588, Appoint, 577-6587. Inpatient, Weed Army Comm Hosp, Fort Irwin, Barstow, Comm: 714-256-1786, EX-2266. Chapels serving all faiths, EX-6533.

RECREATION: Auto Hobby, EX-6441, Bowling, EX-6264, Ceramics Hobby, EX-6228, Equip Issue, EX-6551, Golf, EX-6431, Gym, EX-6392/6697, Library, EX-6395, Spc Svc Ofc, EX-6541, Wood & Rock Hobby, EX-6469, Skeet Range, EX-6789, Stables, EX-7302. Big Bear Rec Center, 1 hr away, see El Toro MCAS listing.

SPACE-A: See Norton AFB, Comm: 714-382-5354, ATVN: 876-5354.

ATTRACTIONS: Calico Ghost Town, 8 mi E, Solar One, 7 mi E, Lake Delores, 13 mi E.

BEALE AIR FORCE BASE (CA47R4)

Beale Air Force Base, CA 95903-5000

Comm: 916-634-3000
ATVN: 368-1110

LOCATION: From CA-70 N exit S of Marysville, to N Beale Rd, continue for 10 mi until Rd deadends at main gate of AFB. RM: p-11, NK/10. NMC: Sacramento, 35 mi SW.

GENERAL INFORMATION: Strategic Air Command Base, Missile Warning Squadron, Information Systems Squadron, aerial tankers, training.

TEMPORARY MILITARY LODGING: Biltg Ofc, Bldg 2156, Warren Shingle Blvd, 24 hrs dly, Comm: 916-788-0216, ATVN: 368-2084. All ranks. DV/VIP, 788-0216.

LOGISTICAL SUPPORT: Complete spt fac avail. Key phone numbers follow.

BX-788-0221	Caf-788-1070	Chap-634-2111
Chld C-634-2254	Commiss-634-2704	Conv S-788-1271
EM C-634-2559	Fam Svc-634-2294	Gas S-788-0215
Legal-634-2815	Loc-634-3000	Med-634-4158
NCO-634-4364	NCO C-634-2559	Off C-788-0292
Pkg S-788-0293	Pol-634-2131	

HEALTH & WELFARE: USAF Hospital, clinics and depts, 30 bed, Emerg, EX-2992, Central Appoint, EX-2941. Chapels serving all faiths, EX-2111.

RECREATION: Rec Svc, EX-2367, Arts & Crafts, EX-2289, Rec Ctr, EX-2616, Bowling, EX-2478, Swimming at club pools, Gym, EX-4549, Golf, EX-2374. Fam-Camp on Base, Spec Svc, Comm: 916-634-4222, ATVN: 368-4222, 27 camper spaces w/comp hookups.

SPACE-A: Pax Term across st from Base Ops, Bldg 1060, 0730-1630 hrs dly, Comm: 916-634-2700, ATVN: 368-2700, flts CONUS and overseas via KC-10/135.

ATTRACTIONS: Sacramento, Sutter's Fort, Zoo, Old Sacramento 2 hours drive from Reno, Lake Tahoe; 2 1/2 hours drive from San Francisco.

CAMP PENDLETON MARINE CORPS BASE (CA30R4)

Camp Pendleton Marine Corps Base, CA 92055-5001

Comm: 619-725-4110
ATVN: 365-0111
FTS: 892-4110

Camp Pendleton MC Base, Cont'd

LOCATION: On I-5 which is adjacent to Camp Pendleton main gate. Take Camp Pendleton Only off ramp I-5 at Oceanside. RM: p-13, SN/16. NMC: Oceanside, adjacent to Base SE.

GENERAL INFORMATION: The major Marine Corps Amphibious Training Base on the West Coast.

TEMPORARY MILITARY LODGING: Biltg Ofc, Bldg 1341 (Mainside), 24 hrs dly, Comm: 619-725-3451, ATVN: 365-3718/3451. All ranks. DV/VIP, 619-725-6618.

LOGISTICAL SUPPORT: Complete spt fac avail. Key phone numbers follow.

MX-725-6233	**Chap-725-4202**	**Chld C-725-6112**
Commiss-725-4012	**Conv S-725-6233**	**EM C-725-3656**
Fam Svc-725-5361	**Fire-725-3333**	**Gas S-725-4398**
Legal-725-6172	**Loc-725-4111**	**McDonald's-725-6233**
Med-725-1288	**Off C-725-3450**	**Pkg S-725-4788**
Pol-725-3888	**Snack B-725-6233**	**Theat-725-4408**

HEALTH & WELFARE: Camp Pendleton Naval Hospital, comp clinics/depts, 600 bed, Emer, 725-3258, Appoint, 725-3951. Chapels serving all faiths, 725-4202.

RECREATION: Spec Svcs, EX-6288, Swimming Pool, EX-4344/5084, Auto Hobby Shop, EX-5963, Bowling Ctr, EX-5945, Youth Act, EX-6288, Field Archery, EX-6288, Flying Club, EX-4910, Golf Course, EX-4704, Hobby Shop, EX-4880, Physical Fitness Ctr, EX-5984/6394/2951, Horseback Riding Stables, EX-4905, Ticket/Resv Ofc, EX-5675/5416, Library, EX-5104/5669, Skeet/Trap, EX-4832. Beach info, EX-2463/7935, rec equip check-out, EX-5296. Operates 3 camping fac on base: Lake O'Neill, 60 camper spaces, w/water only; Del Mar Rec Beach, 12 cottages, 32 camper spaces w/hookups; San Onofre, 10 mobile homes, 24 spaces w/hookups, 80 spaces w/water only. Joint Public Affairs Office, Comm Relations, Bldg 1160, Comm: 619-725-5566, ATVN: 365-5566, FTS: 892-5566.

SPACE-A: Marine Corps Air Station, El Toro, Santa Ana, Bldg 624, Comm: 714-651-3920/3936, ATVN: 952-3920/3936. Flts to CONUS loc via C-9 aircraft.

ATTRACTIONS: Sandy beaches, race tracks at Hollywood Park, Del Mar, Santa Anita or Los Alamitos, Old California Missions, Disneyland, Knotts Berry Farm, Magic Mountain, Sea World, Wild Animal Park, Lion Country Safari, San Diego Zoo, Balboa Park, Maritime Museum, Carlsbad Raceway, Palomar Observatory.

CASTLE AIR FORCE BASE (CA10R4)

Castle Air Force Base, CA 95342-5000

Comm: 209-726-4636
ATVN: 347-1110

LOCATION: From Sacramento, take CA-99 S to Atwater, take Buhach exit to AFB. RM: p-11, NQ/12. NMC: Fresno, 60 mi S.

GENERAL INFORMATION: Strategic Air Command Base. SAC B-52, KC-135 Crew Training and Fighter Interceptor Squadron.

TEMPORARY MILITARY LODGING: Biltg Ofc, Bldg 1108, C St, 24 hrs dly. Comm: 209-726-2531, ATVN: 347-2531. All ranks. DV/VIP-same as Biltg.

LOGISTICAL SUPPORT: Complete spt fac avail. Key phone numbers follow.

BX-722-8717	**Caf-722-4919**	**Chap-726-2201**
Chld C-726-2289	**Commiss-726-2535**	**Conv S-358-8331**
Fam Svc-726-2131	**Gas S-723-5641**	**Legal-726-2421**

Castle AFB, Cont'd

Loc-726-2011	Med-726-2686	NCO C-726-2259/ 723-4317
Off C-726-2769	Pkg S-723-5110	
Pol-726-2111	Snack B-722-4919	Theat-723-6581

HEALTH & WELFARE: USAF Hospital, clinics and depts, 25 bed. Emer, EX-2686, Appoint, EX-2276. Chapels serving all faiths, EX-2201.

RECREATION: Auto Hobby, EX-2694, Arts/Crafts, EX-2694, Library, EX-2630, Gym, EX-2359, Bowling, EX-2189, Rod/Gun, EX-4300, Rec Ctr, EX-2177, MWR Supply EX-2715, Swimming, EX-4481. Fam-Camp on Base, yr round, 6 camper spaces, w/ hookups, 36 spaces w/o hookups, EX-6031. Also operates fac at Lake McClure (Castle Marina), 38 mi NE of Base, 6 campers, EX-6031.

Space-A: Ltd flts via KC-10/135 aircraft to CONUS and overseas, EX-2611.

ATTRACTIONS: Yosemite National Park, 45 mi E. Sacramento, 100 mi N.

CHINA LAKE NAVAL WEAPONS CENTER (CA34R4)

Comm: 619-939-2303
ATVN: 437-2303

China Lake Naval Weapons Center, CA 93555-6001

LOCATION: From US-395 or US-14, take CA-178 to Ridgecrest and the main gate. RM: p-13, SF/15. NMC: Los Angeles, 150 mi S.

GENERAL INFORMATION: A major Navy Research, Development, and Test Ctr for Naval Weapons Systems principally for air warfare. In the upper Mojave Desert.

TEMPORARY MILITARY LODGING: Biltg Ofc, BOQ A, 24 hrs dly, Comm: 619-939-2383, ATVN: 437-2383. All ranks. VIP, 619-939-1366.

LOGISTICAL SUPPORT: Complete spt fac avail. Key phone numbers follow.

NX-446-2586	Chap-939-3506	Chld C-939-2735
Commiss-939-3281	CPO C-939-3633	EM C-939-2581
Fam Svc-446-5566	Gas S-446-5044	Legal-939-2203
Med-939-2911	Off C-939-3105	Pkg S-446-2203
Pol-939-3323	Qtr Dk-939-2303	Snack B-446-4160
Theat-939-2411		

HEALTH & WELFARE: Outpatient only. USAF Hosp, Edwards AFB, 45 mi S, Comm: 805-277-2330, ATVN: 939-350-2330. Chapels serving all faiths, 619-939-3506.

RECREATION: Camping Equip Rental, EX-2080, Community Rec, EX-2010, Golf, EX-2990, Gym/Swimming, EX-2334, Bowling, EX-3471, Auto Hobby, EX-3107, Ceramics Hobby, EX-2942, Electronics Hobby, EX-2080, Wood-working Hobby, EX-3252, Library, EX-2595, Youth Ctr, EX-2259/2909, Spc Svc Off, EX-3791.

SPACE-A: Armitage Air Field at Center for test flts. Very limited Space-A, Comm: 619-939-2303.

ATTRACTIONS: Desert type attractions and recreation, isolated area.

CONCORD NAVAL WEAPONS STATION (CA58R4)

Comm: 415-671-5066
ATVN: 253-5066

Concord Naval Weapons Station, CA 94520-5000

LOCATION: From I-680 take exit in Concord to CA-24/4, Weapons Station is clearly marked. RM: p-11, NO/8. NMC: San Francisco, 35 mi SW.

Concord NWS, Cont'd

GENERAL INFORMATION: Department of Defense Ammunition Transshipment and Naval Weapons Support for Pacific Fleet.

TEMPORARY MILITARY LODGING: None. Ck listings, San Francisco area for TML.

LOGISTICAL SUPPORT: Limited spt fac avail. Key phone numbers follow.

NX-671-5446	**Chap-671-5749**	**Conv S-671-5446**
Fam Svc-671-5448	**Loc-671-5130/5741**	**Med-671-5864**
Off C-671-5676/5255	**Pkg S-671-5088**	**Pol-671-5130**

HEALTH & WELFARE: Dispensary/Emerg, EX-5864. Inpatient, Oakland Naval Hosp, Comm: 415-633-5441, ATVN: 855-5441. Chapels serving all faiths, EX-5749.

RECREATION: Spec Svc, EX-5336/5448, Tickets, EX-5336, Gym, EX-5735, Auto Hobby, EX-5236, Wood Hobby, EX-5574, Library, EX-5448, Rec Gear Issue, EX-5001, Swimming, EX-5676.

ATTRACTIONS: The Weapons Station is a wild life preserve - deer, elk, eagles, quail, foxes, to name a few. Beautiful San Francisco, 35 mi SW.

CORONADO NAVAL AMPHIBIOUS BASE (CA38R4)

Comm: 619-522-4969
ATVN: 987-4969

Coronado Naval Amphibious Base, San Diego, CA 92155-5000

LOCATION: From San Diego, I-5 S to Palm Ave (CA-75) W 10 mi. Follow sign US Amphibious Base, Coronado. RM: p-15, O/23. NMC: San Diego, 5 mi N.

GENERAL INFORMATION: West Coast Naval Amphibious Training Base, Naval Surface and Marine Corps Landing Forces, and Special Warfare Group.

TEMPORARY MILITARY LODGING: Biltg Ofc, Bldg 500, Tulagi St, 24 hrs dly, Comm: 619-522-4902, ATVN: 987-9902. Off, DV/VIP, Comm: 619-522-4902, ATVN: 987-4902, Enl, Comm: 619-522-4838/4839, ATVN: 987-4838.

LOGISTICAL SUPPORT: No commissary, limited exchange, full fac at NAS North Island, 1 mi NW. Key commercial (619) phone numbers follow.

NX-435-0181	**Chap-522-4819**	**Commiss-522-4882**
Conv S-435-2845	**EM C-522-4983**	**Gas S-437-7029**
Legal-522-4887	**Loc-437-6496**	**McDonald's-435-0074**
Med-437-2375	**Pol-437-6383**	**Theat-437-2925**

HEALTH & WELFARE: Outpatient clinic only. Inpatient, at San Diego Naval Hosp, Comm: 619-233-2417, ATVN: 987-2417. Chapels serving all faiths, EX-2471.

RECREATION: Bowling, Theater, Tennis Courts, Marina, Gym, Swimming, Beaches, and Picnic Area. No Rec Area. Call main number and ask for facility.

SPACE-A: North Island NAS, Bldg 700, 24 hrs, Comm: 619-437-6936/6935, ATVN: 951-6936/6935. Flts to CONUS NAS and USAF Bases.

ATTRACTIONS: San Diego Bay Area, Seaport Village, and San Diego Zoo.

EDWARDS AIR FORCE BASE (CA48R4)

Edwards Air Force Base, CA 93523-5000

Comm: 805-277-1110
ATVN: 527-1110

Edwards AFB, Cont'd

LOCATION: Eighteen mi E of Rosamond, and 30 mi NE of Lancaster, off CA-14. Also, 10 mi SW of Boron, off CA-58. RM: p-13, SI/14. NMC: Los Angeles, 90 mi SW.

GENERAL INFORMATION: Air Force Systems Command Base. Air Force Flight Test Center, USAF Test Pilot School, primary and secondary landing site for space shuttle landings.

TEMPORARY MILITARY LODGING: Biltg Ofc, Bldg 5602, 24 hrs dly, Comm: 805-277-3302, ATVN: 350-3302. All ranks. DV/VIP, EX-3302.

LOGISTICAL SUPPORT: Complete spt fac avail. Key phone numbers follow.

BX-258-6573	**Caf-258-6678**	**Chap-277-2110**
Chld C-277-4310	**Commiss-277-2336**	**Fam Svc-277-3039**
Gas S-258-5037	**Legal-277-4310**	**Loc-277-2777**
Med-277-4427	**NCO-277-3040**	**NCO C-277-3230**
Off C-277-2830	**Pkg S-277-3768**	**Pol-277-3340**
Snack B-277-5987	**Theat-277-4178**	

HEALTH & WELFARE: USAF Hospital, clinics and depts, 25 bed, Bldg 5500, Emer, EX-2331, Appoint, EX-3730. Chapels serving all faiths, EX-2110.

RECREATION: Bowling, EX-2590, Golf, EX-3469, Gym, EX-2374/3077, Hobby Shops, Auto, EX-4275, Ceramics, EX-4171, Lapidary, EX-4262, Photo, EX-4504, Wood work, EX-3139, Rec Supply, EX-2895, Teen Ctr, EX-2217, Library, EX-2375. Fam-Camp on Base, Comm: 805-277-3394, ATVN: 527-3394, year round, 6 camper spaces w/hookups. Also operates Lake Isabella Rec Camp, 102 mi N of Edwards AFB off US-395, year round, Comm: 805-277-2895, ATVN: 527-2895, 7 camper spaces w/hookups, 14 mobile homes w/equip kit. Full fac and lake activities.

SPACE-A: Pax Ops and Lounge, Bldg 1200, 0600-2200 hrs dly. Comm: 805-277-2222/3808, ATVN: 350-2222/3808. Ltd flight opportunities CONUS and overseas.

ATTRACTIONS: Mojave Desert area. Adjacent to mountain and winter sports areas; 90 min from beaches; historic mining and ghost towns nearby.

EL CENTRO NAVAL AIR FACILITY (CA09R4)

Comm: 619-339-2555
ATVN: 958-8555

El Centro Naval Air Facility, CA 92243-5000

LOCATION: Take I-8, 2 mi W of El Centro, to Forrester Rd exit, 1.5 mi to Evan Hewes Hwy left (W) for 4 mi, right on Bennet Rd to main gate. RM: p-13, SP/21. NMC: El Centro, 7 mi E.

GENERAL INFORMATION: Operates bombing and gunnery ranges in support of fleet operations. Ideal flying conditions.

TEMPORARY MILITARY LODGING: Biltg Ofc, Bldg 270, B & 2nd Sts, 24 hrs dly, Comm: 619-339-2535, ATVN: 958-8535. All ranks. DV/VIP - same as Biltg. Four 2-bdrm mobile homes, thru Navy Exchange, 339-2478.

LOGISTICAL SUPPORT: Complete spt fac avail. Key phone numbers follow.

NX-339-2478	**Caf-339-2472**	**Chap-339-2457**
Chld C-399-2560	**Commiss-339-2483**	**Conv S-339-2670**
EM C-339-2570	**Gas S-339-2670**	**Legal-339-2519**
Loc-339-2555	**Med-339-2675/2666**	**Off C-399-2491**
Pkg S-339-2640	**Pol-339-2555**	**Snack B-339-2570**
Theat-339-2662		

El Centro NAF, Cont'd

HEALTH & WELFARE: Outpatient care only, EX-2675/2666. Inpatient, San Diego Naval Hosp, Comm: 619-233-2417, ATVN: 957-2417. Chapels serving all faiths, EX-2457.

RECREATION: Rec Svc, EX-2481, for all rec fac, 10 self-contained rec vehicles available for rent, EX-2481.

SPACE-A: Limited to no Space-A, Comm: 619-339-2555.

ATTRACTIONS: Winter home of the Navy Blue Angels (Jan-Mar). Imperial Valley, near Mexico.

EL TORO MARINE CORPS AIR STATION (CA22R4)

Comm: 714-651-3736
ATVN: 997-3736

El Toro Marine Corps Air Station, CA 92709-5000

LOCATION: Off I-5, take the Sand Canyon Rd exit. Follow signs to the MCAS. RM: p-15, L/15. NMC: Los Angeles, 40 mi NW.

GENERAL INFORMATION: Marine Corps Air Bases, Western Area, Hq 3rd Marine Aircraft Wing, and supporting units.

TEMPORARY MILITARY LODGING: Biltg Ofc, Bldg 276, N 6th St & 8th St, 0730-2130 hrs dly. Comm: 714-651-2484, ATVN: 997-2484. All ranks. DV/VIP, EX-3814.

LOGISTICAL SUPPORT: Complete spt fac avail. Key phone numbers follow.

MX-651-3340	**Caf-651-3340**	**Chap-651-3824**
Chld C-651-2487	**Commiss-651-3747**	**Conv S-651-3340**
EM C-651-2477	**Fam Svc-651-2771**	**Gas S-651-3340**
Legal-651-3609	**Loc-651-3736**	**Med-651-2731**
NCO-651-3901	**Off C-651-2464**	**Pkg S-651-2491**
Pol-651-3527	**Snack B-651-3340**	**SNCO C-651-2566**

HEALTH & WELFARE: Dispensary/Emerg, EX-3172, Appoint, EX-2731. Inpatient, Long Beach Naval Hosp, Comm: 213-420-5201. Chapels serving all faiths, EX-3824/3825.

RECREATION: Aero Club, EX-2480/2463, Library, EX-2569, Auto Hobby Shop, EX-2481, Stables, 651-2911, Gym, EX-3714, Bowling, EX-2567, Golf, EX-2577, Swimming, EX-2469. Operates Big Bear Lake Rec Fac, Spec Svc, Ticket Ofc, Comm: 714-651-2626, ATVN: 997-2626, Rec Area, Comm: 714-866-3965, 120 mi E, CA-330 to lake, 8 cottages, 5 camper sites w/hookups, 5 camper sites w/o hookups, full rec fac and all activities.

SPACE-A: Pax Term and Lounge, Bldg 624, 0600-2200 hrs M-F, Sat-Sun, 0600-1800, Comm: 714-651-3920/3936, ATVN: 997-3920/3936, CONUS and overseas flts.

ATTRACTIONS: Orange County and greater Los Angeles area, Disneyland in Anaheim, Anaheim Stadium (home of the California Angels baseball team and Los Angeles Rams football team), Knott's Berry Farm in Buena Park, Marineland located at the tip of the Palo Verdes peninsula, Movieland Wax Museum in Buena Park, and Crystal Cathedral in Garden Grove.

FORT HUNTER LIGGETT (CA37R4)

Comm: 408-385-2544
ATVN: 949-2544

Fort Hunter Liggett, CA 93928-5000

Fort Hunter Liggett, Cont'd

LOCATION: From US-101 S exit at King City to CA-G-14, S to main gate. RM: p-13, SE/5. NMC: San Luis Obispo, 60 mi S.

GENERAL INFORMATION: Sub-Installation of Fort Ord. Another primary training area for 7th Infantry Division (LIGHT), 40th Infantry Division (MECH) CALGD, and testing area for USACEDEC.

TEMPORARY MILITARY LODGING: Biltg Ofc, Bldg T-105, 0800-1630 hrs dly, Comm: 408-385-2511, ATVN: 949-2511. All ranks. DV/VIP, EX-2511/2558.

LOGISTICAL SUPPORT: Limited spt fac avail. Key phone numbers follow.

PX-385-2537	**Gas S-385-2080**	**Med-385-2516/2570**

HEALTH & WELFARE: Outpatient care only, Emerg, EX-2516. Inpatient, Silas B. Hays Army Community Hosp, Fort Ord, Monterey, Comm: 408-242-2020, ATVN: 929-2020. Chapels serving all faiths, EX-2465.

RECREATION: Bowling/Golf/Gym/Craft Shops/Rec Ctr, EX-2612. No rec area.

SPACE-A: None. Try, Lemoore NAS, Bldg 180, Comm: 209-998-3283, ATVN: 949-3183, flts to CONUS NAS's and other locations.

ATTRACTIONS: Famous for yearly return of swallows - like Capistrano. Old CA, hacienda formerly owned by the Hearst family. San Antonio Mission.

FORT IRWIN NATIONAL TRAINING CENTER (CA01R4)

Comm: 619-386-4111
ATVN: 470-4111

Fort Irwin National Training Center, CA 92310-5000

LOCATION: Take I-15 E from Los Angeles for 125 mi or I-15 W from Las Vegas, NV for 150 mi. Fort is N of I-15 near Barstow. Watch for signs. RM: p-13, SH/18. NMC: San Bernardino, 60 mi SW.

GENERAL INFORMATION: The Army National Training Center.

TEMPORARY MILITARY LODGING: Biltg Ofc, Bldg 212, 0800-1630 hrs dly, Comm: 619-386-3222, ATVN: 470-3222. All ranks. DV/VIP, EX-3000. Very limited (new guest house available next year).

LOGISTICAL SUPPORT: Complete spt fac avail. Key phone numbers follow.

PX-386-3427	**ACS-386-3513**	**Caf-386-2331**
Chap-386-3562	**Chld C-386-1029**	**Commiss-386-3423**
Conv S-386-1523	**EM C-386-3491**	**Gas S-386-1088**
Legal-386-3257	**Loc-386-3369**	**Med-386-3128**
NCO C-386-3206	**Off C-386-3018**	**Pkg S-386-3208**
Pol-386-3474	**SDO-386-3530/3600**	**Theat-386-3430**

HEALTH & WELFARE: Weed Army Community Hospital, clinics & depts, 25 bed, Emer, EX-3115/3128/3116, Appoint, EX-3124. Chapels serving all faiths, EX-3562/3440.

RECREATION: Bowling, EX-4249, Swimming, EX-4951-3046, Gym, EX-3457, Riding Stables, EX-3348, ITT Ofce, EX-3583, Auto/Photo Craft, EX-3531, Library, EX-3462, Arts/Crafts, EX-3431/3432, Outdoor Equip, EX-3434, Mini-Golf Course/ Batting, EX-4653, Youth Act, EX-3696, Rec Center, EX-3585, Teen Center, EX-3696, Pre-teen Center, EX-3732, Photo Craft, EX-4054, Morale Support Hq, EX-3582. No Rec Areas.

Fort Irwin NTC, Cont'd

SPACE-A: None on post. Try George AFB, Bldg 707, 0700-2400 hrs dly, Comm: 619-269-2700, ATVN: 353-2700, flts to CONUS AFBs.

ATTRACTIONS: Goldstone Deep Space Station. Desert and isolated area.

FORT ORD (CA36R4)

Fort Ord, CA 93941-5000

Comm: 408-242-4200
ATVN: 929-4200

LOCATION: From San Francisco, S for 100 mi on US-101, right onto CA-156 for 10 mi to main gate of Post. RM: p-13, SC/11. NMC: Monterey, 7 mi S.

GENERAL INFORMATION: Hq, 7th Infantry Division (Light) and Combat Development and Experimentation Center.

TEMPORARY MILITARY LODGING: Biltg Ofc, Martinez Hall, Bldg 2798, 2nd Ave & 12th St, 0800-1600 hrs dly, Comm: 408-242-3242/4307, ATVN: 929-3242/4307. All ranks. DV/VIP, EX-5088/3492.

LOGISTICAL SUPPORT: Complete spt fac avail. Key phone numbers follow.

PX-899-2336	**Caf-899-1790**	**Chap-242-4873**
Chld C-899-4791	**Commiss-899-3663**	**Conv S-899-1980/3550**
EM C-242-2563	**Fam Svc-242-6687/5713**	**3 Gas S-394-8219/ 2443/7011**
Legal-242-7861	**Loc-242-2271**	
Med-242-2020	**NCO C-899-2373**	**Off C-899-3445**
2 Pkg S-899-1833/1870	**Pol-242-7851**	**Presid/Monterey C- 649-6677**
SDO-242-3432/4209	**4 Seasons-899-2339**	
Snack B-899-1766	**Theat Info-242-5566**	

HEALTH & WELFARE: Silas B. Hays Army Community Hospital, clinics & depts, 450 bed, Emerg, EX-7631/2020, Appoint, EX-7681. Chapels serving all faiths, EX-3700/4873.

RECREATION: Sports Arena, EX-4077, Gym, EX-5283, Golf, EX-3268, Stilwell Rec Ctr, EX-5283, Swimming Pool, EX-3426, Arts & Crafts, EX-3707, Multi-Crafts Ctr, EX-3584, Photo Craft, EX-2496, Auto Craft, EX-4124, Ceramics, EX-3983, Wood, EX-4366, Outdoor Rec, EX-7466, Youth Act, EX-6303, Fort Ord Youth Ctr, EX-3265, Libraries, EX-3421/3420, ITT, EX-3092. East Garrison Travel Camp, on Post, year round, Comm: 408-242-3486, ATVN: 929-3486, 38 camper spaces w/hookups, 27 camper spaces w/o hookups, comp fac and activities.

SPACE-A: Try Oakland IAP, Comm: 415-635-8482, ATVN: 837-5789.

ATTRACTIONS: Monterey, Pebble Beach, Carmel nearby. Great area.

GEORGE AIR FORCE BASE (CA49R4)

George Air Force Base, CA 92394-5000

Comm: 619-269-1110
ATVN: 353-1110

LOCATION: From I-15 take D St exit and follow signs. From US-395 at Adelanto exit E on Air Base Rd, 2 mi on left. RM: p-13, SJ/16. NMC: Barstow, 30 mi NE.

GENERAL INFORMATION: Tactical Air Command Base. Fighter crew training including German Air Force. Home of the F-4 Wild Weasel Mission.

TEMPORARY MILITARY LODGING: Biltg Ofc, Bldg 454, 2nd St, 24 hrs dly, Comm: 619-269-2086, ATVN: 353-2514. All ranks. DV/VIP, EX-2620/2086.

George AFB, Cont'd

LOGISTICAL SUPPORT: Complete spt fac avail. Key phone numbers follow.

BX-246-5433	Caf-246-7662	Chap-269-2630
Chld C-269-3056	Commiss-246-8686	Conv S-246-6125
Fam Svc-269-3375	Gas S-246-5828	Legal-269-3630
Loc-269-3208	Med-269-3069	NCO-269-2800
NCO C-246-2506	Off C-269-2501	Pkg S-246-4141
Pol-269-3510	Theat-246-2888	

HEALTH & WELFARE: USAF Hospital, clinics and depts, 35 bed, Emerg, EX-3069, Appoint, EX-3830. Chapels serving all faiths, EX-2630.

RECREATION: Arts & Crafts, EX-2552, Auto Hobby, EX-3116, Golf, EX-2613, Gym, EX-3550, Library, EX-3228, Skeet Range, Swimming, EX-3255, Off C, 269-2501, NCO C, 246-2506, Wood Hobby, EX-2552, Youth Act/Ctr, 269-2254, Rec Dir, EX-2916, Rec Supply, 269-3775. Fam-Camp on Base, 7 spaces w/hookups. Also operates Lake Isabella Rec Area, off Base, take CA-179 to lake, year round, 7 mobile homes. Complete fac and full activities on lake. For both, Comm: 619-269-3233/3480, ATVN: 353-3233/3480.

SPACE-A: Pax Term and Lounge, Bldg 707, Comm: 619-269-2700, ATVN: 353-2700, limited flts to CONUS and overseas (rarely).

ATTRACTIONS: Mountain Recreational Playground. Snow skiing and other mountain sports.

LEMOORE NAVAL AIR STATION (CA06R4)

Lemoore Naval Air Station, CA 93245-5001

Comm: 209-998-2211
ATVN: 949-2211

LOCATION: On CA-198, 24 mi E of I-5, 30 mi W of CA-99 in the south central part of the state. RM: p-13, SD/9. NMC: Fresno, 40 mi NNE.

GENERAL INFORMATION: Tactical Naval Aircraft Base. Carrier Air Wing, and other Tac Air Units.

TEMPORARY MILITARY LODGING: Biltg Ofc, Bks 6, Bldg 843, Hancock Circle, 24 hrs dly, Comm: 209-998-3266, ATVN: 949-3266. All ranks and Navy Lodge. DV/VIP, EX-3121.

LOGISTICAL SUPPORT: Complete spt fac avail. Key phone numbers follow.

NX-998-3082	Caf-998-3084	Chap-998-3496
Chld C-998-3044	Commiss-998-3467	Conv S-998-3681
CPO C-998-3130	EM C-998-3331	Fam Svc-998-3456
Gas S-998-3088	Legal-998-3251	Loc-998-3059
Med-998-4256	Off C-998-3550	Pkg S-998-3292
Pol-998-3362	Snack B-998-3479	Theat-998-3235

HEALTH & WELFARE: Naval Hospital, clinics and depts, 20 bed, Emerg, EX-4256/4257/3131, Appoint, EX-6801/6831. Chapels serving all faiths, EX-3360/3496.

RECREATION: Gym, EX-3450, Arts/Crafts, EX-3137, Auto Hobby, EX-3217, Electronics Hobby, EX-3217, Flying Club, EX-3081, Bowling, EX-3502, Gear Locker, EX-3075, Library, EX-3144, Rec Ctr, EX-3163, Skeet & Trap, EX-2426, Youth Ctr, EX-3175.

SPACE-A: Pax Term and Lounge, Bldg 180, Comm: 209-998-3183, ATVN: 949-3183, flts to CONUS NAS's and other locations.

ATTRACTIONS: Sequoia National Park, 70 mi E of NASL. Ski areas nearby.

LOS ANGELES AREA

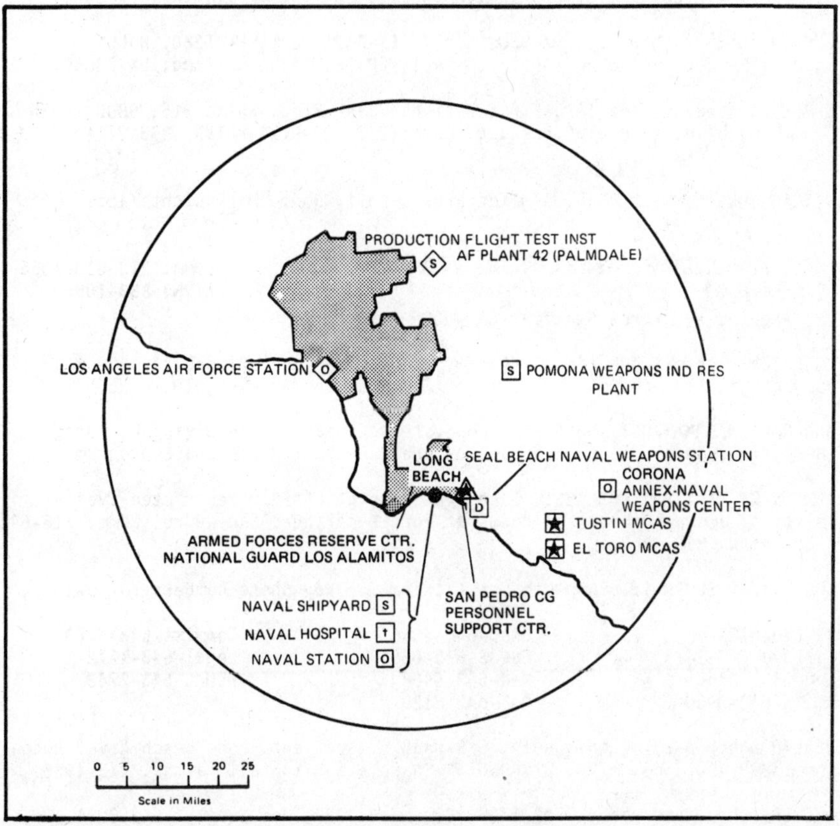

LONG BEACH NAVAL STATION (CA55R4)

Long Beach Naval Station, CA 90822-5000

Comm: 213-547-6202
ATVN: 360-0111

LOCATION: Take Long Beach Freeway, CA-7, to Terminal Island exit to Naval Station. Clearly marked. RM: p-15, L/9. NMC: Long Beach, 2 mi E.

GENERAL INFORMATION: Home port for Pacific fleet ships, Destroyer Squadron, and Naval Regional Dental Center.

TEMPORARY MILITARY LODGING: Try Bltg, EM-360-7924/7928, Off-360-6060.

LOGISTICAL SUPPORT: Most spt fac avail. Key phone numbers follow.

NX-519-5100	**Caf**-547-6295	**Chap**-547-6981
Chld C-547-6295	**Commiss**-547-7281	**Conv S**-519-5100
EM C-547-7007	**Fam Svc**-547-6805	**Gas S**-519-5220
Legal-547-6031	**Loc**-547-6202	**Med**-420-5201
NCO C-547-8206	**Off C**-547-7209	**Pkg S**-547-6087
Pol-547-7731	**Snack B**-519-5100	**Theat**-547-6351

Long Beach NS, Cont'd

HEALTH & WELFARE: Long Beach Naval Hospital, 7500 E Carson St, Emerg, Comm: 213-420-4741/5201, ATVN: 873-9201. Chapels serving all faiths, EX-6981.

RECREATION: Auto Hobby, EX-8302, Golf, EX-6311, Gym, EX-6380, Hobby Shops, EX-7284, Swimming - Off, EX-7401, EM, EX-7400, Sailing, EX-7285.

SPACE-A: Los Angeles IAP, First National Bank Bldg, Suite 316, 9800 Sepulveda Blvd, 0730-2100 hrs dly, Comm: 213-643-0714, ATVN: 833-0714, flts to CONUS and overseas.

ATTRACTIONS: Greater Los Angeles area & Hollywood. Many sights/act.

LOS ANGELES AIR FORCE STATION (CA75R4)

Comm: 213-643-1056
ATVN: 833-1056

Los Angeles Air Force Station, CA 90009-2960

LOCATION: From I-405 (San Diego Fwy) take El Segundo Blvd exit, 200 N Douglas St, El Segundo. RM: p-15, I/7. NMC: Los Angeles, 10 mi NE.

GENERAL INFORMATION: Hq Air Force Systems Command Space Division, manages the design, development, acquisition and launch of DOD's space program.

TEMPORARY MILITARY LODGING: None available at LAAFS. Restricted availability of a small number of rooms at Fort MacArthur, San Pedro, Comm: 213-643-0260, ATVN: 833-0260.

LOGISTICAL SUPPORT: Complete spt fac avail. Key phone numbers follow.

BX-640-0129	**Chap-643-1956**	**Commiss-643-2768**
Fam Svc-643-1475	**Gas S-615-0295**	**Legal-643-1272**
Loc-643-1876	**Med-643-0468**	**NCO C-643-2743**
Off C-643-0960	**Pol-643-2123**	

HEALTH & WELFARE: Clinic only, EX3-0468. Inpatient, Long Beach Naval Hosp, Comm: 213-420-5503, ATVN: 873-9503. Chapels serving all faiths, EX3-1956.

RECREATION: Auto Hobby, EX3-1705, Gym, EX3-0759, Rec Supply, EX3-2081. Discount tickets for some local attractions from SATO office, EX3-1064.

SPACE-A: Administrative headquarters: First Nationwide Bank Building, Suite 316, 9800 Sepulveda Blvd, Los Angeles, 1.5 mi S from Los Angeles IAP, 0800-1700 hrs dly, Comm: 213-643-0715/6, ATVN: 833-0715/6. Pax Term in Tom Bradley Terminal at LAX. Comm: 213-643-1296, 0730-2230. Ltd number of flights to CONUS base and overseas via Air Force contract commercial aircraft.

ATTRACTIONS: Greater Los Angeles area, Hollywood, TV shows, beaches, unltd sights and events.

MARCH AIR FORCE BASE (CA08R4)

Comm: 714-655-1110
ATVN: 947-1110

March Air Force Base, CA 92518-5000

LOCATION: Off CA-60 & on I-215 which bisects AFB. RM: p-15, I/23. NMC: Riverside, 7 mi NW.

GENERAL INFORMATION: Strategic Air Command Base. Hq 15th Air Force and Hq 26th Air Division. Air Refueling Wing, ANG and AFRES units.

TEMPORARY MILITARY LODGING: Biltg Ofc, Bldg 100, Myers & O Sts, 24 hrs dly, Comm: 714-655-5241, ATVN: 947-5241. All ranks. DV/VIP, EX-4764.

March AFB, Cont'd

LOGISTICAL SUPPORT: Complete spt fac avail. Key phone numbers follow.

BX-653-3111	**Caf-655-3656**	**Chap-655-4105**
Chld C-655-2521	**Commiss-655-4311**	**Conv S-653-0711**
EM C-653-1153	**Fam Svc-655-4795**	**Gas S-653-8118**
Legal-655-4455	**Loc-655-3192**	**Med-655-5381**
NCO C-653-1153	**Off C-653-2121**	**Pol-655-2981**
Theat-655-2060	**SDO-655-3192**	

HEALTH & WELFARE: USAF Hospital, clinics and depts, 125 bed, Emer, EX-4266, Appoint, EX-5381. Chapels serving all faiths, EX-4105.

RECREATION: Auto Hobby, EX-2553, Library, EX-2203, Rec Ctr, EX-4123, Youth Ctr, EX-2803, Bowling, EX-2257, Golf, EX-653-2693, Physical Cond, EX-3017, Gym, EX-2292/3213, Swimming, EX-3691.

SPACE-A: Pax Term and Lounge, Graeber St near ATC tower, M-F 0730-1630, Comm: 714-655-2913, ATVN: 947-2913, flts to CONUS bases plus Hawaii and Guam via KC-10/135 aircraft.

ATTRACTIONS: Greater Los Angeles area & near Palm Springs resort.

MARE ISLAND NAVAL COMPLEX (CA56R4)

Comm: 707-646-1111
Mare Island Naval Complex, CA 94592-5000
ATVN: 253-0111

LOCATION: Exit I-80 N or S at Tennessee St, or CA-37 and follow signs to Mare Island Naval Station. RM: p-11, NO/7. NMC: San Francisco, 35 mi SW.

GENERAL INFORMATION: Naval Station Mare Island, Mare Island Naval Shipyard, Combat Systems Technical Schools Command, and more than 20 tenant activities and commands are all part of the Mare Island Naval Complex. Supports Naval units in the San Francisco Bay Area.

TEMPORARY MILITARY LODGING: Try Enlisted, 646-4174, Off, 646-2324, Naval Sta Mare Island. Alameda NAS, Navy Lodge, Comm: 415-523-4917, Oakland Naval Supply Ctr, Navy Lodge, Comm: 415-466-6187. Check other SF area base listings.

LOGISTICAL SUPPORT: Complete spt fac avail. Key phone numbers follow.

NX-554-5815	**Caf-554-5868**	**Chap-646-4251**
Chld C-642-2373	**Commiss-642-8942**	**Conv S-552-4544**
CPO C-646-4232	**EM C-642-4243**	**Fam Svc-646-2527**
Gas S-554-5863	**Legal-646-2181**	**Med-646-2391**
Off C-644-2339	**Pkg S-642-0134**	**Pol-646-2491**
Ships' Info-646-2355	**Theat-646-3301**	

HEALTH & WELFARE: Clinic only, Emer, 646-4444. Inpatient, Oakland Naval Hosp, Comm: 415-633-5441, ATVN: 855-5441. Chapels serving all faiths, 646-4251.

RECREATON: Athletic Gear, 646-2760, Bowling, 646-3219, Camping Gear, 646-3486, Picnic, 646-4289, Fishing Boats, 646-4345, Golf, 644-3888, Trailer Rentals, 646-3486, Saddle Club, 646-3436, Model Shop, 646-4345, Swimming, 646-2760, Tennis, 646-3436, Gym, 646-2760, Hobby Shop, 646-4345, Library, 646-3338, Racquetball, 646-2760, Rifle Range, 646-2411, Youth Act, 646-3301/3436.

SPACE-A: Oakland IAP, main lobby, 24 hrs dly, Comm: 415-635-8482, ATVN: 837-5789, CONUS & overseas flts. Travis AFB, Comm: 707-438-2311, ATVN: 837-2311.

ATTRACTIONS: Near Napa Valley wine growing region, San Francisco Bay Area.

MATHER AIR FORCE BASE (CA12R4)

Mather Air Force Base, CA 95655-5000

Comm: 916-364-1110
ATVN: 828-1110

LOCATION: In Rancho Cordova, US-50 to Mather Field Rd direct to main gate. RM: p-11, NM/10. NMC: Sacramento, 12 mi W.

GENERAL INFORMATION: Air Training Command Base. DOD exclusive manager for navigator training, SAC and AFRES units.

TEMPORARY MILITARY LODGING: Biltg Ofc, Bldg 2750, "A" Av at Gilbert St, 24 hrs dly, Comm: 916-364-2457, ATVN: 828-2932/2457. All ranks. DV/VIP, EX-2460.

LOGISTICAL SUPPORT: Complete spt fac avail. Key phone numbers follow.

BX-363-7363	Caf-362-4796	Chap-364-4611
Chld C-364-4253	Commiss-364-3471	Fam Svc-364-4382
Gas S-363-7389	Legal-364-4391	Loc-364-2597
Med-364-4692	NCO-364-2957	NCO C-362-8030
Off C-362-3257	Pkg S-363-6598	Pol-364-2200
Snack B-362-4391	Theat-364-4504	

HEALTH & WELFARE: USAF Hospital, clinics and depts, 75 bed, Emerg, EX-3212, Appoint, EX-4411. Chapels serving all faiths, EX-4611.

RECREATION: Auto Hobby, EX-3536, Bowling, EX-4776, Ceramics/Crafts, EX-4837, Rec Ctr, EX-4945, Golf, EX-3100, Gym, EX-4786, Hobby Sales Store, EX-4648, Library, EX-4759, Rec Supply, EX-2623, Swimming - Off/EX-2462 and Airmen/EX-2702, Housing Area EX-2700, Youth Ctr, EX-3113.

SPACE-A: Pax Term and Lounge, Bldg 4579, Comm: 916-364-2579/2510, ATVN: 828-2579/2510, CONUS and overseas flts via KC-10/135 aircraft.

ATTRACTIONS: Silver Wings Museum, Planetarium (on Base), Old Sacramento, Sutters Fort, Sacramento Railroad Museum, Folsom Lake.

McCLELLAN AIR FORCE BASE (CA35R4)

McClellan Air Force Base, CA 95652-5000

Comm: 916-643-2111
ATVN: 633-1110

LOCATION: Off I-80 N. From I-80 take Madison Ave exit. Clearly marked. RM: p-11, NK/25. NMC: Sacramento, 10 mi SW.

GENERAL INFORMATION: Air Force Logistics Command Base. Hq Sacramento Air Logistics Center. US Coast Guard Station McClellan. Also MAC, TAC AFCC, and AFRES units.

TEMPORARY MILITARY LODGING: Biltg Ofc, Bldg 89, Palm & 30th Sts, 24 hrs dly, Comm: 916-643-6223, ATVN: 633-6223. All ranks.

LOGISTICAL SUPPORT: Complete spt fac avail. Key phone numbers follow.

BX-920-0537	Caf-929-9295	Chap-643-6021
Chld C-643-3611	Commiss-925-8541	Conv S-332-3872
EM C-643-2259	Fam Svc-643-3815	Gas S-922-8868
Legal-643-3150	Loc-643-4113	Med-643-4733
NCO C-922-9657	Off C-927-5013	Pkg S-920-4009
Pol-643-6160	Snack B-920-9394	Theat-922-8706

HEALTH & WELFARE: Dispensary/Clinics/Emerg, 643-6736 Appoint, EX-4733. Inpatient, USAF Hospital, Mather AFB, Comm: 916-364-3213, ATVN: 828-3213. Chapels serving all faiths, EX-6021/6023.

RECREATION: Gym, 643-2596, Bowling, EX-5752, Hobby Shop, EX-3004, Golf,

McClellan AFB, Cont'd

EX-3313, Library, EX-4640, Rec News, EX-5311, Rec Supply, 643-2738, Service Club, EX-2259, Youth Ctr, EX-2074, Picnic Area, 643-3596, Swimming, Rec Center, 643-3304.

SPACE-A: Pax Term, Bldg 251, 0600-1615 hrs dly, Comm: 916-643-3944, ATVN: 633-3944, flts to CONUS and overseas. Also, try the Coast Guard Air Station Ops at the AFB.

ATTRACTIONS: Lake Tahoe and Reno, NV nearby, 2 1/2 hr drive E. San Francisco Bay Area, 2 hrs drive W.

MIRAMAR NAVAL AIR STATION (CA14R4)

Miramar Naval Air Station, CA 92145-5000

Comm: 619-271-3011
ATVN: 959-3011

LOCATION: Fifteen mi N of San Diego, off I-15. Take Miramar Way exit. RM: p-15, K/23. NMC: San Diego, 15 mi SW.

GENERAL INFORMATION: Hq Fighter Early Warning Wing. Major Navy Fighter/ Early Warning Aircraft Base.

TEMPORARY MILITARY LODGING: Biltg Ofc, Bldg 516, 24 hrs dly, Comm: 619-271-2142, ATVN: 959-2142. All ranks. DV/VIP, EX-2142.

LOGISTICAL SUPPORT: Complete spt fac avail. Key phone numbers follow.

NX-695-7200	**Caf-695-7278**	**Chap-271-3411**
Chld C-271-2280	**Commiss-696-6114**	**Conv S-695-7293**
EM C-271-2270	**Fam Svc-271-2746**	**Gas S-695-7253**
Legal-271-3601	**Loc-271-3500**	**Med-271-3687**
NCO C-271-2270	**Off C-271-2171**	**Pkg S-271-3535**
Pol-271-2451/2/3	**Snack B-695-7275/8**	**Theat-271-3619**

HEALTH & WELFARE: Medical/Dental Clinic, Emerg, EX-2233, Appoint, EX-2711/27712. Inpatient, San Diego Naval Hosp, Comm: 619-233-2417, ATVN: 957-2417. Chapels serving all faiths, EX-3411.

RECREATION: Tickets, EX-3567, Rec Svc, EX-3526, Gym, EX-3558, Arts/Crafts/ Woods, EX-3615/2530, Bowling, EX-3548, Golf, EX-3431, Swimming, EX-3538/2461, Picnic (Mills Park), EX-3526/7, Amateur Radio, EX-3620, Youth Ctr, EX-3397, Auto Hobby, EX-3553, Camp Gear/Equipment Rental, EX-3600, Library, EX-3557. Rec Vehicle Park, on Base, yr round, Comm: 619-271-3600, ATVN: 959-3600, 18 camper spaces w/hookups, 6 camper spaces w/o hookups, full fac & all act.

SPACE-A: Pax Term, Lounge, Bldg 476, 0700-2400 hrs dly, Comm: 619-271-2330, ATVN: 959-2330. Flts to CONUS, overseas NAS's, and other bases.

ATTRACTIONS: Old Town San Diego, Sea World (clean, great show), San Diego Zoo, Wild Animal Park, beautiful beaches and boardwalk, Mission Blvd.

MOFFETT FIELD NAVAL AIR STATION (CA15R4)

Comm: 415-966-5411
ATVN: 462-5411

Moffett Field Naval Air Station, CA 94035-5000

LOCATION: On Bayshore Freeway, US-101, 35 mi S of San Francisco. RM: p-11, NK/25. NMC: San Jose, 7 mi S.

GENERAL INFORMATION: Patrol Wing, Pacific Fleet Units, 129th ARRS CANG, and Nav Air Res Units.

Moffett Field NAS, Cont'd

TEMPORARY MILITARY LODGING: Biltg Ofc, Bldg 555, 0800-1600 hrs dly, Comm: 415-966-5158, ATVN: 462-5158. New Navy Lodge 1-800 NAVY INN. All ranks. DV/VIP, EX-5411.

LOGISTICAL SUPPORT: Complete spt fac avail. Key phone numbers follow.

NX-940-6200	**Caf-940-6200**	**Chap-966-5020**
Chld C-966-5604	**Commiss-969-5173**	**Conv S-940-6200**
CPO C-966-5471	**EM C-966-5471**	**Fam Svc-966-4911**
Gas S-940-6262	**Legal-966-5006**	**Loc-966-5061**
Med-966-5111	**Off C-966-5306**	**Pkg S-940-6200**
Pol-966-5141	**Snack B-940-6200**	**Theat-966-5464**

HEALTH & WELFARE: Clinic/Emerg, EX-5111, Appoint, EX-5406. Inpatient, Oakland Naval Hosp, Comm: 415-633-5441, ATVN: 855-5441. Chapels serving all faiths, EX-5326/5020.

RECREATION: Auto Hobby, EX-5385, Bowling, EX-5793, Camping Gear, EX-5074, Picnic, EX-5206, Crafts Hobby, EX-5159, Handball, EX-5363, Golf, EX-5332, Gym, EX-5206, Library, EX-5455, Pistol Range, EX-5086, Rec Svc Ofc, EX-5619, Stables, EX-6124, Swimming, EX-5133, Ticket Ofc, EX-5464, Youth Act, EX-4980.

SPACE-A: No central flt info service. Comm: 415-966-5233/5231, ATVN: 462-5233/5231, for info on flts by each unit on Base.

ATTRACTIONS: Located in the heart of the "Silicon Valley", San Jose and Santa Cruz are nearby. San Francisco is 35 miles North.

MONTEREY NAVAL POSTGRADUATE SCHOOL (CA16R4)

Comm: 408-646-0111
ATVN: 878-0111

Monterey Naval Postgraduate School, CA 93943-5000

LOCATION: Take CA-1 to central Monterey exit, right at light onto Camino Aguajito. Right at stop sign onto 3rd St and to Base entrance. Or N on CA-1, Aguajito Rd exit to Mark Thomas Dr, to left on Sloat Ave, right at 3d St gate. RM: p-13, SB/22. NMC: Monterey, in city limits.

GENERAL INFORMATION: Naval Postgraduate School, Research & Defense Resource Management Education Center.

TEMPORARY MILITARY LODGING: Biltg Ofc, Bldg 220, Herrmann Hall, Middle Rd, 24 hrs dly, Comm: 408-646-2060/2069, ATVN: 878-2060/2069. Officer only. DV/VIP, EX-2513.

LOGISTICAL SUPPORT: Most spt fac avail. Key phone numbers follow.

NX-373-0557/7252	**Chap-646-2241**	**Chld C-646-2734**
Conv S-375-0959	**EM C-646-2358**	**Fam Svc-2441**
Gas S-373-7271	**Legal-646-2506**	**Loc-646-2441**
Med-647/5664/5663	**NCO-646-2441/878-2441**	**Off C-646-2170**
Pkg S-373-7511	**Pol-646-2555**	**Snack B-375-5933**

HEALTH & WELFARE: Clinic/Emerg, EX-242-8614, Appoint, 647-5664. Inpatient, Silas B. Hays Army Community Hosp, Fort Ord, Monterey, Comm: 408-242-7631/2020, ATVN: 929-7631/2063. Chapels serving Prot/Cath, EX-2242/2241.

RECREATION: Rec Ofc, EX-2466/2467, Golf, EX-2167, Gym & Gear Issue, EX-2167, Swimming, EX-2275.

Monterey NPG Sch, Cont'd

SPACE-A: Try NAS Lemoore, Bldg 180, Comm: 209-998-3183, ATVN: 949-3183.

ATTRACTIONS: Beaches, seafood, Carmel Valley, Carmel-By-The-Sea, & Salinas Valley. Beautiful area.

NORTH ISLAND NAVAL AIR STATION (CA43R4)

Comm: 619-437-6011
ATVN: 951-0111

North Island Naval Air Station, CA 92135-5000

LOCATION: From I-5 N or S exit at Coronado Bridge (toll). Also, from CA-75 N to CA-282 to Base. In Coronado. RM: p-15, N/22. NMC: San Diego, 4 mi NE.

GENERAL INFORMATION: Largest Naval Industrial Complex on the West Coast.

TEMPORARY MILITARY LODGING: Biltg Ofc, Bldg 1402, Navy Lodge, 24 hrs dly, Comm: 619-435-0191, ATVN: 951-0191. All ranks.

LOGISTICAL SUPPORT: Complete spt fac avail. Key phone numbers follow.

NX-522-7215	**Chap-437-7269**	**Chld C-437-5769**
Commiss-437-6997	**Conv S-522-7277**	**CPO C-437-7372**
EM C-437-5659	**Fam Svc-437-6693**	**Gas S-435-7029**
Legal-437-7621	**Loc-437-7151**	**McDonalds 435-0074**
Med-437-6263	**Off C-437-7651**	**Pkg S-435-2285**
Pol-437-6383	**Theat-435-7657**	

HEALTH & WELFARE: Dispensary/Emer, EX-6278, Appoint, EX-6273/74/75. Inpatient, San Diego Naval Hosp, Emer, Comm: 619-233-2417, ATVN: 957-2417. Chapels serving all faiths except Jewish, EX-7269.

RECREATION: MWR, EX-7642, Golf, EX-7577, Racquet Sports, EX-7634, Ball Fields, EX-7634, Gym, EX-7634, Bowling, EX-7585, EM Rec Ctr, 435-9344, Rec Social Ctr, EX-7584, Skeet, EX-5521, Tickets, EX-7659, Hobby Shops, EX-6897, Library, EX-7041.

SPACE-A: Pax Term and Lounge, Bldg 700, Comm: 619-437-6936/6489, ATVN: 951-6936/6489. Flts to CONUS NAS's and other installations. Some overseas.

ATTRACTIONS: Beaches, sportfishing, Old San Diego, Sea World, Bay Area, and Mexico nearby.

NORTON AIR FORCE BASE (CA17R4)

Comm: 714-382-1110
ATVN: 876-1110

Norton Air Force Base, CA 92409-5000

LOCATION: From I-10 in Loma Linda take Tippecanoe exit N to Base, approx 1.5 mi. RM: p-15, E/24. NMC: San Bernardino, 3 mi NW.

GENERAL INFORMATION: Military Airlift Command Base, Ballistic Missile Office (MX-Program Manager), AFRES units, Aerospace Audiovisual Agency, AF Inspection and Safety Ctr, and AF Audit Agency.

TEMPORARY MILITARY LODGING: Biltg Ofc, Bldg 512, 3rd St, 24 hrs dly, Comm: 714-382-5531, ATVN: 876-5531. All ranks. DV/VIP, EX-7617.

LOGISTICAL SUPPORT: Complete spt fac avail. Key phone numbers follow.

BX-884-4807	**Caf-885-4741**	**Chap-382-4087**
Chld C-382-7773	**Commiss-382-2127**	**EM C-382-5769**

Norton AFB, Cont'd

Fam Svc-382-7137
Loc-382-5381
Off C-889-4451/884-8701
Theat-382-6787
Gas S (Mil)-382-3060
Med-382-7818
Pkg S-885-6005
Legal-382-6084
NCO C-889-0635
SDO-382-3012

HEALTH & WELFARE: Clinic/Emerg, EX-2333, Appoint, EX-3633. Inpatient, USAF Regional Hosp, March AFB, Comm: 714-653-4266, ATVN: 947-4266. Chapels serving all faiths, Prot/Cath, EX-4087, Jewish by resv, EX-4087.

RECREATION: Hobby, EX-3788, Bowling, EX-2070, Youth Ctr, EX-3236, Gym, EX-7334, Rec Svc, EX-5769, Golf, EX-3421, Library, EX-7119. Rec Vehicle Park, on Base, year round, Comm: 714-382-5531, ATVN: 876-5531, 7 camper spaces w/ hookups and 25 camper spaces w/o hookups.

SPACE-A: Pax Term and Lounge, Bldg 673, 24 hrs dly, Comm: 714-382-5354, ATVN: 876-5354, flts CONUS and overseas, Pacific, and AU/NZ areas.

ATTRACTIONS: Snow skiing N of Norton AFB, greater Los Angeles area, Palm Springs, resort city nearby.

SAN FRANCISCO AREA

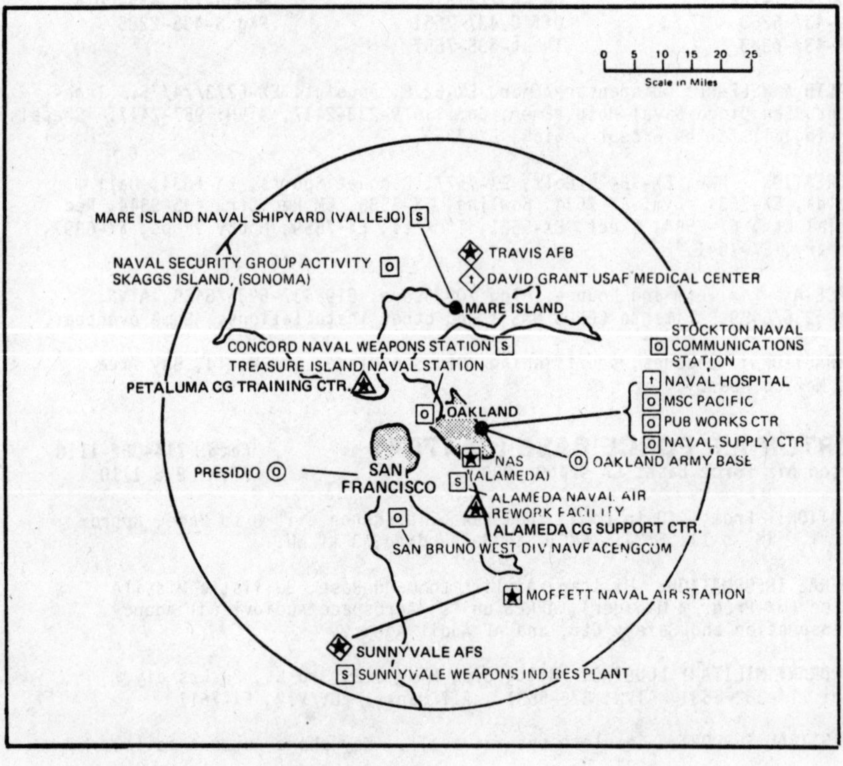

OAKLAND ARMY BASE (CA18R4)

Oakland Army Base, CA 94626-5000

Comm: 415-466-9111
ATVN: 859-9111

Oakland Army Base, Cont'd

LOCATION: Near junction of I-80 and CA-17, S of San Francisco-Oakland Bay Bridge. RM: p-11, NC/22. NMC: Oakland, 2 mi SE.

GENERAL INFORMATION: Western Hq for Military Traffic Management Command and other western area Army support units.

TEMPORARY MILITARY LODGING: Biltg Ofc, Bldg 560, 12th St, ATTN: MTW-PES-S, 24 hrs dly, Comm: 415-466-3113, ATVN: 859-3113/3067. All ranks. DV/VIP call 859-3113/9111. Controls billeting at Fort Mason, on Franklin St, 4 blocks N of US-101 (Lombard St) in San Francisco. Also, contact Inst Club Manager, Bldg 780, Comm: 415-441-7700, ATVN: 859-7700.

LOGISTICAL SUPPORT: Complete spt fac avail. Key phone numbers follow.

PX-836-4277	**Caf-465-3175**	**Chap-466-2617**
Commiss-466-2680	**EM C-466-2791**	**Fam Svc-466-3457**
Gas S-465-3176	**Legal-466-2921**	**Med-466-2245**
NCO C-466-2791	**Off C-441-7700**	**Off C(Ft Mason)-414-7700**
Pkg S-451-2743	**Pol-466-3333**	**SDO-466-3131**
Theat-466-3680		

HEALTH & WELFARE: Dispensary/Emer, EX-2161/2245, Appoint, EX-2918. Inpatient, Oakland Naval Hosp, Comm: 415-633-5441, ATVN: 855-5441. Chapels, serving all faiths, EX-2616.

RECREATION: Bowling, EX-3020, Handball, Tennis, Gym, EX-2246, Hobby Shops, EX-3541, Service Club, EX-2727, Tickets, EX-2727, Youth Ctr, EX-2727.

SPACE-A: Det 1, 60 APS, #1 Airport Drive, Box 44, Oakland IAP, CA 94621, Pax Term, Lounge, main lobby of IAP, 0700-2300 hrs dly, Comm: 415-635-8515, ATVN: 837-5789. Commercial contr flts to CONUS and overseas (Pacific area).

ATTRACTIONS: San Francisco and Bay Area. Cable cars in operation now. For great seafood visit Tadich Grill, 240 CA St, 415-391-2373, tell Chris (Manager) you were sent by Military Living.

OAKLAND NAVAL HOSPITAL (CA41R4)

Oakland Naval Hospital, CA 94627-5000

Comm: 415-633-6200
(Aft dty hrs EX-6200)
ATVN: 855-6200

LOCATION: Off I-580 S from San Francisco-Oakland Bridge, take either Keller Ave or Golf Links Rd exits. Clearly marked. RM: p-11, ND/24. NMC: Oakland, 38 mi NW.

GENERAL INFORMATION: Naval Medical Command, Northwest Region.

TEMPORARY MILITARY LODGING: Biltg Ofc, Navy Lodge, 24 hrs dly, Comm: 415-633-6096, ATVN: 855-6096. DV/VIP, EX-6200.

LOGISTICAL SUPPORT: Limited spt fac avail. Key phone numbers follow.

NX-633-6090	**Gas S-633-6090**	**Med-633-5440**

Caf, Snack B, Off C, CPO C, EM C, Pkg S, Theat - EX-5440 for numbers.

HEALTH & WELFARE: Naval Hospital, all clinics and depts, 497 bed, Emer, EX-5190/5440, Appoint, EX-2087. Chapels serving all faiths, EX-2251.

RECREATION: Bowling, EX-6016, Spec Svc, EX-6014, Hobby Shops, EX-6013, Library, EX-6127, Tennis, EX-6013.

Oakland Nav Hosp, Cont'd

SPACE-A: See Oakland Army Base listing.

ATTRACTIONS: San Francisco, Oakland and Bay Area.

OAKLAND NAVAL SUPPLY CENTER (CA42R4)

Comm: 415-466-0112
ATVN: 836-0111

Oakland Naval Supply Center, CA 94625-5000

LOCATION: Can be reached from CA-17 N or S and CA-24 from E. Clearly marked. RM: p-11, NC/22. NMC: Oakland, 4 mi E.

GENERAL INFORMATION: Major Naval West Coast Supply Base, Support Activity for Navy Resale System.

TEMPORARY MILITARY LODGING: Biltg Ofc, Bldg 521, Navy Lodge, 0800-1630 hrs M-F, 0900-1700 hrs other days, Comm: 415-466-6187, ATVN: 836-6187. All ranks. DV/VIP, SDO.

LOGISTICAL SUPPORT: Limited spt fac avail. Key phone numbers follow.

NX-466-6227/6228	**Gas S-466-5678**	**Loc-466-0112**
SDO-466-5713 (aft 1600 hrs)		

HEALTH & WELFARE: No med fac on base. Inpatient, Oakland Naval Hosp, Comm: 415-633-5441, ATVN: 855-5441. Chapels avail at Oakland Army Base.

RECREATION: Spec Svc, EX-5496, Tickets, EX-6778.

SPACE-A: See Oakland Army Base listing.

ATTRACTIONS: San Francisco, Oakland and bay area.

PETALUMA COAST GUARD TRAINING CENTER (CA23R4)

Comm: 707-7657211
FTS: 623-7382

Petaluma Coast Guard Training Center, CA 94952-5000

LOCATION: Exit US-101 N to Washington Ave, 9 mi W to Coast Guard Training Center. RM: p-11, NN/6. NMC: San Francisco, 31 mi S.

GENERAL INFORMATION: The major West Coast Coast Guard Training Center.

TEMPORARY MILITARY LODGING: Biltg Ofc, Bldg 140, Pennsylvania Ave, 0730-1600 hrs dly, Comm: 707-765-2382, ATVN: 623-2382. All ranks. DV/VIP, EX-2215.

LOGISTICAL SUPPORT: Complete spt fac avail. Key phone numbers follow.

CGX-765-2300	**Caf-765-2287**	**Chap-765-2220**
Chld C-765-2278	**Consol C-765-2287**	**Fam Svc-765-2317**
Gas S-765-2336	**Loc-765-2215**	**Med-765-2250**
Pkg S-765-2356	**Pol-765-2215**	**Snack B-765-2287**
Theat-765-2259		

HEALTH & WELFARE: Dispensary/Emerg, EX-2250. Inpatient, Oakland Naval Hosp, Comm: 415-633-5441, ATVN: 855-5441. Chapels serving all faiths, EX-2220.

RECREATION: Arts & Crafts, Bowling, Gym, Library, Trap/Archery. Rec Area on Base, year round, 10 camper spaces w/water only, most fac, 765-7348.

SPACE-A: Try Oakland IAP, Space-A desk in lobby, Comm: 415-635-8482, ATVN: 837-8482. Flts via contr comm aircraft to CONUS and overseas (Pacific area).

Petaluma CG Tng Ctr, Cont'd

ATTRACTIONS: The Base is a federal game preserve, scenic atmosphere. North of San Francisco, Bay Area (31 mi).

POINT MUGU PACIFIC MISSILE TEST CENTER (CA40R4)

Comm: 805-989-1110
ATVN: 351-1110

Point Mugu, CA 93042-5000

LOCATION: Eight mi S of Oxnard and 40 mi N of Santa Monica, on Coastal Highway, CA-1. RM: p-13, SL/11. NMC: Los Angeles, 50 mi SE.

GENERAL INFORMATION: Pac Missile Test Ctr, NAS, Navy Astronautics Group, VXE-6, VX-4, VAQ-34, Naval Air Reserve. No general visits permitted.

TEMPORARY MILITARY LODGING: Biltg Ofc, Bldg 241, 24 hrs dly, Comm: 805-989-7617/7378, ATVN: 351-7317/7378. All ranks. DV/VIP, EX-8672

LOGISTICAL SUPPORT: Complete spt fac avail. Key phone numbers follow.

NX-989-7271	**Caf-989-7661**	**Chap-989-4421/4571**
Chld C-9898375	**Commiss-989-7993**	**Conv S-989-7993**
CPO C-989-7517	**EM C-989-8420**	**Fam Svc-989-7271**
Gas S-989-7791	**Legal-989-8233**	**Loc-989-7938**
Med-989-3331	**NCO-989-7209/7294**	**Off C-989-8407**
Pkg S-989-7496	**Pol-989-7907**	**Snack B-989-7637**
Theat-989-7606		

HEALTH & WELFARE: Dispensary/Emerg, EX-8821/3331, Appoint, EX-4501. Inpatient, Long Beach Naval Hosp, Comm: 213-420-5201, ATVN: 873-9201. Chapels serving all faiths, EX-4421/4358/5276.

RECREATION: Arts & Crafts, EX-5278/4693, Auto Hobby, EX-4478, Billiard Ctr, EX-4874, Bowling, EX-4617, Golf, EX-4620, Gym, EX-5173, Hobby Shops, EX-4294, Library, EX-4411, Skating, EX-5156, Swimming, EX-4409, Tickets, EX-4103, Youth Ctr, EX-4218. Point Mugu Rec Area, on Base, year round, Comm: 805-989-8770, ATVN: 351-8770. 22 camper spaces w/full hookups, 11 camper spaces w/o hookups, 6 cabins, 12 motel units, comp fac and all beach activities.

SPACE-A: Pax Term, Lounge, Base Ops Bldg, Ops hrs, Comm: 805-989-8094, ATVN: 351-8094, flts (USAF C-141, and Navy LC-130 aircraft) in spt of "Oper Deep Freeze" (US Naval Support Force, Antarctica (VXE-6), winter mo here-summer mo there (AU/NZ) and other flts to CONUS and overseas locations.

ATTRACTIONS: Beautiful beaches and Los Angeles area nearby. Disneyland, Marineland, Movieland, Universal Studios, Queen Mary and Spruce Goose, Lion Country Safari and more.

POINT HUENEME NAVAL CONSTRUCTION BATTALION CENTER (CA32R4)

Comm: 805-982-4711
ATVN: 360-4711
FTS: 799-4711

Port Hueneme, CA 93043-5000

LOCATION: Seven mi W of US-101. Take Victoria Ave exit in Ventura to Channel Islands Blvd (turn left), to Ventura Road (turn right), to Pleasant Valley Road (turn right). Enter at Pleasant Valley Gate. RM: p-13, SL/11. NMC: Los Angeles, 50 mi SE.

GENERAL INFORMATION: Naval Mobile Construction Battalions and spt units.

Port Hueneme NCBC, Cont'd

TEMPORARY MILITARY LODGING: Biltg Ofc, ATTN: Code 33, Bldg 1182, 32nd Ave & Harris St, 24 hrs dly, Comm: 805-982-4804, ATVN: 360-4804, Navy Lodge. All ranks. DV/VIP, EX-4401.

LOGISTICAL SUPPORT: Complete spt fac avail. Key phone numbers follow.

NX-982-5212	**Caf-982-4435**	**Chap-982-4358**
Chld C-982-4567	**Commiss-982-4939**	**Conv S-982-4939**
CPO C-982-4473	**EM C-982-5293**	**Fam Svc-982-4885**
Gas S-982-5123	**Legal-982-4567**	**Loc-982-4711**
Med-982-6301	**Off C-982-4380**	**Pkg S-982-5981**
Pol-982-4591	**Snack B-982-5611**	**Theat-982-5491**

HEALTH & WELFARE: Clinic/Emer, 982-6301-3332, Appoint, 984-8433. Inpatient, Long Beach Naval Hosp, Comm: 213-420-5201, ATVN: 873-9201. Chapels serving all faiths, EX-4421/4358.

RECREATION: Arts/Crafts, EX-4693, Auto Hobby, EX-4478, Billiard, EX-4874, Bowling, EX-4617, Comm Ctr, EX-4035, Gear Issue, EX-5116, Golf, EX-4620, Gym, EX-5173, Seabee Pk, EX-4035, Skating Rnk, EX-5156, Swimming, EX-4409, Tickets, EX-4103, Wood Hobby, EX-4294, Youth Ctr, EX-4218. Pt Mugu Rec Area, 5 mi SE.

SPACE-A: Base Ops, Point Mugu NAS (Pac Missile Test Center), Comm: 805-989-8094, ATVN: 351-8094, flts to CONUS and overseas. See Point Mugu listing.

ATTRACTIONS: CEC/SEABEE Museum at Ctr. Beaches, Channel Isls, two marinas.

PRESIDIO OF SAN FRANCISCO (CA19R4)

Presidio of San Francisco, CA 94129-5000

Comm: 415-561-2211
ATVN: 586-1110

LOCATION: At the S end of the Golden Gate Bridge. From US-101 N exit onto Lombard St. From US-101 S exit onto Merchant Road, follow signs to Lombard St and Presidio. RM: p-11, ND/19. NMC: San Francisco, S & E in city limits.

GENERAL INFORMATION: Hq, Sixth US Army, Letterman Army Medical Hospital.

TEMPORARY MILITARY LODGING: Billeting Ofc, Bldg 42, Lincoln Blvd. ATTN: AFZM-DEH-HB, 24 hrs dly, Comm: 415-561-2096, ATVN: 586-2096. DV/VIP, EX-3950/2540. **MARINES' MEMORIAL CLUB, 609 Sutter St, San Francisco, CA 94102, Comm: 415-673-6672.** NOT "military lodging" a club/hotel for uniformed personnel, AD & Ret, their families and guests. Rooms, suites, and club fac, library, museum, swimming, gym, launderette, valet, exch store, pkg store, lounges, restaurant overlooking the city. Moderate rates. Cable car, 1 block at Powell Street.

LOGISTICAL SUPPORT: Complete spt fac avail. Key phone numbers follow.

PX-922-4592	**Caf-922-3546**	**Chap-561-4719/4711**
Chld C-561-4712	**Commiss-561-4855**	**Conv S-922-6933**
Fam Svc-561-5155	**Gas S-922-5728**	**Legal-561-2925**
Loc-561-4431	**Med-561-5656**	**NCO C-921-6896**
NCO/EMC 921-1896/1897	**Off C-921-1612**	**Pkg S-921-1634**
Pol-561-2251/2252	**SDO-561-2045/3601**	**Snack B-922-3546**
Theat-921-3852		

HEALTH & WELFARE: Letterman Army Medical Center, clinics and depts, 400 bed, Emer, EX-5771, Appoint, EX-5324. Chapels serving all faiths, EX-4719/4711.

RECREATION: Arts & Crafts, EX-4497/5915, Wood Shop, EX-4497/5915, Photo

Presidio of SF, Cont'd

Shop, EX-3445, Auto Shop, EX-5521, Golf, 751-0562, Gym, EX-5032, Library, EX-3448, Riding Stables, 332-9712, Recreation Ctr, EX-2000, Rec Supply, EX-5141, Swimming, EX-3673, Bowling, EX-5042. Operates Lake Tahoe Army Rec Fac, 4 town-house-type units, Comm: 415-561-4324.

SPACE-A: Oakland IAP, main lobby, 24 hrs dly, Comm: 415-635-8515, ATVN: 837-5789. Comm contract flts to CONUS and overseas (Pacific area). Coast Guard Air Station at San Francisco, IAP, North Access Rd next to Flying Tigers, Comm: 415-876-2900, ATVN: 859-6900, may have very infrequent flts on transient aircraft, only helicopters stationed here.

ATTRACTIONS: Golden Gate Park, cable cars, Fishermans Wharf. Best Seafood in the world is at Tadich Grill, 240 Calif St, 415-391-2373, ask for Chris, the manager, tell him Military Living sent you there. Great Chinese food at Yuet Lee, 1300 Stockton St, 415-982-6020.

SACRAMENTO ARMY DEPOT (CA53R4)

Sacramento, CA 95813-5012

Comm: 916-388-2211
ATVN: 839-1110

LOCATION: From CA-99 N or S take Fruitridge Rd exit E to Depot at intersection of Fruitridge Rd & Florin Perkins Rd. RM: p-11, NO/25. NMC: Sacramento, in the city.

GENERAL INFORMATION: Major Communications-Electronics Depot of Army Materiel Command.

TEMPORARY MILITARY LODGING: None reported.

LOGISTICAL SUPPORT: Limited spt fac avail. Key phone numbers follow.

PX-383-3690	**Caf-388-2904**	**Chap-388-3129**
Fam Svc-388-2059	**Gas S-383-5652**	**Legal-388-3274**
Loc-388-3324	**Med-388-2225**	**NCO C-388-2692**
Off C-388-2587	**Pol-388-2910**	

HEALTH & WELFARE: Dispensary/Emerg, EX-2225. Inpatient, USAF Hosp, Mather AFB, Sacramento, Comm: 916-364-2114643-2334. Chapels serving all faiths, EX-3129.

RECREATION: Rec Svc, EX-2950, Auto Hobby, EX-2041, Swimming, EX-2950, Library, EX-2660.

SPACE-A: Mather AFB, Bldg 4579, 24 hrs dly, Comm: 916-364-2579/2510, ATVN: 828-2579/2510, flts to CONUS and overseas.

ATTRACTIONS: Old Sacramento, China Tn, 90 mi SW to San Francisco/Bay Area.

SAN DIEGO AREA

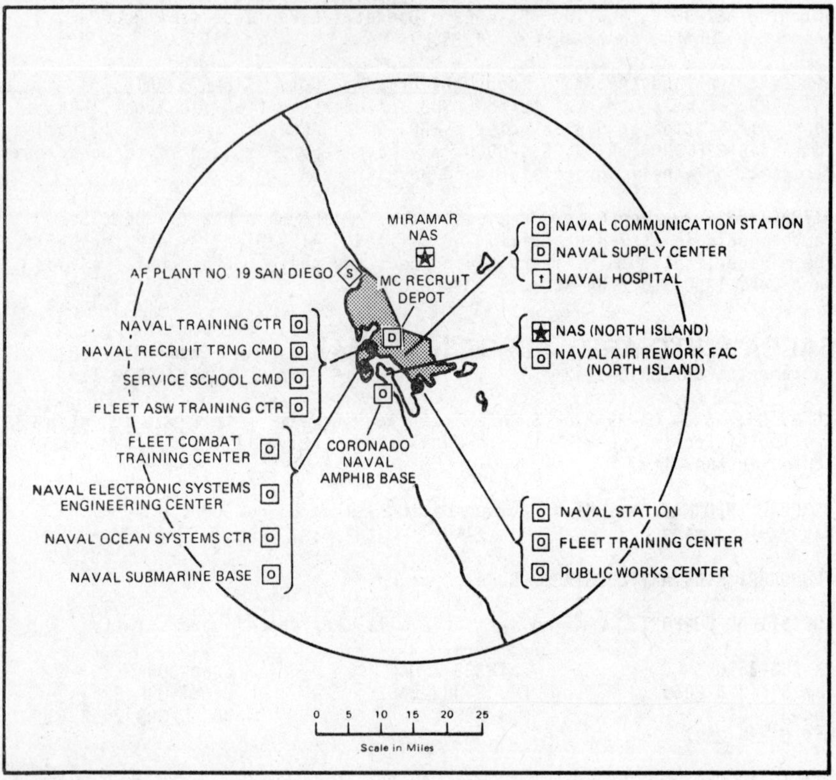

SAN DIEGO MARINE CORPS RECRUIT DEPOT (CA57R4)

Comm: 619-225-4011
ATVN: 957-0111

San Diego, CA 92140-5000

LOCATION: From I-5 N or S exit to San Diego IAP (Lindbergh Field). Adjacent to and on the west side of IAP. RM: p-15, N/22. NMC: San Diego, 2 mi NE.

GENERAL INFORMATION: USMC/Commanding General/Marine Corps Recruit Depot - Western Recruiting Region/San Diego CA 92140-5000.

TEMPORARY MILITARY LODGING: Ltd TML for unaccompanied Enlisted, EX-3571. Navy Lodges: San Diego Nav Stn & North Is NAS, see these listings for details.

LOGISTICAL SUPPORT: Most spt fac avail. Key phone numbers follow.

MX-297-2500	Chap-225-4647	Chld C-225/5284/4496
EM C-225-4158	Fam Svc-225-4356	Gas S-225-5719-EX-246
Legal-225-3165	Loc-225-4520	Med-225-3501
NCO-225-4713	NCO C-225-5521	Off C-225-5640
Pkg S-225-3850	Pol-225-3601	Snack B-297-2500-EX-315/225-3960
Theat-225-4389		

San Diego MCRD, Cont'd

HEALTH & WELFARE: Dispensary only. Inpatient, San Diego Naval Hosp, Emerg, 233-2417/2841. Outpatient, 233-2946. Chapels serving all faiths, EX-4647.

RECREATION: Bowling, EX-3757/3960, Camping Equip, EX-4573, Fitness Ctr, EX-4495/5313, Boats, EX-4166, Tickets, EX-5593, Rec Svc Ofc, EX-4611.

SPACE-A: North Island NAS, Bldg 700, 24 hrs dly, Comm: 619-437-6936, ATVN: 951-6936, flts to CONUS and overseas.

ATTRACTIONS: Sea World, San Diego Zoo, Balboa Park.

SAN DIEGO NAVAL HOSPITAL (CA59R4)

San Diego, CA 92134-5000

Comm: 619-233-2417
ATVN: 987-2417

LOCATION: Exit fr I-5 to CA-163 and Park Blvd. RM: p-15, N/23. NMC: San Diego, 3 mi NE.

GENERAL INFORMATION: Large Naval medical fac serving the San Diego area.

TEMPORARY MILITARY LODGING: Try for Enlisted only, EX-2748. Navy Lodges at Naval Station, 234-6142, North Island NAS, 435-0191, Miramar NAS, 271-7111. BOQ Naval Training Center, 226-0444.

LOGISTICAL SUPPORT: Limited spt fac avail. Key phone numbers follow.

NX-232-9286	Chap-233-2311/2417	SDO-233-2417

HEALTH & WELFARE: Emer, 233-2417/2841. Outpatient Admin, 233-2946, 600 bed Inpatient, San Diego Naval Hosp, Comm: 619-233-2411, ATVN: 987-4201.

RECREATION: Library, EX-2816, Rec Svc Ofc, EX-2245/2456, Athletic Equip, EX-2245, Swimming, EX-2245, Tennis, EX-2245, Tickets, EX-2749/2949, Tours, EX-2749, Weight Room, EX-2245.

SPACE-A: See North Island NAS listing.

ATTRACTIONS: Sea World, Zoo, Balboa Park, city of San Diego, Wild Animal Park.

SAN DIEGO NAVAL STATION (CA26R4)

San Diego Naval Station, CA 92136-5000

Comm: 619-235-2011
ATVN: 958-2011

LOCATION: Off I-5, 7 mi S of San Diego. Take 28th St exit. NAVSTA is at 28th & Main Streets. RM: p-15, O/23. NMC: San Diego, in the city.

GENERAL INFORMATION: Pacific Fleet Training Center, Public Works Center, and home port for major ships of the Pacific Fleet.

TEMPORARY MILITARY LODGING: No central Biltg Ofc. Navy Lodge near Harbor Dr, 24 hrs dly, Comm: 619-234-6142, ATVN: 958-6142. Resv: 1-800-628-9466. All ranks. DV/VIP, EX-1548.

LOGISTICAL SUPPORT: Complete spt fac avail. Key phone numbers follow.

NX-239-9761	Caf-239-9761	Chap-235-1306
Chld C-235-0861	Commiss-696-6091	Conv S-239-9761
CPO C-235-1550	EM C-235-1428	Fam Svc-235-4706
Gas S-239-9761	Legal-696-6121	Med-696-6286
Off C-696-6156	Pkg S-236-9548	Pol-235-1260

San Diego NS, Cont'd

Snack B-239-9761	Theat-235-1135

HEALTH & WELFARE: Dispensary/Emerg, 696-6286. Inpatient, San Diego Naval Hosp, Emerg, Comm: 619-233-2417, info for Central Appoint, Comm: 619-233-2205. Chapels serving all faiths, EX-1306.

RECREATION: Adm Robinson Ctr, EX-1981, Rec Equip, EX-1530, Gym, EX-1455, Tennis, EX-1455, Golf, EX-1503, Swimming, EX-1456, Field House, EX-1353, Rec Lounge, EX-1150, Sauna, EX-1456, Tickets, EX-2919, Courts, EX-1353, Indoor Courts, EX-4788, Hobby Shop, EX-3303, Sport Fishing, EX-2287. Mission Gorge Rec Ctr, off Base, year round, Comm: 619-282-8481, ATVN: 958-1135, 24 camper spaces w/hookups, 4 camper spaces w/o hookups, comp fac and many act. Also, Golf Crest RV Park, near campground, Comm: 619-281-0432, 14 spaces for self-contained units only.

SPACE-A: North Island NAS, Bldg 700, 24 hrs dly, Comm: 619-437-6936, ATVN: 951-6489, flts to CONUS locations and some overseas.

ATTRACTIONS: San Diego and Mission Bays, Mexico nearby, Sea World, zoo and museum.

SAN DIEGO NAVAL TRAINING CENTER (CA54R4)

Comm: 619-225-4011
ATVN: 957-4011

San Diego, CA 92133-5000

LOCATION: From I-8 W take Rosecrans exit to Lytton, left on Lytton St to NTC gate #1. From I-5 N exit to Harbor Dr to Rosecrans, right on Rosecrans to Lytton, right on Lytton to gate #1. RM: p-15, N/22. NMC: San Diego, 4 mi E.

GENERAL INFORMATION: Major West Coast training facility, including both advanced and basic training. Component commands include Recruit Training Command, Service School Command, Naval Training Station, and Personnel Support Activity.

TEMPORARY MILITARY LODGING: Try NAVADMINCOM/NTC, BEQ-225-4141, BOX-225-4508.

LOGISTICAL SUPPORT: Complete spt fac avail. Key phone numbers follow.

NX-221-2000	Caf-222-1022	Chap-225-5501
Chld C-225-4463	Commiss-225-4191	Conv S-221-1074
CPO C-225-3231	EM C-225-3449	Fam Svc-235-4706
Gas S-221-1049	Legal-225-5146	Loc-225-4011
Med-233-2205	Off C-223-9839	Pkg S-222-5580
Pol-225-4011	Snack B-225-9651	Theat-222-6916

HEALTH & WELFARE: Naval Hospital, Emerg, 223-2417, Appoint, 223-2946. Chapels serving all faiths, 225-5501.

RECREATION: Library, EX-5470, Gym, EX-5370, Bowling, EX-5386, Golf, EX-5450, Sailing, EX-4551, Rec Veh Park, EX-5211, Game Room, EX-5386, Tickets, EX-5214, Swimming, EX-5270, Rec Equip Loan, EX-4197, Marina, EX-4227, Auto Hobby, EX-3300, Crafts Hobby, EX-5160, Rec Svc Ofc, EX-5214.

SPACE-A: North Island NAS, Bldg 700, 24 hrs dly, Comm: 619-437-6936, ATVN: 951-6936, CONUS and overseas flts.

ATTRACTIONS: Zoo, Sea World, Baja Cal, Mexico.

SAN PEDRO COAST GUARD PERSONNEL SUPPORT CENTER (CA25R4)

San Pedro, CA 90731-5000

Comm: 213-548-2843
FTS: 796-2843

LOCATION: On Coast Guard Base, Terminal Island, San Pedro CA, 6 mi W of Long Beach. RM: p-15, L/9. NMC: Long Beach, 2 mi E.

GENERAL INFORMATION: The CG Support Center in the Los Angeles area.

TEMPORARY MILITARY LODGING: Local Housing Authority, PCS, Base PO Box 8, Terminal Island Station, San Pedro, 0700-1600 hrs dly, Comm: 213-548-2482. All ranks. DV/VIP-same number.

LOGISTICAL SUPPORT: Limited spt fac avail. Exchange and Clubs.

HEALTH & WELFARE: No medical facility at this Base. Inpatient, Long Beach Naval Hosp, Comm: 213-420-5201, ATVN: 873-9201.

RECREATION: On the beach in Los Angeles area.

SPACE-A: Los Angeles IAP, see other LA listings.

ATTRACTIONS: Greater Los Angeles area.

SHARPE ARMY DEPOT (CA52R4)

Lathrop, CA 95331-5122

Comm: 209-982-2011
ATVN: 462-2011

LOCATION: Off I-5 at Roth Road, or from CA-99 N, left on CA-120, N on Airport Way, left on Roth Road to Depot. RM: p-11, NO/11. NMC: Stockton, 12 mi N.

GENERAL INFORMATION: Area oriented Spare Parts Depot, Depot Systems Command, Army Materiel Command.

TEMPORARY MILITARY LODGING: None reported.

LOGISTICAL SUPPORT: Limited spt fac avail. Key phone numbers follow.

PX-982-5108	**Caf-982-2544**	**Chap-982-2530**
Consol C-982-2523	**Fam Svc-982-2415**	**Legal-982-2443**
Loc-982-2633	**Med-982-2303**	**Pol-982-2566**
Theat-982-3608		

HEALTH & WELFARE: Dispensary/Emer, EX-2586, Appoint, EX-2303. Inpatient, USAF Hosp, Mather AFB, Sacramento, 55 mi N, Comm: 916-364-2114. Chapels serving all faiths, EX-2530.

RECREATION: Spec Svc, EX-2419, Arts & Crafts, Library, Swimming, Racquet Courts, Youth Ctr. El Descanso (The Rest) Army Trailer Camp, on post, year round, Comm: 209-982-2419, ATVN: 462-2419/2420, 12 camper spaces w/hookups, most fac and activities.

SPACE-A: Oakland IAP, main lobby, 24 hrs, Comm: 415-635-8482, ATVN: 837-5789, flts to CONUS and overseas.

ATTRACTIONS: San Francisco and Bay Area to west, nearby.

SIERRA ARMY DEPOT (CA44R4)

Herlong, CA 96113-5000

Comm: 916-827-4544
ATVN: 830-9910

Sierra Army Depot, Cont'd

LOCATION: Fifty-five mi N of Reno, NV, off US-395. Right on CA-A26 from Reno. When traveling S on US-395, left on CA-A25. RM: p-11, NH/13. NMC: Reno, 55 mi SE.

GENERAL INFORMATION: Major Army Materiel Depot, maint/storage activities.

TEMPORARY MILITARY LODGING: Biltg Ofc, Bldg T-75, dty hrs, after hrs, Bldg P-2, Comm: 916-827-9544, ATVN: 830-9544. All ranks. DV/VIP, EX-4345/9910.

LOGISTICAL SUPPORT: Most spt fac avail. Key phone numbers follow.

PX-827-2017/8716	**Caf-827-6166**	**Chap-827-9128/9**
Commiss-827-4480/6161	**Conv S-827-8716**	**Fam Svc-827-4425/9425**
Gas S-827-2787/5151	**Legal-827-4544**	**Loc-827-4544/9910**
Med-827-9221/4575	**NCO C-827-2504/5241**	**Off C-827-2771/5231**
Pkg S-827-2504/5241	**Pol-827-4345**	**Snack B-827-6166**
Theat-827-7421		

HEALTH & WELFARE: Army Clinic, Emerg, EX-9221, Appoint, EX-9422. Inpatient, Letterman Army Medical Center, San Francisco, Emerg, Comm: 415-561-5656, ATVN: 586-5656, Appoint, EX-4822. Chapels serving Cath/Prot, EX-9128/9.

RECREATION: Auto Hobby, EX-9198, Bowling, EX-8481, Ceramics Shop, EX-9464, Photo, EX-9464, Gym, EX-9345, Swimming, EX-5231, Rec Svc, EX-9464.

SPACE-A: Travis AFB, Bldg P-3, 24 hrs, Comm: 707-438-2311, ATVN: 837-2311, flts to CONUS and overseas.

ATTRACTIONS: Reno for entertainment, casinos, Lake Tahoe 1.5 hr drive S.

STOCKTON NAVAL COMMUNICATIONS STATION (CA51R4)

Comm: 209-944-0284
ATVN: 466-7284

Stockton, CA 95203-5000

LOCATION: From I-5 N exit at Rough & Ready Island, right at Fresno St to Washington St to Station. RM: p-11, NO/10. NMC: Stockton, in the city.

GENERAL INFORMATION: Major Nav Communications Station on the West Coast.

TEMPORARY MILITARY LODGING: Biltg Ofc, EX-343, 24 hrs. Ask for directions at gate, Comm: 209-466-7343 for Enl, 209-466-7330 for Off. DV/VIP, Biltg Ofc.

LOGISTICAL SUPPORT: Most spt fac avail. Key phone numbers follow.

NX-944-0332	**Chaplain C-944-0282**	**Commiss-944-0224**
CPO C-944-0552	**EM C-944-0522**	**Gas S-944-0479**
Loc-944-7284	**Med-944-0445**	**Off C-944-0330**
Pkg S-944-0400	**Pol-944-0336**	

HEALTH & WELFARE: Clinic/Emerg, EX-0445. Inpatient, Oakland Naval Hosp, Comm: 415-633-5441, ATVN: 855-5441.

RECREATION: Bowling, EX-0325, Golf, EX-0442, Marina, EX-0364, Rec Svc, EX-0582.

SPACE-A: Oakland IAP, main lobby, 24 hrs, Comm: 415-635-8482, ATVN: 837-5789, flts to CONUS and overseas.

Stockton NCS, Cont'd

ATTRACTIONS: Equal distance to Sacramento and San Francisco. Great sights and shopping.

TRAVIS AIR FORCE BASE (CA50R4)

Travis Air Force Base, CA 94535-5000

Comm: 707-438-4011
ATVN: 837-1110

LOCATION: Off I-80 N, take Travis AFB Parkway exit. Clearly marked. RM: p-11, NN/8. NMC: San Francisco, 45 mi SW.

GENERAL INFORMATION: Mil Airlift Command Base. AFRES units stationed here.

TEMPORARY MILITARY LODGING: Biltg Ofc, Bldg 404, Sevedge Dr, 24 hrs dly, Comm: 707-437-4661, ATVN: 837-4661. All ranks. DV/VIP, EX-3185.

LOGISTICAL SUPPORT: Complete spt fac avail. Key phone numbers follow.

BX-437-4634	Caf-437-2092	Chap-438-2652
Chld C-438-3041	Commiss-438-3757	Conv S-437-6606
Fam Svc-437-2774	Gas S-437-2678	Legal-438-3251
Loc-438-2026	Med-428-8034	NCO C-438-5071
Off C-437-3368	Pkg S-438-3126	Pol-438-3293
Theat-438-3855		

HEALTH & WELFARE: David Grant USAF Medical Center, clinics and depts, 290 bed, Emerg, 438-8034, Appoint, 425-1160. Chapels serving all faiths, EX-2562/2427.

RECREATION: Bowling, EX-5048, Golf, 438-5997, Rec Ctr, EX-3076, Gym, EX-5680, Hobby Shops, EX-2929, Library, EX-5254, Youth Ctr, EX-5392, Aero Club, 437-3470, Saddle Club, 437-9060, Rod & Gun Club, 437-2095. Travis Fam-Camp on Base, year round, through Rec Center Comm: 707-438-3076, ATVN: 837-3076, 24 camper spaces w/hookups, 12 camper spaces w/o hookups, comp fac.

SPACE-A: Pax Term and Lounge, Bldg P-3, 24 hrs dly, Comm: 707-438-2311. ATVN: 837-2311, flts to CONUS and overseas locations.

ATTRACTIONS: Short drive to both Sacramento and San Francisco. Many sights and activities. Vacaville, 7 mi N, is the onion capital of CA.

TREASURE ISLAND NAVAL STATION (CA21R4)

San Francisco, CA 94130-5000

Comm: 415-765-6411
(0730-1600 M-F)
415-765-6232/4(Oth)
ATVN: 869-6232

LOCATION: On Treasure Island in San Francisco Bay off I-80 (Oakland Bay Bridge), take exit from left lane. RM: p-11, NC/20. NMC: San Francisco, 3 mi SW.

GENERAL INFORMATION: Naval Technical Training Center, Hq 12 Marine Corps District, Navy and Marine Corps Reserve Units.

TEMPORARY MILITARY LODGING: None available.

LOGISTICAL SUPPORT: Complete spt fac avail. Key phone numbers follow.

NX-765-6966	Caf-765-6087	Chap-765-6823
Chld C-765-5765	Commiss-765-6694/5	CPO C-765-6284

Treasure Island NS, Cont'd

EM C-765-5121	Fam Svc-765-5789	Gas S-765-5874
Legal-986-6166	Med-765-5223	Off C-765-5445
Pkg S-765-5291	Theat-765-5995	

HEALTH & WELFARE: Clinic/Emer EX-5222. Inpatient care, Oakland Naval Hosp, Comm: 415-633-5441, ATVN: 855-5441. Chapels serving all faiths, EX-6823/6824.

RECREATION: Bowling, EX-6337, Crafts/Hobby, EX-5771, Library, EX-5809, Marina & Sailing, EX-5065, Rod & Gun, EX-5778, Youth Ctr, EX-5012, Auto Hobby, EX-5451, Sport Fishing, EX-5778.

SPACE-A: Oakland IAP, main lobby, 24 hrs dly, Comm: 415-635-8482, ATVN: 837-5789, flts to CONUS and overseas locations.

ATTRACTIONS: San Francisco, Fisherman's Wharf, China Town, cable cars.

TWENTYNINE PALMS MARINE CORPS AIR/GROUND COMBAT CENTER (CA27R4)

Twentynine Palms, CA 92278-5000

Comm: 619-368-6000
ATVN: 952-6000

LOCATION: From W on I-10 exit on CA-62 NE to Base. From E on I-40 exit S at Amboy. RM: p-13, SK/20. NMC: Palm Springs, 60 mi SW.

GENERAL INFORMATION: The world's largest Marine Corps Base, training and maintenance of the Corps.

TEMPORARY MILITARY LODGING: Biltg Ofc, 5th St & Griffin Rd behind hospital, 24 hrs dly, Comm: 619-368-6642/7375, ATVN: 952-7375. All ranks. Hostess House, 24 rm. EX 6583-6573. DV/VIP, EX-6109.

LOGISTICAL SUPPORT: Complete spt fac avail. Key phone numbers follow.

MX-368-6163	Caf-368-6989	Chap-368-6464
Chld C-368-6660	Commiss-368-6606	Conv S-368-6138
EM C-368-6696	Fam Svc-368-6344	Gas S-368-6693
Legal-368-6782	Loc-368-7369	Med-368-6819
Off C-368-6610	Pkg S-368-6860	Pol-368-6297
S NCO C-368-6608	Theat-368-7269	

HEALTH & WELFARE: Branch Naval Hospital, clinics and depts, 40 bed, Emerg, EX-7354, Appoint, EX-3585. Chapels serving all faiths, EX-6456.

RECREATION: Spec Svc Ofc, EX-6597, Rec Lounge, Gym, Hobby Shops, Youth Act, Physical Fitness Ctr, Golf, Riding Stables, Skeet Range, Hunting, Swimming, and more. The Marine Corps Rec Fac at Big Bear Lake, operated by MCAS, El Toro is in San Bernardino Mountains off I-15, Comm: 714-651-2626, ATVN: 997-2626. See MCAS El Toro listing for details.

SPACE-A: Norton AFB, Bldg 673, 24 hrs dly, Comm: 714-382-5354, ATVN: 876-5354, flts to CONUS and overseas (Pacific and AU/NZ).

ATTRACTIONS: Joshua Tree National Monument, Palm Springs, 60 mi SW.

VANDENBERG AIR FORCE BASE (CA29R4)

Vandenberg Air Force Base, CA 93437-5000

Comm: 805-866-1611
ATVN: 276-1611

LOCATION: From S on US-101, W on CA-246, N on CA-S20 to AFB. From N on US-101, W on US-1 from Gaviota, N on CA-S20 to AFB. RM: p-7, SJ/7. NMC: Lompoc, 6 mi S.

Vandenberg AFB, Cont'd

GENERAL INFORMATION: Strategic Air Command Base. Space and Ballistic Missile Training and Test. Only base to test launch Intercontinental Ballistic Missiles (ICBM).

TEMPORARY MILITARY LODGING: Biltg Ofc, ATTN: 4392-AEROSG/SVH, Bldg 13005, Oregon at L St, 24 hrs dly, Comm: 805-866-8913, ATVN: 276-8913. All ranks. DV/VIP, EX-3711.

LOGISTICAL SUPPORT: Complete spt fac avail. Key phone numbers follow.

BX-734-4341	**Caf-734-1488**	**Chap-866-6655**
Chld C-866-3237	**Commiss-866-5181**	**Conv S-734-2439**
EM C-866-3330	**Fam Svc-866-5484**	**Gas S-734-2185**
Legal-866-5823	**Loc-866-1841**	**Med-866-1847**
NCO-866-1611	**NCO C-866-3330**	**Off C-734-4311**
Pol-866-3911	**Snack B-734-3832**	**Theat-734-1315**

HEALTH & WELFARE: USAF Hospital, clinics and depts, 45 bed, Emer 866-6206, Appoint, 866-1847. Chapels serving all faiths, EX-6655.

RECREATION: Aero, 734-2733, Arts & Crafts, EX-6438, Auto Hobby, EX-6014, Bowling, 734-1310, Ceramics, EX-5209, Golf, EX-6262, Gym, EX-3832, Library, EX-6414, Picnic Area, EX-8415, Rec Ctr, EX-7976, MWR Supply, EX-5908, Rod & Gun Club, EX-4560, Swimming, EX-3581, Wood Hobby, EX-4567, Youth Ctr, EX-6898. Fam-Camp, on Base, year round, Comm: 805-866-8579, ATVN: 276-8579, 20 camper spaces w/hookups, 30 camper spaces w/w & e only, 19 camper spaces w/o hookups, an RV parts store.

SPACE-A: Pax Term and Lounge, Bldg 1749, 0800-1700 hrs dly, Comm: 805-866-1854, ATVN: 276-1854, limited flts to CONUS installations.

ATTRACTIONS: Great beaches, historic missions, & Hearst Castle, N of AFB.

Small Installations Cont'd from page 10

North Bay Air Force Station
Canadian Forces Base
Ontario CN 00100-5000

Long Island Sound Coast Guard
Group, 120 Woodward Ave
New Haven CT 06512-5000

Wilmington Air National
Guard, Bldg 1504
New Castle DE 19720-5000

Mayport Coast Guard Base
PO Box 385
Mayport FL 32267-5000

Panama City
Naval Coastal Systems Center
Panama City FL 32407-5000

Cape Canaveral Air Force
Station, Bldg 1733
Canaveral Sp Ctr FL 32920-5000

Miami Coast Guard Air Station
Opa Locka Airport
Opa Locka FL 33054-5000

Miami Beach Coast Guard Base
100 MacArthur Causeway
Miami Beach FL 33139

Clearwater Coast Guard
Air Station
Clearwater FL 33520-5000

St Petersburg Coast Guard Gp
600 Eighth Ave, SE
St Petersburg FL 33701-5000

Barking Sands Pacific Missile
Range Facility, Kekaha
Kauai HI 96752-5000

Honolulu Coast Guard Base
Sand Island Access Rd
Honolulu HI 96819-5000

Cont'd page 67

COLORADO

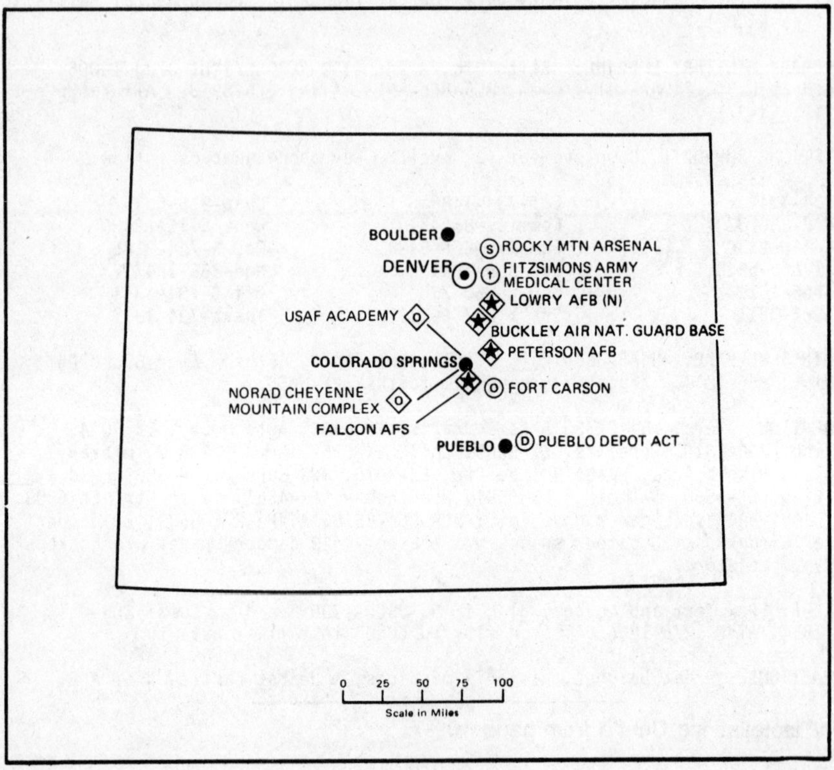

FITZSIMONS ARMY MEDICAL CENTER (CO10R3)

Aurora, CO 80045-5000

Comm: 303-361-8241
ATVN: 943-8241
FTS: 337-8241

LOCATION: From I-70 take Peoria Ave (281), exit S on Peoria Ave, 1 mi to left on Colfax Ave, Center on left. From I-25 take I-225 N to Colfax Ave, W on Colfax Ave to Center on right. RM: p-17, C/26. NMC: Denver, 8 mi E.

GENERAL INFORMATION: One of the US Army's largest Medical Research, Teaching, and Treatment Facilities.

TEMPORARY MILITARY LODGING: Biltg Ofc, Bldg 400, W Bruns Ave, 0715-2130 hrs dly, Comm: 303-361-8903, ATVN: 943-8903. All ranks. DV/VIP, EX-8824.

LOGISTICAL SUPPORT: Complete spt fac avail. Key phone numbers follow.

PX-361-3060	**Chap**-361-8121	**Chld C**-361-8126
Commiss-361-8270	**Conv S**-361-8071	**Fam Svc**-361-8925
Gas S-361-8663	**Legal**-361-8262	**Loc**-361-8223
Med-361-8350	**NCO C**-361-8961	**Off C**-361-8061
Pol-361-3791	**Snack B**-361-8475	**Theat**-361-8658

Fitzsimons AMC, Cont'd

HEALTH & WELFARE: Fitzsimons Army Medical Center, over 50 clinics and depts, Emer, EX-8350, Appoint/Info, EX-8223, 506 bed. Chapels serving all faiths, EX-8121.

RECREATION: Arts & Crafts, EX-8689, Auto Hobby, EX-8480, Bowling, EX-8379, Golf, EX-2222/3172, Gym, EX-2205, Library, EX-8749, Theater, EX-8658, Tickets, EX-2135. Rec Area at Rocky Mountain Arsenal Camping Area, in NE metro Denver area, on Base, year round, Comm: 303-289-0367, ATVN: 556-2367, 8 camper spaces w/o hookups, wilderness area.

SPACE-A: Try Buckley Air National Guard Base, 8 mi E of Denver, Bldg 809, 24 hrs, Comm: 303-366-5363-EX-9662, ATVN: 877-9011, provides MEDEVAC for FAMC flts to CONUS locations.

ATTRACTIONS: Many snow ski areas nearby. Denver.

FORT CARSON (CO02R3)

Fort Carson, CO 80913-5000

Comm: 719-579-5811
ATVN: 691-5811

LOCATION: From Colorado Springs, take I-25 or CO-115 S, clearly marked. RM: p-17, J/24. NMC: Colorado Springs, 6 mi N.

GENERAL INFORMATION: Hq Fourth Infantry Division, 43rd Support Group.

TEMPORARY MILITARY LODGING: Biltg Ofc, Bldg 6227, Prussman St, Ivy Inn, 24 hrs dly, Comm: 719-579-5090, ATVN: 691-2551/4832/2527. All ranks. DV/VIP, EX-4832.

LOGISTICAL SUPPORT: Complete spt fac avail. Key phone numbers follow.

PX-576-6313/6372	**Caf-576-6670**	**Chap-579-3393/5229**
Chld C-576-6042	**Commiss-579-5644**	**Conv S-576-7923**
EM C-579-2212/4937/4163	**Fam Svc-579-4590**	**Gas S-576-4096**
Legal-579-5572	**Loc-579-3341/3342**	**Med-579-7111**
NCO C-576-7540	**Off C-576-6646**	**Pkg S-576-6531**
Pol-579-2333	**Theat-576-3719**	

HEALTH & WELFARE: Evans US Army Community Hospital, Emerg, 579-7111, Appoint, 579-7700, 154 bed. This is a new hospital opened June 1986. Chapels serving all faiths, EX-3393.

RECREATION: Morale Spt Act Ofc, EX-3712/2432, Arts & Crafts, EX-3020, Auto Craft, EX-2147, Wood Craft, EX-3487, Bowling, EX-5542, Youth Act, EX-2680/3546, Golf, EX-4122, Library, EX-2350/2842, Little Theater, EX-3179, Outdoor Rec, EX-5474/2083, Equip Checkout, EX-2541, Physical Fitness Ctr, EX-2583, Swimming, Indoor, EX-5739, Outdoor, EX-4456. Turkey Creek Ranch, 12 mi S on CO-115, Rec Area and Ranch House, Comm: 303-579-3905.

SPACE-A: Base Ops, Butts Army Airfield, EX-5462, flts via exec aircraft. Try Peterson AFB, Pax Term, Lounge, Bldg 122, 0530-1930 M-F, 0700-1700 Sat-Sun, Comm: 303-554-4521, ATVN: 692-4707/4521, flts to CONUS locations.

ATTRACTIONS: USAF Academy, guided tours, Colorado Springs.

LOWRY TECHNICAL TRAINING CENTER (CO05R3)

Lowry Air Force Base, CO 80230-5000

Comm: 303-370-1110
ATVN: 926-1110

Lowry TTC, Cont'd

LOCATION: From I-70 take I-225 S to 6th Ave, W 3 mi to "E" gate of Base at 6th Ave & Dayton St. RM: p-17, D/25. NMC: Denver, 6 mi W.

GENERAL INFORMATION: Air Training Command Base. Large Air Force Technical Training Facility, Air Force Accounting & Finance Center, Air Reserve Personnel Center.

TEMPORARY MILITARY LODGING: Biltg Ofc, 3415 ABG/SVH, ATTN: Resv Clerk, Bldg 1400, 24 hrs dly, Comm: 303-370-7255, ATVN: 926-7255. All ranks. DV/VIP, EX-2261.

LOGISTICAL SUPPORT: Complete spt fac avail. Key phone numbers follow.

BX-388-4625	Chap-370-4536	Chld C-370-2683
Commiss-370-2647	EM C-388-0827	Fam Svc-370-2992
Gas S-333-5847	Legal-370-2470	Loc-370-2950
Med-370-4107	NCO C-320-4483	Off C-320-4002
Pkg S-370-3078	Pol-370-2881	

HEALTH & WELFARE: Dispensary/Emerg, EX-2333. Inpatient, Fitzsimons Army Medical Center, Aurora, Comm: 303-361-8241, ATVN: 943-8241. Chapels serving all faiths, EX-4536.

RECREATION: Rec Ctr, EX-2642, Tickets/Tours, EX-2640, Auto Hobby, EX-2585, Crafts, EX-2870, Library, EX-3836, Gym, EX-4306, Bowling, EX-4448, Golf, EX-3426, Aero Club, 340-9474, Rec Supply, EX-2892. Fam-Camp, on Base, yr round, Comm: 303-370-2888, ATVN: 926-2888, 20 camper spaces w/hookups, other fac and activities. Also, Dillon Rec Area, off Base, 80 mi W of Denver on Dillon Reservoir, Comm: 303-370-2888, ATVN: 926-2888, 33 camper spaces, w/o hookups, restrooms, fishing, boating, other activities.

SPACE-A: Buckley Air National Guard Base, Bldg 809, 24 hrs, Comm: 303-340-9662, ATVN: 877-9662, flts to CONUS locations.

ATTRACTIONS: Great snow ski opportunities nearby. Denver cultural and sports activities are numerous.

PETERSON AIR FORCE BASE (CO06R3)

Peterson Air Force Base, CO 80914-5000

Comm: 719-554-7321
ATVN: 692-7011

LOCATION: Off US-24 (Platte Ave) E of Colorado Springs. Clearly marked. RM: p-17, I/26. NMC: Colorado Springs, 6 mi NW.

GENERAL INFORMATION: Space Command Base. Hq Space Command, North American Air Defense Command, Air Force Hq. 1st Space Wing.

TEMPORARY MILITARY LODGING: Biltg Ofc, Bldg 1164, Stewart Ave, 24 hrs, Comm: 719-554-7851/4513, ATVN: 692-7851/4513. All ranks. DV/VIP, EX-3011.

LOGISTICAL SUPPORT: Complete spt fac avail. Key phone numbers follow.

BX-596-7270	Caf-554-7769	Chap-554-4442
Chld C-554-7880	Commiss-554-4215	Conv S-554-0300
Fam Svc-554-7614	Gas S-597-0350	Legal-554-4871
Loc-554-4020	Med-554-4333	NCO C-597-7875
Off C-574-4100	Pkg S-596-1715	Pol-554-4805
Theat-554-7171		

HEALTH & WELFARE: Clinic/Emer, EX-4333, Appoint, EX-7425. Inpatient, US Air

Peterson AFB, Cont'd

Force Academy Hosp, Colorado Springs, Comm: 303-472-3110, ATVN: 259-3110. Chapels serving all faiths, EX-4444/5311.

RECREATION: Aero Club, EX-4310, Bowling, EX-4607, Golf, EX-7233, Gym, EX-4475, Hobby Shops, EX-7941/8841, Lib, EX-7462, Swimming, EX-4608, Rec Ctr, EX-4488, Skeet, EX-4434, Ski Shop, EX-7547, Youth Ctr, EX-7220, Rod/Gun, EX-4343.

SPACE-A: Pax Term and Lounge, Bldg 122, 0530-1930 M-F, 0700-1700 Sat-Sun, Comm: 303-554-4707/4521, ATVN: 692-4707/4521, flts to CONUS locations.

ATTRACTIONS: Ski resorts, mountain sports, US Olympic Training Center.

PUEBLO DEPOT ACTIVITY (CO03R3)

Pueblo, CO 81001-5000

Comm: 719-549-4111
ATVN: 877-4111

LOCATION: Take I-25 NW from Pueblo, exit 100A, US-50 E for 18 mi to Depot. RM: p-17, K/15. NMC: Pueblo, 18 mi W.

GENERAL INFORMATION: A depot activity of the Army Materiel Command.

TEMPORARY MILITARY LODGING: None Reported.

LOGISTICAL SUPPORT: Limited spt fac avail. Key phone numbers follow.

Comm C-549-4146	**Loc-549-4135**	**Med-549-4176**
NCO-549-4236	**Pol-549-4236**	**Theat-549-4144**

HEALTH & WELFARE: Dispensary, 549-4176. Inpatient, US Army Community Hosp, Fort Carson, Comm: 719-579-3842/4040. No chapels at the Depot, services in Pueblo.

RECREATION: MWR Office, EX-4144, fishing at Linda-Ann Reservoir at the Depot, Community Club, EX-4764

SPACE-A: See listing for Peterson AFB.

ATTRACTIONS: Sangre de Christo Arts and Conference Center, Pueblo Reservoir (fishing, boating, water skiing), Cuchara Ski Area (65 mi S), Conquistador Ski Area (60 mi W), Nature Center (within city), inter city bike and hike trails adjoining reservoir, dog track, and Pueblo Symphony.

UNITED STATES AIR FORCE ACADEMY (CO07R3)

United States Air Force Academy, CO 80840-5151

Comm: 719-472-1818
ATVN: 259-3110

LOCATION: West of I-25 N from Colorado Springs. Two gates, about 5 mi apart, provide access from I-25 and are clearly marked. RM: p-17, F/24. NMC: Colorado Springs, 8 mi S.

GENERAL INFORMATION: US Air Force Academy, education of future Air Force officers, and other tenant units.

TEMPORARY MILITARY LODGING: Biltg Ofc, Bldg 3130, Academy Dr, Comm: 719-472-4910. All ranks. DV/VIP, EX-3540.

LOGISTICAL SUPPORT: Complete spt fac avail. Key phone numbers follow.

BX-472-0861	**Chap-472-3300**	**Chld C-472-4733**

US Air Force Academy, Cont'd

Conv S-472-0861	EM C-472-4475	Fam Svc-472-3595
Gas S-472-0395	Legal-472-3642	Loc-472-4262
Med-472-3333	NCO C-472-4377	Off C-472-4677
Pkg S-472-0554	Pol-472-2000	Theat-472-4582

HEALTH & WELFARE: USAF Academy Hospital, Emerg, EX-3333, Appoint, EX-5030, 85 bed. Chapels serving all faiths, EX-3300/2636.

RECREATION: Arts & Crafts, EX-4538, Auto Hobby, EX-4752, Bowling, EX-4709, Golf, EX-3456, Gym, EX-4522, Library, EX-4665, Pool, EX-4522, Rec Svc, EX-2410, Stables, EX-4607, Sports Tickets, 472-1895, Visitors Center, Self-guided tours, Ex-2555. Farish Memorial Fam Camp, off Base, 6 mi NE of Woodland Park, 1 May-30 Sep, 24 camper spaces w/elec, 11 camper spaces w/o hookups, 4 bdrm lodge, full fac and many activities, Comm: 303-687-9098.

SPACE-A: Peterson AFB, Bldg 122, 0530-1930 M-F, 0700-1700 Sat-Sun, Comm: 303-554-4707/4521, ATVN: 692-4707/4521, flts to CONUS locations.

ATTRACTIONS: Pikes Peak, snow skiing, ghost towns, Colorado Springs.

CONNECTICUT

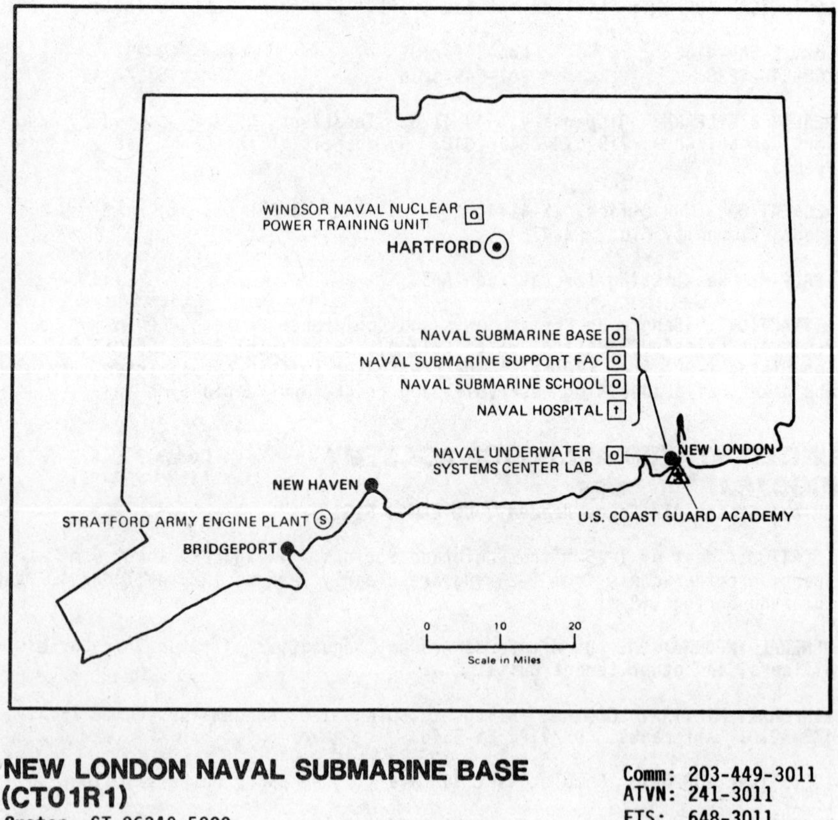

NEW LONDON NAVAL SUBMARINE BASE (CT01R1)

Groton, CT 06349-5000

Comm: 203-449-3011
ATVN: 241-3011
FTS: 648-3011

New London NSB, Cont'd

LOCATION: From I-95 N take exit 86 to CT-12 & Crystal Lake Rd. Base clearly marked. RM: p-19, I/17. NMC: Hartford, 50 mi NW.

GENERAL INFORMATION: Naval Submarine School, Naval Underwater Medical Institute, Naval Hospital, submarine units.

TEMPORARY MILITARY LODGING: No Biltg Ofc, Navy Lodge, 77 Dewey Ave, 0700-2300 dly, Comm: 203-446-1160, ATVN: 241-1160. All ranks. DV/VIP, EX-3416.

LOGISTICAL SUPPORT: Complete spt fac avail. Key phone numbers follow.

NX-449-3811	**Caf-449-3679**	**Chap-449-3232**
Chld C-449-3500	**Commiss-449-3912**	**Conv S-446-8593**
CPO C-449-3237	**EM C-449-3050**	**Fam Svc-449-4847**
Gas S-449-3626	**Legal-449-3741**	**Loc-449-3709**
Med-449-3666	**Off C-449-3352**	**Pkg S-449-3535**
Pol-449-3224	**Theat-449-3358**	

HEALTH & WELFARE: Naval Hospital, clinics and depts. Emerg, EX-3666, Appoint, EX-4460, 74 bed. Chapels serving all faiths, EX-3232.

RECREATION: Auto Shop, EX-3582, Bowling, EX-3477/3454, Golf, EX-3763, Gym, EX-3521/4205, Hobby, EX-3217, Marina, EX-3164, Library, EX-3723, Swimming, EX-3562, Rec Svc, EX-3687.

SPACE-A: Try Bradley IAP, N of Hartford, Comm: 203-623-8291, ATVN: 636-8310, ANG units and sometimes MAC flts.

ATTRACTIONS: US Coast Guard Academy, Mystic Seaport, Ocean Beach Park, Nautilus Memorial and Submarine Force Library and Museum and Croker Submarine Memorial.

UNITED STATES COAST GUARD ACADEMY (CT02R1)

Comm: 203-444-8270
FTS: 642-8270

New London, CT 06320-4195

LOCATION: From I-95 N take exit 83 in New London. Clearly marked. RM: p-19, J/17. NMC: New London, in the city.

GENERAL INFORMATION: The Officer Education and Training Academy for the US Coast Guard. The nation's oldest seagoing service and agency for maritime law enforcement and marine safety.

TEMPORARY MILITARY LODGING: None.

LOGISTICAL SUPPORT: Limited spt fac avail. Key phone numbers follow.

CGX-444-8488	**Chap-444-8480**	**Conv S-444-8492**
CPO C-444-8457	**Cr Un-446-8200**	**EM C-444-8455**
Gas S-444-8494	**Hous Ofc-444-8211**	**Legal-444-8253**
Main Gate-444-8614	**Med-444-8400**	**Off C-444-8141**
Pkg S-444-8491	**Pol-444-8597**	**Pers Sup Ctr-444-8689**

HEALTH & WELFARE: Base OOD, 444-8450, fire and emergencies, 444-8555. Inpatient, New London Naval Hospital, Groton, Comm: 203-444-8400.

SPACE-A: See the New London Naval Submarine Base listing.

RECREATION: Bowling, 444-8470, Library, 444-8510, Visitors Center, 444-

US Coast Guard Academy, Cont'd

8611, Gym, 444-8600, Musical Activities, 444-8471, Hobby Shop, 444-8489, Acad Events Recording, 442-1092, and Theater Ticket Office, 444-8154.

ATTRACTIONS: Visitors are invited to take a self-guided tour of the Academy daily 9 a.m. to 5 p.m.. The Museum is open from 8 a.m. to 4 p.m. weekdays throughout the year. May through October, the Visitor's Pavilion and Museum are open daily, weekends, and holidays from 10 a.m. to 5 p.m. When in port, the barque EAGLE is open to the public on Sundays from 1 p.m. to 4 p.m. and during the Spring and Fall, formal reviews by the corp of cadets are held on the Washington parade ground.

DELAWARE

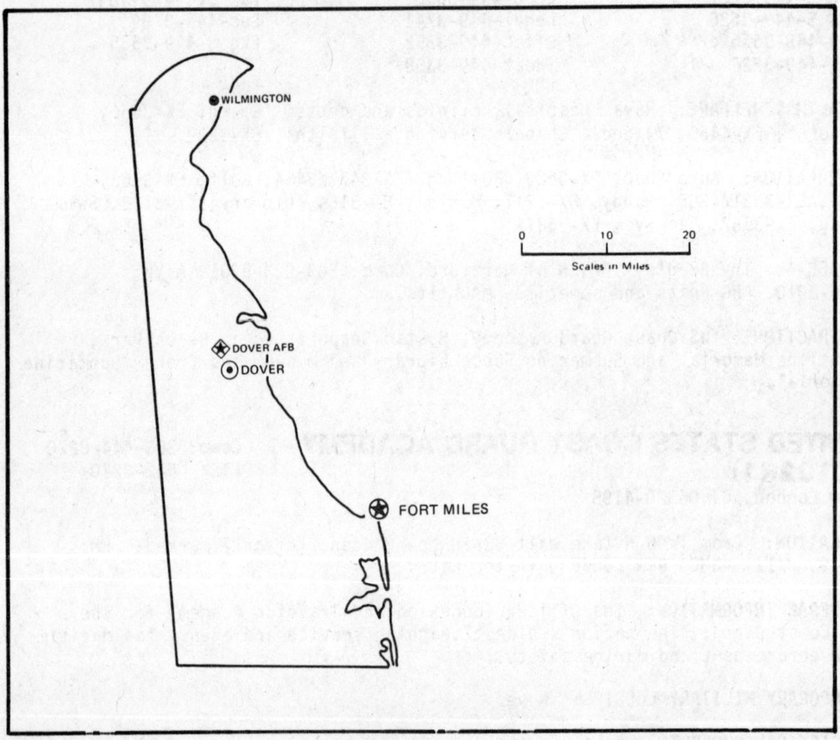

DOVER AIR FORCE BASE (DE01R1)

Dover Air Force Base, DE 19902-5000

COMM: 302-678-7011
ATVN: 455-6881

LOCATION: Off US-113. Clearly marked. RM: p-41,F/24. NMC: Dover, 5 mi NW.

GENERAL INFORMATION: Military Airlift Command Base. Largest air cargo port on East Coast. AFRES units.

TEMPORARY MILITARY LODGING: Biltg Ofc, Bldg 805, 14th St, 24 hrs dly, Comm: 302-678-6329, ATVN: 455-6329. All ranks. DV/VIP, EX-6095.

LOGISTICAL SUPPORT: Complete spt fac avail. Key phone numbers follow.

Dover AFB, Cont'd

BX-678-4862	Caf Term-674-3380	Chap-678-6341/6417
Chld C-674-2994	Commiss-674-4189	Conv S-674-3551
EM C-678-2600	Fam Svc-678-6370	Gas S-674-4228
Legal-678-6526	Loc-678-6333	NCO C-678-6307
Off C-678-6364	Pol-678-6113	Snack B-674-2241
Theat-678-8711(off base)	Theat-678-7575(on base)	

HEALTH & WELFARE: USAF Hospital, clinics and depts, Emerg, 735-2600, Appoint, 735-2500, 40 bed. Chapels serving all faiths, 678-6341/6342/6343.

RECREATION: Auto Hobby, EX-6205, Beach, 674-2736, Bowling, EX-6304, Golf, EX-6067, Gym, EX-6345, Hobby, EX-6205, Library, EX-6246, Swim, EX-6345, Rec Ctr Ex-6513, Rec Fac, EX-6066, Tennis, EX-6345, Youth Act, EX-6252, Skeet, EX-6060.

SPACE-A: Bldg 500, 24 hrs dly, Comm: 302-678-6982, ATVN: 455-6982/3, flts to CONUS (mid/West Coast) overseas (Europe/Middle-East). Also, try Greater Wilmington Airport, Bldg 2812, 0800-1630, M-F, Comm: 302-322-2261 EX-204, ATVN: 455-9204, flts to CONUS locations. ANG units.

ATTRACTIONS: State capital, great beaches and the Delaware Bay.

DISTRICT OF COLUMBIA

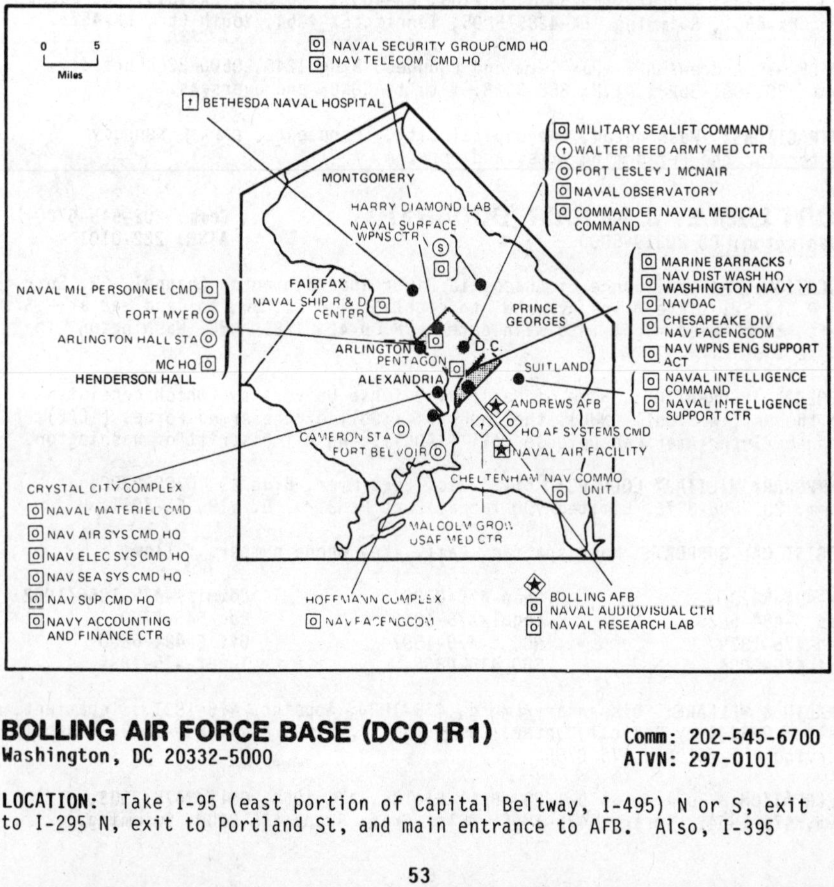

BOLLING AIR FORCE BASE (DC01R1)

Washington, DC 20332-5000

Comm: 202-545-6700
ATVN: 297-0101

LOCATION: Take I-95 (east portion of Capital Beltway, I-495) N or S, exit to I-295 N, exit to Portland St, and main entrance to AFB. Also, I-395

Bolling AFB, Cont'd

(Southeast Freeway) N, exit South Capitol St, cross Anacostia River on South Capitol St, main entrance to AFB on right. Clearly marked. RM: p-42, I/8. NMC: Washington, in SE section of the city.

GENERAL INFORMATION: Air Force District of Washington. Support for Air Force activities of the National Capital Region.

TEMPORARY MILITARY LODGING: Biltg Ofc, Bldg 602, Theisen St, 24 hrs dly, Comm: 202-767-5878/5771, ATVN: 297-5878/5771. All ranks. DV/VIP, EX-4195.

LOGISTICAL SUPPORT: Complete spt fac avail. Key phone numbers follow.

BX-562-3000	**Caf-562-4419**	**Chap-767-5900**
Chld C-767-2890	**Commiss-767-4044**	**Conv S-562-3105**
Fam Svc-767-4464	**Gas S-561-0484**	**Legal-767-4439**
Loc-767-4522	**Med-767-4253/5520**	**NCO C-563-1400**
Snack B-562-4419	**Theat-767-4608**	

HEALTH & WELFARE: Clinic/Emerg, EX-4253/5520, Appoint, EX-4253. Inpatient, Malcolm Grow USAF Medical Center, Andrews AFB, MD, Comm: 301-981-6867, ATVN: 858-6867. Chapels serving all faiths, EX-5900.

RECREATION: Arts & Crafts, EX-4422, Auto Hobby, EX-4564, Bowling, EX-1701, Gym, EX-5269, Library, EX-5578, Marina, EX-4651, Rec Ctr, EX-5315, Rec Equip Ctr, EX-4651, Swimming, EX-4286/5895, Tennis, EX-4354, Youth Ctr, EX-4575.

SPACE-A: Andrews AFB, Pax Term and Lounges, Bldg 1245, 0600-2200 hrs dly, Comm: 301-981-3528, ATVN: 858-3528, flts to CONUS and overseas.

ATTRACTIONS: Washington, the capital city. Monuments, parks, Kennedy Center for the Performing Arts, and more.

FORT LESLEY J. McNAIR (DC05R1)

Washington, DC 20319-5000

Comm: 202-545-6700
ATVN: 222-0101

LOCATION: At confluence of Anacostia River and Washington Channel, SW. Entr on P St, SW. Take Maine Ave, SW, to right on 4th St, SW, to dead end at P St. Left, then immed right into main gate. RM: p-42, I/8. NMC: Washington, in SW sect of city.

GENERAL INFORMATION: Home of National Defense University (which consists of the Natl War Coll (NWC), the Industrial Coll of the Armed Forces (ICAF), and the Inter-American Defense Coll (IADC)). Hq Mil District of Washington.

TEMPORARY MILITARY LODGING: Biltg Ofc, Fort Myer, Bldg 49, 0700-2300, Comm: 202-696-3576. Limited VOQ fac at Fort McNair. DV/VIP, EX-7051.

LOGISTICAL SUPPORT: Most spt fac avail. Key phone numbers follow.

PX-484-5823	**Chap-475-0706**	**Commiss-475-1966/1962**
Gas S-484-5825	**Legal-475-1069**	**Loc-545-6700**
Med-475-1829	**NCO C-475-1597**	**Off C-484-5800**
Pol-475-2004	**SDO-475-0866**	**Theat-475-1999**

HEALTH & WELFARE: Dispensary/Emerg, 475-1829, Appoint, 475-1831. Inpatient, Walter Reed Army Medical Center, Comm: 202-576-1199, ATVN: 291-1199. Chapels serving all faiths, 475-0706.

RECREATION: Bowling, 475-1729, Ball Fields, 475-1964, Golf, 475-2003, Gym, 475-1964, Library, 475-1995, Multi-Craft Shop, 475-2000, Swimming,

Fort Lesley J McNair, Cont'd

475-0504, Tennis, 475-1964, Youth Ctr, 475-1262.

SPACE-A: See Andrews AFB, MD, listing for details.

ATTRACTIONS: Very historic fort, site of the trial and execution of President Lincoln conspirators, Washington waterfront along Maine Ave, restaurants, seafood markets, monuments, and Potomac River.

PENTAGON (VA41R1)

Washington, DC 20310-5000

Comm: 202-545-6700
ATVN: 227-0101

LOCATION: From I-395 S, exit to South Parking or from Columbia Pike (VA-244) exit to South Parking. North Parking entr can be reached from Boundary Drive, exit from I-395 S. Visitor parking entr is from Boundary Dr and is paid park. Due to tightened security the River and Mall entrances are accessible only by walking from other parking areas. Bus, METRO, and taxi service fr Concourse. Concourse and all entrances have directions posted to support activities/ facilities. RM: p-42, H/7.

GENERAL INFORMATION: Hq Department of Defense, Joint Chiefs of Staff and the military departments. The world's largest office building, 17.5 mi of corridors, employing about 27,000 people.

TEMPORARY MILITARY LODGING: See nearby Fort Myer, VA, listing.

LOGISTICAL SUPPORT: Very ltd. Most are commercial activities. Military Living book titles are on sale at book store in Concourse, 695-0870/5433.

Chap-695-3336	**Legal-AR-697-3170, AF-695-2450, NA-695-1332**
Med-697-4778/4598	**Off C-521-5020**
Snack B-2 per floor	**Caf-4 + Dining Rm 3C1063**

HEALTH & WELFARE: Military and civilian health clinic - mil: Rm 1E356 (Concourse area), 697-4778/4598, civ: Rm 2A750, 697-3375. Inpatient, Walter Reed Army Medical Center and other installations in the area. See the resp listings in this book. Chapel, Rm 5A1070, 695-3336. Armed Forces Hostesses Association, Rm 1A736, 697-6857/3180, volunteers, info on the area.

RECREATION: Athletic Center, adj to River Entrance, Membership Club, 521-5020, has work-out rooms, jogging, racquet sports, swimming, sauna and health rooms. Library, 697-4301, Rec Svc - USA: Rm 3A146, 697-3816, USAF: Rm 5E367, 697-9866, OSD Welfare & Rec: Rm 3C1055, 695-5338, Ski Club, 697-9866, GELCO Travel, 892-9311.

SPACE-A: See Andrews AFB, MD, listing.

ATTRACTIONS: The Pentagon guided tour is an exciting tourist attraction, (office on Concourse). Free tours leave hourly. Great military art on the corridor walls and the Hall of Heroes, Medal of Honor winners.

WALTER REED ARMY MEDICAL CENTER (DC03R1)

Washington, DC 20307-5000

Comm: 202-576-3501
ATVN: 291-3501

LOCATION: 6900 Georgia Ave, NW. From I-495 (Capital Beltway) take Georgia Ave/Silver Spring exit south to Center, enter first or second gate. To reach the Forest Glen support facilities from Georgia Ave, south, right turn on to Linden Lane, cross over B&O railroad bridge, support fac on left (.75 miles

Walter Reed AMC, Cont'd

from Georgia Ave). RM: p-42, F/8. NMC: Washington, DC, in city limits.

GENERAL INFORMATION: The major Army Medical Research, Teaching, and Treatment Facility.

TEMPORARY MILITARY LODGING: Biltg Ofc, ATTN: HSHL-LF-UPH, Bldg 18, Georgia Ave & Butternut St, 24 hrs dly, Comm: 202-576-2076/2096, ATVN: 291-2076/2096. All ranks. DV/VIP, EX-3501.

LOGISTICAL SUPPORT: Complete spt fac avail. Key phone numbers follow.

PX-588-4790	Chap-576-1305	Chld C-576-3516
Commiss-427-5013	Conv S-829-1303	EM C-829-8932
Fam Svc-576-3412	Gas S-291-7819	Legal-576-2531
Loc-576-3501	Med-576-1199	NCO C-829-8932
Off C-829-0067	Pol-576-2511	Snack B-565-0900/ EX-109/110
SNCO-576-1000		

HEALTH & WELFARE: Major medical center, clinics and depts, Emerg, EX-1199, Appoint, EX-3501, 1280 bed. Chapels serving all faiths, EX-1305.

RECREATION: Auto Hobby, EX-2433, Bowling, EX-5436, Gym, EX-2324/5500, Hobby Shops, 427-5205, Library, EX-1314, Swimming, EX-2324, Tennis, EX-2324, Rec Ctr, EX-3032, Tickets, EX-3032.

SPACE-A: See Andrews AFB listing for details.

ATTRACTIONS: Washington, DC, White House, Capitol, Skyline Drive, Ocean City, MD, PA Dutch Country.

WASHINGTON, DC, AREA CLUBS

ARMY & NAVY CLUB (DCO7R1)

17th and I Streets, NW
Washington, DC 20006-5000

Comm: 202-628-8400
ATVN: None

LOCATION: At 17th and I Streets, NW, in the city. RM: p-42, G/7.

GENERAL INFORMATION: A private club for members (active and retired officers of the uniformed services) and military associated club members. Undergoing extensive renovation and reconstruction, 12 stories with two levels of parking, scheduled for completion in 1986.

TEMPORARY LODGING: Accommodations for transient members and guest of members. 29 guest suites, check at front desk. Comm: 202-628-8400.

SUPPORT AND RECREATION: Bar, dining and banquet facilities and temporary lodging. Library and lounges. Extensive private office space in new building.

USO WORLD HQ'S & USO OF METROPOLITAN WASHINGTON (DCO6R1)

601 Indiana Avenue NW
Washington, DC 20004-5000

Comm: 202-783-8117
ATVN: None

LOCATION: In the Bob Hope USO Building at 601 Indiana Avenue, NW. The USO of Metropolitan Washington's Center is located in the basement. The USO lobby is on the first floor, the USO of Metropolitan Washington's administrative offices are on the second floor, while the offices of USO World Headquarters

Wash DC Area Clubs, Cont'd

are on the third and fourth floors. RM: p-42, G/7.

RECREATION: Lounge, snacks, reading room, showers, etc. Free tickets for shows and sporting events. Ticket and information hot line: 202-879-4700. There are nine Outreach Locations of the USO in the greater Washington, DC, area serving uniformed personnel and their families. See the USO appendix in this book for addresses and telephone numbers.

WASHINGTON NAVY YARD (DC02R1)

Washington, DC 20374-5000

Comm: 202-545-6700
ATVN: 222-6700

LOCATION: Main entrance is on M St, SE, at 9th St. Exit I-395/I-695 N at 11th St, SE, right turn on M St, west to entr at 9th St on left side of M St. Also, exit I-395/I-695 at South Capitol St, SE, stay right, left at 2d traffic light onto M St, SE. Entr on right at 9th St. RM: p-42, H/8. NMC: Washington, DC, in SE section.

GENERAL INFORMATION: Hq Naval Dist of Washington. Also, **USMC BARRACKS (DC04R1), at 8th and I Streets SE near the Washington Navy Yard. The oldest post of the Corps. Sunset parades in summer. Hq's USMC Band. Limited spt fac avail, 979-8420.**

TEMPORARY MILITARY LODGING: Biltg Ofc, ATTN: Anacostia CMAA: NAVSTA in the Anacostia complex adjacent to Bolling AFB. From I-95 (Beltway East) take I-295 N exit at South Capitol St. NDW Anacostia is clearly marked. Comm: 202-433-3232, ATVN: 288-3232. New Navy Lodge in Bellevue Navy Housing area, Bolling AFB, 0700-2300 hrs dly, Comm: 202-563-6950.

LOGISTICAL SUPPORT: Most spt fac avail. Key phone numbers follow.

NX-889-4909	**Caf-889-7910**	**Chap-433-2679/2057**
Chld C- 981-3035	**CPO C-433-2523**	**EM C-433-2048**
Fam Svc-433-6150	**Gas S-433-2710**	**Legal-433-4331**
Loc-694-3155	**Med-433-3407/3750**	**Off C-433-3041**
Pol-433-2411	**Snack B-889-7910**	**SNCO-433-2217/2607**

HEALTH & WELFARE: Dispensary/Emerg, EX-3407/3269. Inpatient, Naval Medical Command National Capital Region, clinics and depts, Comm: 202-295-5385, ATVN: 295-5385. Chapels serving all faiths, 433-2679.

RECREATION: Library, 433-2386, Auto Hobby, 433-2066, Rec Ctr, 433-2061, Rec Equip, 433-2269, Tennis, 433-2062, Swimming, 433-2061. NDW Marina at the Anacostia Naval Station offers boat rentals, 472-3750.

SPACE-A: Washington NAF at Andrews AFB, MD, Pax Term and Lounge, Bldg 3198, 0600-2400 dly, Comm: 301-981-2740/2744, ATVN: 858-2740/2744, flts to CONUS locations. Also, Andrews AFB, Pax Term and Lounge, Bldg 1245, 0600-2200 hrs dly, Comm: 301-981-3528, ATVN: 858-3528, flts to CONUS and overseas.

ATTRACTIONS: Navy Memorial Museum, Combat Art Gallery, Navy Historical Center, Display Ship Barry (DD-933), Washington, DC, monuments, Capitol, White House, cherry blossoms in April.

FLORIDA

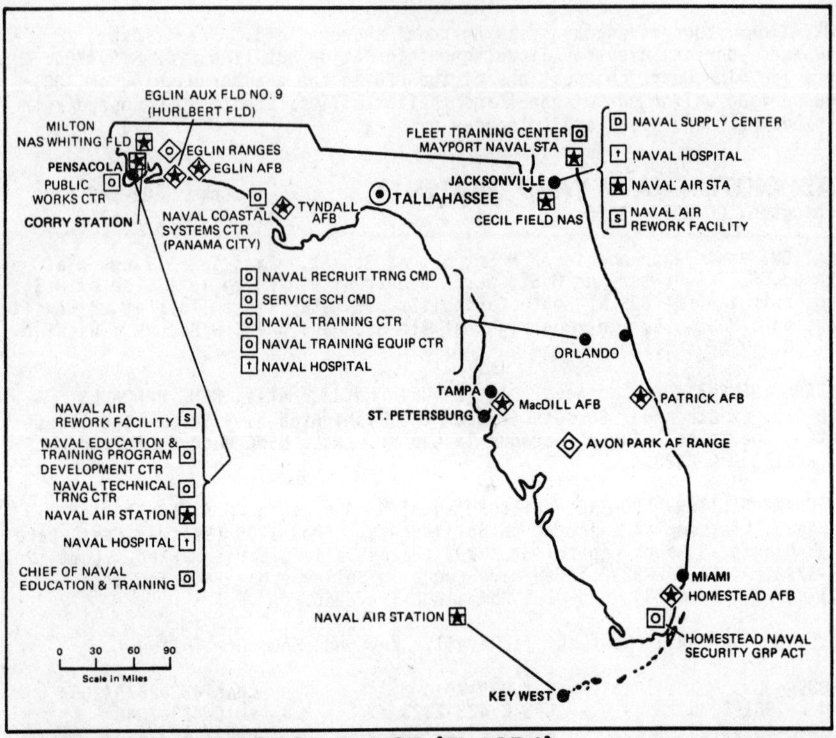

AVON PARK AIR FORCE RANGE (FL16R1)

Avon Park, FL 33825-5000

Comm: 813-452-1110
ATVN: 968-1110-E114

LOCATION: Off FL-64, 10.5 mi E of Avon Park, 15.5 mi NE of Sebring. RM: p-21, N/13. NMC: Orlando, 60 mi N.

GENERAL INFORMATION: Bombing and Gunnery Range.

TEMPORARY MILITARY LODGING: Biltg Ofc, Bldg 236, 0730-1630 hrs dly, Comm: 813-452-4128/4114, ATVN: 968-4128/4114. All ranks. DV/VIP, Biltg Ofc.

LOGISTICAL SUPPORT/RECREATON: Limited spt fac and rec avail. Key phone numbers follow.

BX-452-4123	Gas S-452-4153

CECIL FIELD NAVAL AIR STATION (FL06R1)

Cecil Field Naval Air Station, FL 32215-5000

Comm: 904-778-5675
ATVN: 860-5675

LOCATION: Take Normandy exit W off I-295 and follow Normandy (FL-228) to main gate. RM: p-21, E/11. NMC: Jacksonville, 13 mi E.

GENERAL INFORMATION: Naval Combat Aircraft Support and Training Station.

TEMPORARY MILITARY LODGING: Navy Lodge, Bldg 822, 24 hrs dly, Comm: 904-772-0947.

LOGISTICAL SUPPORT: Complete spt fac avail. Key phone numbers follow.

Cecil Field NAS, Cont'd

NX-778-3571	Caf-778-5225	Chap-778-5340
Chld C-778-5638	Commiss-778-5485	Conv S-778-0251
EM C-778-6154	Fam Svc-778-6160	Gas S-771-0778
Legal-778-5208	Loc-778-5240	Med-772-2241
NCO C-778-6154	Off C-778-5255	Pkg S-778-5555
Pol-778-5381	Theat-778-6113	

HEALTH & WELFARE: Dispensary/Emerg, EX-2241. Inpatient, Jacksonville Naval Hosp, Comm: 904-772-2241, ATVN: 942-2241. Chapels serving all faiths, EX-5340.

RECREATION: Golf, EX-5245, Rod & Gun, EX-5181, Skeet, EX-5181, Tickets, EX-6112, Ceramics Shop, EX-6112, Auto Hobby, EX-6112, Gym, EX-5498, Spec Svc Ofc, EX-6112, Spec Svc Rental, EX-6112, Woodwork, EX-6112, Youth Act, EX-6112, Swimming, EX-5281. Lake Newman, on Base, year round, club house and picnic area, EX-6112. Lake Fretwell Rec Area, on Base, year round, 4 camper spaces w/hookups, comp fac and all rec activities.

SPACE-A: Pax Term and Lounge, Bldg at end of "D" Ave, 24 hrs dly, Comm: 904-778-5536/5481, ATVN: 860-5536/5481, flts to NASs in CONUS.

ATTRACTIONS: Beaches, golf, Busch Gardens, Disney World (Orlando Area), Cypress Gardens, and St Augustine, America's oldest city.

CORRY NAVAL STATION (FL19R1)

Pensacola, FL 32511-5000 **Comm: 904-452-6512**
ATVN: 922-6512

LOCATION: Off US-98, 3 mi N of Pensacola NAS. RM: p-21, X/3. NMC: Pensacola, 5 mi NE.

GENERAL INFORMATION: Home of the Naval Technical Training Center.

TEMPORARY MILITARY LODGING: None. Navy Lodges at Pensacola and Whiting Field NAS's, see those listings.

LOGISTICAL SUPPORT: Most spt fac avail. Key phone numbers follow.

NX-453-5311	Chap-452-6561	Conv S-453-5311
EM C-452-6347	Loc-452-6512	Med-452-6326
CPO C-452-6330	Off C-452-6300	Pol-452-6155

HEALTH & WELFARE: Dispensary/Emerg, 452-6326. Inpatient, Pensacola Naval Hosp, 452-6611. Chapels serving all faiths, 452-6561.

RECREATION: Archery, Bowling, Hobby Shop, Swimming, Tennis, Theater, Rec Ctr, 452-6553. Rec vehicle park at Pensacola NAS.

SPACE-A: See Pensacola NAS listing.

ATTRACTIONS: Visit the historic section of Pensacola which is undergoing renovation and restoration. Beaches. Naval Aviation Museum at NAS Pensacola.

EGLIN AIR FORCE BASE (FL27R1)

Eglin Air Force Base, FL 32542-5000 **Comm: 904-881-6668**
ATVN: 872-1110

LOCATION: Exit I-10 at Crestview & follow posted direction signs to Niceville and Valparaiso (Eglin AFB). RM: p-21, W/5. NMC: Fort Walton Beach, 7 mi S.

Eglin AFB, Cont'd

GENERAL INFORMATION: Air Force Systems Command Base. Largest AFB in Free World, Armament Div and Lab, Air Warfare Center, AFRES units.

TEMPORARY MILITARY LODGING: Biltg Ofc, Bldg 11001, Boatner Rd, 24 hrs dly, Comm: 904-882-8761, ATVN: 872-8761. All ranks. DV/VIP, EX-3011.

LOGISTICAL SUPPORT: Complete spt fac avail. Key phone numbers follow.

BX-651-2512	**Chap-882-4426**	**Chld C-882-3861**
Commiss-882-3560	**Fam Svc-882-2893**	**Gas S-678-7222**
Legal-882-4611	**Loc-882-4478**	**Med-885-3227**
NCO C-678-5127	**Off C-651-1010**	**Pol-882-2502**
Theat-882-3813		

HEALTH & WELFARE: USAF Regional Hospital, Emerg, 885-2333/3227, Appoint, 885-4131/5001, 160 bed. Chapels serving all faiths, 882-2111.

RECREATION: Golf, 882-2695, Hobby Shops, 882-3570, Library, 882-2460, Youth Act, 882-5074. Also, Swimming, Auto Hobby, Bowling, Gym, Skeet, and large and small Game Hunting and Fishing on the AFB. Fam-Camp, on Base, year round, 22 camper spaces w/hookups, Comm: 904-882-5058. Comp fac and full act.

SPACE-A: Pax Term and Lounge in Bldg 60, 24 hrs dly, Comm: 904-882-5732, ATVN: 872-5732, flts to CONUS locations.

ATTRACTIONS: Fort Walton Beach and beach areas, sport fishing, deep sea fishing, dog races, Pensacola, 35 mi W.

HOMESTEAD AIR FORCE BASE (FL17R1)

Homestead Air Force Base, FL 33039-5000 — **Comm: 305-257-8011** — **ATVN: 791-0111**

LOCATION: Off Florida Turnpike, US-1, take Homestead exit. Clearly marked. RM: p-21, V/15. NMC: Miami, 40 mi NE.

GENERAL INFORMATION: Tactical Air Command Base. Fighter aircraft operations and training. AFRES units stationed here.

TEMPORARY MILITARY LODGING: Biltg Ofc, Bldg 945, St Nazair Street, 24 hrs dly, Comm: 305-257-5831, ATVN: 791-5831. All ranks. DV/VIP, EX-7212.

LOGISTICAL SUPPORT: Complete spt fac avail. Key phone numbers follow.

BX-257-5811	**Caf-257-5117**	**Chap-257-7895**
Chld C-257-8303	**Commiss-257-8717**	**Conv S-257-5851**
EM C-257-8949	**Fam Svc-257-7261**	**Gas S-257-2358**
Legal-257-7206	**Loc-257-8011/0111**	**Med-257-7233/8137**
NCO C-257-5610/5511	**Off C-257-8791**	**Pkg S-257-8293**
Pol-257-7200	**Snack B-257-2826**	**Theat-257-7559**

HEALTH & WELFARE: USAF Regional Hospital, Emerg, EX-8137/7233, Appoint, EX-8081. Chapels serving all faiths, EX-7895.

RECREATION: Auto Shop, EX-2647, Bowling, EX-2247, Golf, EX-2872, Gym, EX-7404, Hobby Shops, EX-5591, Library, EX-8433, Swimming, EX-8384, Rec Ctr, EX-7139, Youth Act, EX-8219. Mystic Lake Fam-Camp, on Base, year round, Comm: 305-257-8949, ATVN: 791-8949, 15 camper spaces w/hookups, full spt fac and comp rec including tickets and tours.

SPACE-A: Pax Term and Lounge, Bldg 701, 24 hrs dly, Comm: 305-257-8659, ATVN: 791-8659, flts to CONUS locations.

Homestead AFB, Cont'd

ATTRACTIONS: Biscayne Bay, deep sea fishing, Everglades National Park, horse and dog racing, boating, and water sports.

HURLBURT FIELD (FL18R1)

Hurlburt Field, FL 32544-5000

Comm: 904-881-6668 (Eglin Base Info)
ATVN: 872-1110

LOCATION: Off US-98, 5 mi W of Fort Walton Beach. Clearly marked. RM: p-21, X/4. NMC: Pensacola, 40 mi W.

GENERAL INFORMATION: Military Airlift Command Base, on the Eglin AFB reservation, Hq 2nd Air Div, First Special Operations Wing, Air Force Air Ground Ops School.

TEMPORARY MILITARY LODGING: Biltg Ofc, Bldg 90509, 24 hrs dly, Comm: 904-884-6245/7115, ATVN: 872-6245/7115. All ranks. DV/VIP, Biltg Ofc.

LOGISTICAL SUPPORT: Most spt fac avail. Key phone numbers follow.

BX-581-0030	Chap-884-7795	Commiss-581-3621
Conv S-581-0030	Gas S-581-2224	Med-884-7882
NCO C-884-6468	Off C-581-3110	Pkg S-884-6370
Pol-884-6423	Theat-884-7648	

HEALTH & WELFARE: Clinic/Emer, 884-7882, Appoint, 884-7882. Inpatient, USAF Regional Hosp, Eglin AFB, Comm: 904-882-4131, ATVN: 872-4131. Chapels serving all faiths, 884-7795.

RECREATION: Auto Hobby, 884-6674, Bowling, 884-6470, Hobby Shops, 884-6492, Library, 884-6947, Marina, 884-6939, Swimming, 884-6866, Youth Ctr, 884-6355. **See** Eglin AFB listing for Rec Areas.

SPACE-A: Try Base Ops at Hurlburt, 884-7806. Eglin AFB, Bldg 60, 24 hrs, Comm: 904-882-5313, ATVN: 872-5313, flts to CONUS installations.

ATTRACTIONS: Sport fishing, water sports, Pensacola (west) and Panama City (east), both short drives.

JACKSONVILLE NAVAL AIR STATION (FL08R1)

Jacksonville Naval Air Station, FL 32212-5000

Comm: 904-772-2338
ATVN: 942-2345

LOCATION: Access from US-17 S (Roosevelt Blvd). On the St Johns River. RM: p-21, E/15. NMC: Jacksonville, 9 mi NE.

GENERAL INFORMATION: Combat aircraft support and training, Naval Hospital.

TEMPORARY MILITARY LODGING: Navy Lodge: 904-772-6000, ATVN: 942-2549 or central Navy Lodge Resv, 800-NAVY-INN.

LOGISTICAL SUPPORT: Complete spt fac avail. Key phone numbers follow.

NX-777-7200	Chap-772-3051	Chld C-772-5434
Commiss-772-3431	Conv S-777-7170	CPO C-772-3461
EM C-772-3521	Fam Svc-772-2766	Legal-772-2941
Loc-2951/2340	Med-777-7300	Off C-772-3041
Pkg S-772-2869	Pol-772-2661	Theat-772-2858

HEALTH & WELFARE: Naval Hospital, Emerg, 777-7340, Appoint, 778-2000,

Jacksonville NAS, Cont'd

310 bed. Chapels serving all faiths, 777-7531.

RECREATION: Athletic Dir, EX-3239, Hobby Shop, EX-3523, Library, EX-3415, Spec Svc, EX-3111/2, Marina, EX-3260, Tickets, EX-3318, Bowling, EX-3493, Golf, EX-3249, Swimming, EX-2930.

SPACE-A: Pax Term and Lounge, Bldg 118, 24 hrs dly, Comm: 904-772-2537, ATVN: 942-2537, flts to CONUS and Caribbean area.

ATTRACTIONS: Beaches, deep sea fishing, St Augustine, Orlando (3 hrs S).

KEY WEST NAVAL AIR STATION (FL15R1)

Key West Naval Air Station, FL 33040-5000

Comm: 305-292-2434
ATVN: 483-2434

LOCATION: Take Florida Turnpike, US-1 S to exit signs for Key West NAS on Boca Chica Key, 7 mi N of Key West. RM: p-21, Z/11. NMC: Miami, 150 mi N.

GENERAL INFORMATION: Tactical aircraft and rescue aircraft operations, Coast Guard, reserve and other services support detachments.

TEMPORARY MILITARY LODGING: Biltg Ofc, Bldg 2076, 24 hrs dly, Comm: 305-294-2112/5571, ATVN: 483-2112/5571. All ranks. DV/VIP, 292-2178.

LOGISTICAL SUPPORT: Complete spt fac avail. All EX through switchboard.

NX-296-5121	Chap-292-2318	Chld C-294-4924
Commiss-296-6628	Conv S-292-2538	CPO C-292-2502
EM C-296-5537	Fam Svc-294-5760	Gas S-296-5745
Legal-EX-8210	Loc-292-2178	Med-294-3500
Off C-296-6431	Pol-296-7706	Theat-EX-2511

HEALTH & WELFARE: Clinic/Emerg, 294-0302. USAF Regional Hospital, Homestead AFB, Comm: 305-257-8011, ATVN: 791-0111. Chapels serving all faiths, 292-2318.

RECREATION: Auto Hobby, 292-2615, Bowling, 292-2130, Flying Club, 292-5081, Gym, 292-8323, Marina, 292-3664, Rec Ctr, 292-2112/2678, Water Rec, 292-2468.

SPACE-A: Pax Term at Control Tower, 0800-1600 hrs dly, Comm: 305-292-2179, ATVN: 483-2179, flts to CONUS installations.

ATTRACTIONS: Boating, sport fishing, water sports, Audubon House, Ernest Hemingway Home and Museum.

MacDILL AIR FORCE BASE (FL02R1)

MacDill Air Force Base, FL 33608-5000

Comm: 813-830-1110
ATVN: 968-1110

LOCATION: Take I-75 S to I-275 S, exit at Dale Mabry Hwy (US-92) S, 5 mi S to MacDill AFB main gate. RM: p-21, 9/4. NMC: Tampa, 5 mi N.

GENERAL INFORMATION: Tactical Air Command Base. Hq US Readiness Command, and US Central Command. F-16 fighter aircraft training.

TEMPORARY MILITARY LODGING: Biltg Ofc, 56 CSG/SVHH/Biltg, PO Box 6826, Bldg 411, corner Garden Dr & Tampa Blvd, 24 hrs dly, Comm: 813-830-4179, ATVN: 968-4179. All ranks. DV/VIP, EX-2056.

LOGISTICAL SUPPORT: Complete spt fac avail. Key phone numbers follow.

MacDill AFB, Cont'd

BX-840-0511	Caf-840-0511	Chap-830-3621
Chld C-830-3332	Commiss-830-3196	Conv S-840-2077
EM C-830-3357	Fam Svc-830-2221	Gas S-839-7089
Legal-830-4421	Loc-830-2444	Med-830-3334
NCO C-830-3357	Off C-837-1031	Pkg S-840-2323
Pol-830-3322	Snack B-840-2211	Theat-830-3973

HEALTH & WELFARE: USAF Regional Hospital, Emerg, 839-3344, Appoint, 839-5711, 75 bed. Chapels serving all faiths, EX-3621.

RECREATION: MWR, EX-2821, Arts & Crafts, EX-4413, Auto Shop, EX-4553, Bowling, EX-3008, Golf, EX-2200, Gym, EX-4648, Library, EX-3607, Marina, EX-2801, Rec Ctr, EX-3594, Youth Act, EX-4244. MacDill Fam-Camp, Base 1-A, Comm: 813-830-3864, ATVN: 968-3864, on Base, year round, 143 camper spaces w/hookups, 12 camper spaces w/o hookups, comp fac and full activities.

SPACE-A: Pax Term and Lounges, Hangar 3, 0730-1630 weekdays, 0730-1230 weekends, Comm: 813-830-2310, ATVN: 968-2310, flts to CONUS locations. Also, USCG Air Station, St Petersburg, Comm: 813-534-1437 EX-219, may have flts.

ATTRACTIONS: Disney World, pro football, sport fishing, water sports.

MAYPORT NAVAL STATION (FL13R1)

Jacksonville, FL 32228-5000

Comm: 904-246-5401
ATVN: 960-5401

LOCATION: From Jacksonville take Atlantic Blvd (FL-10) E to Mayport Rd (FL-A1A), left (N) to Naval Station. RM: p-20, D/12. NMC: Jacksonville, 10 mi W.

GENERAL INFORMATION: Naval Air Fac, Fleet Tng Ctr, home port for 40 ships.

TEMPORARY MILITARY LODGING: Biltg, ATTN: UPO Ofc, Bldg 1394, Bailey Av, 24 hrs dly, Comm: 904-246-5707, ATVN: 960-5707. All rnks. DV/VIP, EX-5401.

LOGISTICAL SUPPORT: Complete spt fac avail. Key phone numbers follow.

NX-246-5292	Chap-246-5212	Chld C-246-5339
Commiss-246-5372	Conv S-246-5548	CPO C-246-5432
EM C-246-5300	Fam Svc-246-5344	Gas S-246-5712
Legal-246-5245	Loc-246-5401	Med-246-5444
Off C-246-5313	Pkg S-246-5566	Pol-246-5583
Snack B-246-5670/5137	Theat-246-5369	

HEALTH & WELFARE: Dispensary/Emer, EX-5444, Appoint, EX-5306/7. Inpatient, Navy Hospital, Jacksonville NAS, Comm: 904-772-5310, ATVN: 942-5310. Chapels serving all faiths, EX-5212/3.

RECREATION: Auto Shop, EX-5392, Beach, EX-5178, Bowling, EX-5377, Golf, EX-5380, Gym, EX-5451, Hobby Shops, EX-5110, Int'l Sports, EX-5451, Library, EX-5393, Swimming, EX-5400, Rec Ctr, EX-5545, Skeet, EX-5228, Tennis, EX-5717, Tickets, EX-5145, Youth Act, EX-5451.

SPACE-A: Jacksonville NAS, Pax Term and Lounge, Bldg 118, 24 hrs dly, Comm: 904-772-2537, ATVN: 942-2537, flts to CONUS and Caribbean.

ATTRACTIONS: Great beaches, deep sea fishing, zoo, and 3 hrs to Orlando.

ORLANDO NAVAL TRAINING CENTER (FL11R1)

Orlando, FL 32813-5005

Comm: 407-646-4111
ATVN: 791-4111

Orlando NTC, Cont'd

LOCATION: On Bennet Road, .5 mi N of FL-50 (Colonial Dr). Bennet Rd is about 3 mi E of I-4 off FL-50. RM: p-21, J/4. NMC: Orlando, in the city.

GENERAL INFORMATION: Training Center, schools, Naval Hospital, Naval Training Systems Center on main Base, Annex has Navy Recruiting, and Audit Team. Annex is 12 mi S adjacent to Orlando IAP, site of the old McCoy AFB.

TEMPORARY MILITARY LODGING: Biltg Ofc, Bldg 2010, Hibiscus St, 0730-1600 hrs dty days, Comm: 407-646-5163/5722. ATVN: 791-5163/5722, Navy Lodge, also handles lodging at Annex. All ranks. DV/VIP, EX-4532.

LOGISTICAL SUPPORT: Complete spt fac avail. Key phone numbers follow.

BX-646-5937	**Caf-646-5826/4330**	**Chap-646-5186**
Chld C-646-5131	**Commiss-857-2350**	**Conv S-646-4171**
CPO C-646-5050	**EM C-646-5205**	**Fam Svc-646-4144/4290**
Gas S-646-5979	**Legal-646-5181**	**Loc-646-4111**
Med-646-4946	**Off C-646-4867**	**Pkg S-646-5733**
Pol-646-4444	**Snack B-646-4330**	**Theat-646-5502**

HEALTH & WELFARE: Naval Branch Medical Clinic, Emerg, EX-4001/2, Appoint, EX-4982/3/4, 104 bed. Chapels serving all faiths, EX-4922.

RECREATION: Auto Hobby, 646-5920, 857-1380, Bowling, 857-2311, 646-5048, Golf, 857-0280, 646-5736, Marina, 857-1380, 646-5850, Sailing, 646-5850, Flying Club, 898-3311, Gym, 646-5161, 857-2660, Library, 857-0690, Swimming, 857-1340, 646-5196, Youth Act, 857-1760/3733. Travel Trailer Park at Annex, Comm: 305-857-2120, ATVN: 791-5164, on Base, year round, 29 camper spaces w/hookups, comp fac and activities.

SPACE-A: Patrick AFB, Bldg 800, 0730-1630 dly, Comm: 305-494-5631, ATVN: 854-5631, flts to CONUS and overseas (Caribbean and South Africa).

ATTRACTIONS: The playground of Florida, Disney World, Busch Gardens, Cypress Garden, water sports, fishing, Sea World, Epcot Center, Church St Station, Friday Recruit Review at Navy Base (free), Wet 'n' Wild, Citrus Circus, Gatorland Zoo, Florida Festival, Elvis Presley Museum, dog racing, Factory Outlet Center, and Jai lai.

PATRICK AIR FORCE BASE (FLO3R1)

Patrick Air Force Base, FL 32925-5152 — **Comm: 407-494-1110** **ATVN: 854-1110**

LOCATION: Take I-95 S to exit 73 (Wickham Rd), 3 mi to FL-404 (Pineda Causeway, toll .20), L on S Patrick Dr to Patrick AFB. RM: p-21, L/15. NMC: Cocoa Beach, 2 mi N.

GENERAL INFORMATION: Air Force Systems Command Base. Eastern Space and Missile Center, support to DOD and NASA, airhead for Cape Canaveral AFS.

TEMPORARY MILITARY LODGING: Biltg Ofc, ATTN: 6550 ABG/SVH, Bldg 400, C St between 1st and 2nd Sts, 24 hrs dly, Comm: 407-494-2075, ATVN: 854-2075. All ranks. DV/VIP, EX-4506.

LOGISTICAL SUPPORT: Complete spt fac avail. Key phone numbers follow.

BX-799-1300	**Dinette-494-4458**	**Chap-494-4073**
Chld C-494-2550	**Commiss-494-5841**	**Conv S-494-4555**
Fam Svc-494-4907	**Gas S-494-2655**	**Legal-494-7356**
Loc-494-4542	**Med-494-8134**	**NCO-494-7001**

Patrick AFB, Cont'd

NCO C-494-7491	Off C-494-4011	Pkg S-494-6686
Pol-494-7777	Snack B-494-2831	Theat-494-2057

HEALTH & WELFARE: USAF Regional Hospital, Emerg, EX-8134, Appoint, EX-8241, 25 bed. Chapels serving all faiths, EX-4073.

RECREATION: MWR, EX-6695, Auto Hobby, EX-2537, Bowling, EX-2958, Aero Club, EX-4356, Golf, EX-7856, Gym, EX-6697, Hobby Shops, EX-2482, Library, EX-6881, Swimming, EX-6796, Rec Ctr, EX-5523, Skeet, EX-4787, Youth Ctr, EX-4747.

SPACE-A: Pax Term and Lounge, Bldg 800, 0730-1630 dly, Comm: 305-494-5631, ATVN: 854-5631, flts to CONUS, Caribbean and South Africa. South Africa requires approval of base commanders.

ATTRACTIONS: Busch Gardens, Disney World, Daytona Beach, Epcot, Spaceport USA, Circus World, Sea World.

PENSACOLA NAVAL AIR STATION (FL14R1)

Pensacola Naval Air Station, FL 32508-5000 — **Comm:** 904-452-2311 **ATVN:** 922-2311

LOCATION: Off US-98, 4 mi S, and 12 mi S of I-10. Take Navy Blvd from US-98 or US-29 directly to NAS. RM: p-21, X/3. NMC: Pensacola, 8 mi N.

GENERAL INFORMATION: Naval Aviation Schools Command, home port of Navy Blue Angels, USS Lexington, and Naval Air Rework Facility.

TEMPORARY MILITARY LODGING: Biltg Ofc, Bldg 600, 24 hrs dly, Comm: 904-452-2255, ATVN: 922-2255. All ranks. Navy Lodge. DV/VIP, EX-4481.

LOGISTICAL SUPPORT: Complete spt fac avail. Key phone numbers follow.

NX-455-0392	Caf-456-9956	Chap-452-2341
Chld C-452-2211	Commiss-452-6891	Conv S-456-5701
CPO C-452-3214	EM C-452-3251	Fam Svc-452-3393
Gas S-456-7447	Legal-452-4321	Loc-452-2353
NCO-452-2353	Off C-455-2276	Pkg S-456-5701
Pol-452-2653	Snack B-455-0969	

HEALTH & WELFARE: Pensacola Naval Hospital, Emerg, EX-6600/6731, Appoint, 432-0258, 138 bed. Chapels for all faiths, EX-2341.

RECREATION: Auto Shop, EX-9424, Beach, EX-4391, Bowling, EX-4630, Marina (Sherman Cove), EX-3369, Golf, EX-2455, Gym, EX-2317, Hobby, EX-2424, Library, EX-4362, Sailing, EX-4152, Tickets, EX-4229, Swimming, EX-4391, Youth Act, EX-4209. Oak Grove Park, on Base, year round, Rec Bldg 632, Comm: 904-452-2535, ATVN: 922-2535, 24 camper spaces w/hookups, 6 apartments, 11 mobile homes.

SPACE-A: Pax Term and Lounge, Bldg 1852, 24 hrs dly, Comm: 904-452-3311, ATVN: 922-3311, flts to CONUS locations.

ATTRACTIONS: Sport fishing, water sports, dog races, Naval Aviation Museum (aboard NASP), annual Fiesta of Five Flags, Ft. Barrancas (aboard NASP), annual Blue Angel Marathon, annual Blue Angels Homecoming Air Show (aboard NASP).

PENSACOLA NAVAL HOSPITAL (FL20R1)

Pensacola, FL 32512-5000 — **Comm:** 904-452-6601 **ATVN:** 922-6601

Pensacola Nav Hosp, Cont'd

LOCATION: On US-98 between Navy Blvd and Fairfield Dr in Pensacola. RM: p-21, X/3. NMC: Pensacola, in the city.

GENERAL INFORMATION: The Naval Hospital, new in 1976.

TEMPORARY MILITARY LODGING: None. Navy Lodges and Biltg at Pensacola NAS and Whiting Field NAS.

LOGISTICAL SUPPORT: Limited spt fac avail. Most spt is at Pensacola NAS.

HEALTH & WELFARE: Complete range of clinics and depts, Emerg, 452-6730/6731, Info Desk, 452-6601. Most clinics 0730-1600 hrs dly.

RECREATION/SPACE-A and ATTRACTIONS: See Pensacola NAS and Corry Naval Station listings.

TYNDALL AIR FORCE BASE (FL04R1)

Tyndall Air Force Base, FL 32403-5000

Comm: 904-283-1110
ATVN: 970-1110

LOCATION: Take I-10, exit to US-231 S to US-98E, signs mark the way to AFB. RM: p-21, Y/8. NMC: Panama City, 10 mi NW.

GENERAL INFORMATION: Tactical Air Command Base. U.S Air Force Defense Weapons Center. 23rd North American Aerospace Defense Command Region and TAC Air Division. Air Force Engineering and Services Center.

TEMPORARY MILITARY LODGING: Biltg Ofc, Bldg 1332, Suwanne and Oak Dr, 24 hrs dly, Comm: 904-283-4210 (reservations EX-2394), ATVN: 970-4210. All ranks. DV/VIP, EX-2232.

LOGISTICAL SUPPORT: Complete spt fac avail. Key phone numbers follow.

BX-286-5804/283-4110	Chap-283-2925	Chld C-283-4747
Commiss-283-4825	Fam Svc-283-2963	Gas S-286-5844
Legal-283-4681	Loc-283-2138/4210	Med-283-7523
NCO C-283-4444	Off C-283-2986	Pol-283-4124
SNCO-283-2965		

HEALTH & WELFARE: USAF Regional Hospital, clinics and depts, Emerg, EX-2333, Appoint, EX-4277/2023, 50 bed. Chapels serving all faiths, EX-2925

RECREATION: Auto Hobby, Beach, Bowling, Flying Club, Golf, Gym, Hobby Shops, Library, Marina, Swimming, Tennis & Theater, EX-2495, Youth Ctr EX-4326. Fam-Camp, on Base, year round, Rec Svc (Marina), Comm: 904-283-2798, ATVN: 970-2798, 20 camper spaces w/hookups, full spt fac and all activities, water sports. No reservations. Also, Panama City Rec Area, Naval Coast Systems Ctr, on US-98, year round, Comm: 904-234-4402, ATVN: 436-4402, 12 camper spaces w/hookups, spt fac and water sports.

SPACE-A: Try Base Ops, EX-4244.

ATTRACTIONS: Panama City and beautiful white sand beaches.

WHITING FIELD NAVAL AIR STATION (FL05R1)

Whiting Field Naval Air Station, FL 32570-5000

Comm: 904-623-7437
ATVN: 868-7437

LOCATION: From US-90E exit, FL-87 N for 8 mi to NAS. RM: p-21, W/4. NMC: Pensacola, 30 mi SW.

Whiting Field NAS, Cont'd

GENERAL INFORMATION: Naval Aviation Training, fixed & rotary wing aircrft.

TEMPORARY MILITARY LODGING: Biltg Ofc, Bldg 2942, Lexington Cir, 24 hrs dly, Comm: 904-623-7605, ATVN: 868-7605. All ranks. Navy Lodge.

LOGISTICAL SUPPORT: Complete spt fac avail. Key phone numbers follow.

NX-623-8066	**Caf-623-5897**	**Chap-623-7211**
Commiss-623-7131	**Conv S-623-6364**	**EM C-623-7591**
Fam Svc-623-7177	**Gas S-623-8088**	**Legal-623-7231**
Loc-623-7437	**Med-623-7509**	**NCO C-623-2339**
Off C-623-7311	**Pkg S-623-7198**	**Pol-623-7387**
Snack B-623-5897	**Theat-623-7471**	

HEALTH & WELFARE: No medical fac at this Base. Inpatient, Pensacola Naval Hosp, Emer, 425-6601/6731, Appoint, 452-6705, 138 bed. Chapels serving all faiths, EX-7211.

RECREATION: Archery, 623-7445, Boat Dock, 623-2383, Bowling, 623-7545, Golf, 623-7348, Gym, 623-2274, Hobby, Library, 623-2171, Rec Ctr, 623 -7293, Skeet, 623-7293. See NAS Pensacola for travel camp.

SPACE-A: Pensacola NAS, Bldg 1852, 24 hrs dly, Comm: 904-452-3311, ATVN: 922-3311, flts to CONUS locations, mostly NAS's.

ATTRACTIONS: Deep sea fishing, water sports, golf, dog races, thieves market each weekend, stock car racing, Pensacola Symphony Orchestra.

Small Installations Cont'd from page 45

Gowen Field, USAF
Bldg T-303
Boise ID 83705-5000

O'Hara Airport, USAF
Mannheim & Higgins Rd, Bldg 10
Chicago IL 60666-5000

Camp Atterbury
Bldg 708
Edinburg IN 46214-5000

Des Moines Site
Rodger Road, Bldg 123
Des Moines IN 50315-5000

Forbes Air National Guard
Bldg 679
Topeka KS 66620-5000

Bluegrass Army Depot
Bldg 229
Lexington KY 40507-5000

New Orleans Coast Guard Base
4640 Urquhart St
New Orleans LA 70117-5000

Natick Laboratories
Bldg P-4
Natick MA 01760-5000

Boston Coast Guard Spt Ctr
427 Commercial St
Boston MA 02109-5000

South Portland Coast Guard
Base, 259 High Street
South Portland ME 04106-5000

Detroit Arsenal
Bldg 718
Warren MI 48090-5000

Detroit Coast Guard Group
Mt Elliott Ave
Detroit MI 48207-5000

Port Austin Air Force Station
Bldg 527
Port Austin MI 48467-5000

Michigan Air National Guard
3367 W Dickman Road
Battle Creek MI 49017-5000

Cont'd page 76

GEORGIA

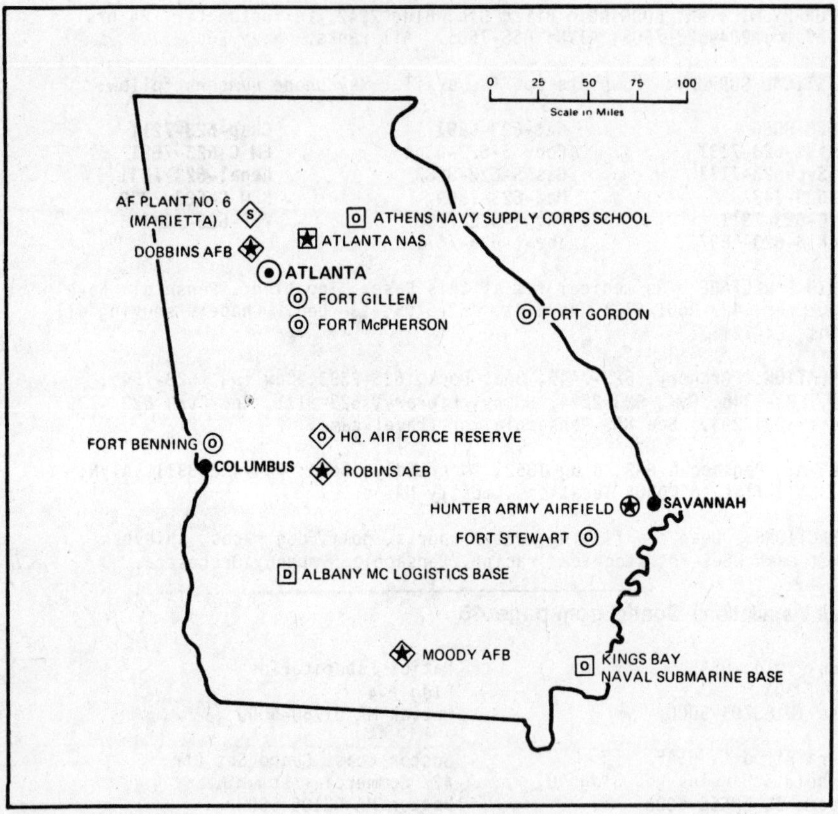

ALBANY MARINE CORPS LOGISTICS BASE (GA17 R1)

Comm: 912-439-5000
ATVN: 460-5000

Albany Marine Corps Logistics Base, GA 31704-5000

LOCATION: Off US-82, take Mock Rd S from 5-Points to Fleming Road. Go east to main gate. RM: p-23, P/6. NMC: Albany, 3 mi W.

GENERAL INFORMATION: The east coast main Marine Corps Supply Base.

TEMPORARY MILITARY LODGING: Four quarters are available for temporary lodging, reservations 439-5234. First come, first served basis.

LOGISTICAL SUPPORT: Most spt fac avail. Key phone numbers follow.

MX-435-1471	Chap-439-5282	Commiss-435-1721
Conv S-435-1471/4271	EM C-439-5267	Fam Svc-439-5276
Gas S-436-8352	Legal-439-5213	Loc-439-5000/5206
Med-435-0806/5984	Nur-439-5285	NCO C-439-5268/5231
Off C-439-5269/5239	Pkg S-439-5267/8/9	Pol-439-5183
Snack B-435-1471	Theat-439-5220	

Albany MC Log Base, Cont'd

HEALTH & WELFARE: Dispensary, Emer, EX-0806, Appoint, EX-5984. Inpatient, USAF Hosp, Moody AFB, Valdosta, 90 mi SE, Comm: 912-333-3232. Chapel serving all faiths, EX-5282.

RECREATION: Skeet Range, EX-5234/5, Golf, EX-5211, Hobby Shops, EX-5226, Bowling, EX-5233, Gym, EX-5246, Library, EX-5242, Swimming, EX-5234, Rod & Gun Club, EX-5000.

SPACE-A: Robins AFB, Bldg 110, 0630-1700 hrs dly, Comm: 912-926-3166, ATVN: 468-3166, flts to CONUS and overseas.

ATTRACTIONS: Drum & Bugle Corps. Tallahassee, FL, easy drive south.

ATHENS NAVAL SUPPLY CORPS SCHOOL (GA12R1)

Comm: 404-354-1500
ATVN: 588-7222

Athens Naval Supply Corps School, GA 30606-5000

LOCATION: From Athens GA take bypass, exit to Prince Ave, continue 1 mi to Base at intersection of Prince & Oglethorpe Aves. RM: p-23, G/8. NMC: Atlanta, 70 mi W.

GENERAL INFORMATION: Navy Supply Corps School, Personnel Support Detach.

TEMPORARY MILITARY LODGING: Biltg Ofc, Brown Hall, 0800-1600 hrs dty dys, Comm: 404-354-7316, ATVN: 588-7316. Off only. DV/VIP, Biltg Ofc.

LOGISTICAL SUPPORT: Limited sup fac avail. Key phone numbers follow.

NX-354-3850	**Commiss-354-7384**	**Gas-354-7350**
NCO-354-7316		

HEALTH & WELFARE: Dispensary, Emerg, 354-7321. Inpatient, Dwight D. Eisenhower Army Med Ctr, Fort Gordon, Augusta, 100 mi SE, Emerg, Comm: 404-791-0110. Chapels serving all faiths, 354-7208.

RECREATION: Gym, 354-7377, Library, 354-7217, Swimming, 354-7300, Tennis, Tickets, 354-7205.

SPACE-A: Atlanta Naval Air Station, Base Ops, Comm: 404-429-5359, ATVN: 925-5359, flts to CONUS locations.

ATTRACTIONS: Athens and the University of Georgia.

ATLANTA NAVAL AIR STATION (GA16R1)

Comm: 404-429-5503
ATVN: 925-5503

Atlanta Naval Air Station, GA 30060-5000

LOCATION: Off I-75 and US-41, exit to GA-280 W to GA-3 S to main gate, adjacent to Dobbins AFB on the west. RM: p-24, B/1. NMC: Atlanta, 15 mi SE.

GENERAL INFORMATION: Navy and Marine Corps tactical aircraft and crew tng.

TEMPORARY MILITARY LODGING: Very ltd. Off and Enl, Comm: 404-429-5393.

LOGISTICAL SUPPORT: Most spt fac avail. Key phone numbers follow.

NX-427-8175/4	**Chap-429-5472**	**Chld C-429-5498**
CPO C-429-5298	**EM C-429-5492**	**Legal-429-5396/7**
Med-429-5302	**Off C-429-5504**	**PAO-429-5406/2**

Atlanta NAS, Cont'd

HEALTH & WELFARE: Dispensary/Emerg, EX-5311, Appoint, EX-5302. Inpatient, USAF Hosp, Robins AFB, Macon, 80 mi S, Emerg, 914-926-3845, Appoint, EX-2841. Chapel serving all faiths, EX-5472/3/4.

RECREATION: Spec Svs, EX-5498, some fac, use Dobbins AFB fac. Operates Lake Allatoona Rec Area, 40 mi N of Atlanta and 20 mi N of NAS, Comm: 404-974-6309, 14 camper spaces w/hookups, 1 camper space w/o hookups, 5 cabins, comp fac, all activities on the lake.

SPACE-A: Try Base Ops at NAS, Comm: 404-429-5359, ATVN: 925-5359 or Dobbins AFB, Comm: 404-429-4903.

ATTRACTIONS: Atlanta is easy drive 15 mi S. Stone Mountain State Park, Six Flags Over Georgia, and other attractions are within 30 mi of NAS.

DOBBINS AIR FORCE BASE (GA13R1)

Dobbins Air Force Base, GA 30069-5000

Comm: 404-421-5000
ATVN: 925-1110

LOCATION: From I-75 N exit to GA-280 W to AFB. Clearly marked. RM: p-24, B/1. NMC: Atlanta, 16 mi SE.

GENERAL INFORMATION: Air Force Reserve Base. Hq 14th Air Force (AFRES), Airlift and Fighter Wings.

TEMPORARY MILITARY LODGING: Biltg Ofc, Bldg 401, Ridenour Dr, 0700-1200 dly, Comm: 404-429-4946, ATVN: 925-4946. All ranks.

LOGISTICAL SUPPORT: Limited spt fac avail. Key phone numbers follow.

BX-428-3054	**Con Ms-429-4594**	**EM C-427-9015**
Gas S-427-4400	**Loc-429-5000**	**Med-429-5302**
Pkg S-429-4594	**Pol-429-4908**	

HEALTH & WELFARE: Dispensary/Emerg, EX-5302. Inpatient, USAF Hosp, Robins AFB, Macon, 85 mi SE, Emerg, Comm: 912-926-3845. Chapels serving all faiths, EX-4955/4956 (Prot/Cath), 873-1731 (Jewish).

RECREATION: Flying Club, Gym, Swimming, Tennis, EX-4872. Lakeside Fam-Camp, on Base, year round, Comm: 404-429-4870, ATVN: 925-4872, 16 camping spaces w/o hookups (hookups avail 9/86), most other fac and activities.

SPACE-A: Pax Term and Lounge, Bldg 737, 0700-2300 hrs dly, Comm: 404-429-4903, ATVN: 925-4903, flts to CONUS, Reserve units.

ATTRACTIONS: Stone Mountain, Six Flags, Underground Atlanta.

FORT BENNING (GA11R1)

Fort Benning, GA 31905-5065

Comm: 404-544-2011
ATVN: 784-2011

LOCATION: Off I-185 and US-27/280. RM: p-23, Y/5. NMC: Columbus, 5 mi NW.

GENERAL INFORMATION: Army Infantry Center and School. Army Infantry Training Center, School of the Americas, 197th Infantry Brigade, 75th Ranger Regiment, 10th Mountain Division, 29th Infantry Regiment, U.S. Army Infantry Board, 36th Engineer Group, 34th Medical Battalion.

TEMPORARY MILITARY LODGING: Biltg Ofc, Bldg 399, 24 hrs dly, Comm: 404-545-1851/7312, ATVN: 835-1851/7312. All ranks. DV/VIP, EX-545-2613.

Fort Benning, Cont'd

LOGISTICAL SUPPORT: Complete spt fac avail. Key phone numbers follow.

PX-687-0384	**Chap-545-2050**	**Chld C-689-8698**
Commiss-544-3663	**Conv S-682-0473**	**EM C-689-0887**
Fam Svc-545-1169	**Gas S-687-6520**	**Legal-545-3281**
Loc-545-5216	**Med-544-2041**	**NCO C-687-0600**
Off C-682-0640	**Pkg S-687-9674**	**Pol-545-5222**
Snack B-687-6013	**Theat-545-1151**	

HEALTH & WELFARE: Martin Army Community Hospital, clinics and depts, Emer, 544-1502, Appoint, 544-2041, 500 bed. Chapels serving all faiths, 545-2050.

RECREATION: Arts & Crafts, Bowling, Flying Club, Golf, Gym, Hobby Shops, Int'l Sports, Library, Pool, Tennis, Rod & Gun Club, Youth Ctr, Rec Ctr, all 545-4436. Uchee Creek Rec Area, on Post, year round, Rec Svc, Comm: 404-545-7238/4053, ATVN: 835-7238/4053, 34 camper spaces w/hookups, all other fac and activities. Also, operates USA Inf Ctr Rec Area, Destin, FL, on Choctawatchee Bay, year round, Comm: 404-545-4155, 20 camper spaces w/hookups, 37 camper spaces w/o hookups, 14 cottages, all other fac and activities.

SPACE-A: Lawson Army Airfield on Post, limited Space-A via executive Army aircraft, 545-2857.

ATTRACTIONS: National Infantry Museum on Post, Springer Opera House, Columbus, and Callaway Gardens, Pine Mountain.

FORT GILLEM (GA21R1)

Fort Gillem, GA 30050-5000

Comm: 404-363-5000
ATVN: 797-1001

LOCATION: From I-75, E on I-285 to US-23 (Moreland Ave/Macon Hwy), S for 3 mi to the main gate. Fort is 5 mi from Hartsfield IAP. RM: p-24, J/5. NMC: Atlanta, 10 mi NW.

GENERAL INFORMATION: Sub-Post of Fort McPherson. Hq 2nd US Army, 2nd US Army Recruiting Brigade (SW), 3rd CID Region, Army CID Lab, Army and Air Force Exchange Distribution Region, and other tenant agencies.

TEMPORARY MILITARY LODGING: Biltg Ofc, Bldg 816, 0730-1600 dty hrs, Comm: 404-363-5431, ATVN: 797-5431. All ranks. DV/VIP, 752-4145.

LOGISTICAL SUPPORT: Most spt fac avail. Key phone numbers follow.

PX-363-5483	**Chap-362-3299**	**Commiss-363-5361**
Conv S-363-5483	**Fam Svc-362-7764**	**NCO C-363-3530**
Off C-363-3830	**Pkg S-363-9047**	**Pol-363-5582**

HEALTH & WELFARE: Dispensary at Fort McPherson, 752-4437. Inpatient, USAF Hosp, Robins AFB, Macon, 80 mi S, Comm: 914-926-3845. Chapels serving all faiths, 362-7395.

RECREATION: Bowling, 363-5460, Golf, Gym, EX-5853, Library, EX-2568, Rec Ctr, EX-5853, Skeet, 363-5316, Tennis, 363-5853, Youth Ctr, 363-5520.

SPACE-A: None. Try Dobbins AFB, Bldg 737, 0700-2300 hrs dly, Comm: 404-429-4903, ATVN: 925-4903, flts to CONUS and some overseas, Reserve units.

ATTRACTIONS: Atlanta Braves, Six Flags, Stone Mountain.

FORT GORDON (GA09R1)

Fort Gordon, GA 30905-5000

Comm: 404-791-0110
ATVN: 780-0110

LOCATION: Between US-78/278 and US-1. Gates are on both US-78 & US-1. RM: p-23, H/13. NMC: Augusta, 12 mi NE.

GENERAL INFORMATION: Army Signal Center & School, Dwight David Eisenhower Army Medical Center.

TEMPORARY MILITARY LODGING: Biltg Ofc, Bldg 250, 24 hrs dly, Comm: 404-791-2277 or 404-793-7160, ATVN: 780-2277. DV/VIP, EX-5376.

LOGISTICAL SUPPORT: Complete spt fac avail. Key phone numbers follow.

PX-793-7171	**Caf-793-8161**	**Chap-791-4683**
Chld C-791-2701	**Commiss-791-3718**	**Conv S-793-1160**
EM C-791-2207	**Fam Svc-791-6967**	**Gas S-793-8363**
Legal-791-7812	**Loc-791-4675**	**Med-791-5811**
NCO C-791-6780	**Off C-791-2205**	**Pkg S-793-5366**
Pol-791-4380	**Snack B-793-9757**	**Theat-791-5425**

HEALTH & WELFARE: Dwight David Eisenhower Army Medical Center, Emerg, 791-5811, Appoint, 791-4521. Chapels serving all faiths, 791-4683/3959.

RECREATION: Bowling, EX-3446, Craft Shops, EX-4476, Gyms, EX-2864/3692, Golf, 791-2433, Library, EX-2449, Riding Stables, EX-4864, Rec Ctr, 541-0063, Swimming, EX-3002, Fish and Wildlife, EX-5025, Youth Act, EX-5104. Clarks Hill Rec Area, 25 mi N of Post at Clarks Hill Reservoir, year round, Rt 1 Box 94, Appling, GA 30802, Comm: 404-791-4517, ATVN: 780-4517, 100 camper spaces w/ hookups, 8 cottages, 8 mobile homes, complete facilities and activities.

SPACE-A: Base Ops, Army Airfield, ltd Space-A via exec aircraft, EX-0110.

ATTRACTIONS: Augusta Masters Golf Tournament, many golf courses, and historical sites in old Augusta.

FORT McPHERSON (GA08R1)

Fort McPherson, GA 30330-5000

Comm: 404-752-3113
ATVN: 588-1110

LOCATION: Off I-75, take Lakewood Freeway (GA-166), exit to US-29 (Main St exit). Main gate is at Main St exit. RM: p-24, H/3. NMC: Atlanta, in city limits.

GENERAL INFORMATION: Hq Army Forces Command and Hq Third Army.

TEMPORARY MILITARY LODGING: Biltg Ofc, Bldg T-22, 0730-1615 dty dys, Comm: 404-752-2253, ATVN: 588-2253. All ranks. DV/VIP, EX-4145.

LOGISTICAL SUPPORT: Complete spt fac avail. Key phone numbers follow.

Mini Mall-753-6258	**Chap-752-2315/3954**	**Chld C-752-3945/2759**
Commiss-752-2231	**Fam Svc-752-2439**	**Gas S-753-2114**
Legal-752-4159	**Loc-752-2743/4174**	**Med-752-3711**
NCO-752-2980/3602	**NCO C-753-4540**	**Off C-753-4520**
Pkg S-753-2444	**Pol-752-3050 or 752-HELP**	**Theat-752-2423**

HEALTH & WELFARE: Dispensary/Emerg, EX-2672, Appoint, EX-2183/3473. Inpatient, USAF Hosp, Robins AFB, Macon, 80 mi S, Comm: 914-926-3845. Chapels serving all faiths, EX-2315/3144.

RECREATION: Arts & Crafts, EX-2221, Auto Hobby, EX-2070, Bowling, EX-2479,

Fort McPherson, Cont'd

Golf, EX-2178/2078, Gym, EX-3034, Library, EX-3045/2518, Rec Ctr, EX-3677, Tickets, EX-3677/4392, Youth Ctr, EX-2763. FORSCOM Rec Area, 40 mi N of Atlanta at Allatoona Lake Reservoir, year round, Comm: 404-974-3413, 24 camping spaces w/hookups, 10 camping spaces w/o hookups, 8 motel units, 10 mobile homes, 10 cabins, all facilities and complete activities.

SPACE-A: Try Atlanta NAS, Marietta, Base Ops, Comm: 404-429-5418, ATVN: 925-5418, limited Space-A.

ATTRACTIONS: 6 Flags, Cyclorama, prof sports, Stone Mountain St Pk.

FORT STEWART (GA15R1)

Fort Stewart, GA 31314-5000

Comm: 912-767-1110
ATVN: 870-1110

LOCATION: On US-82. Accessible from US-17 or I-95. Also GA-119 or GA-144 crosses the Post but may be closed occasionally. RM: p-23, N/14. NMC: Savannah, 35 mi NE.

GENERAL INFORMATION: Hq 24th Infantry Division and Army support units.

TEMPORARY MILITARY LODGING: Biltg Ofc, Bldg 802, 0730-1630 hrs dty days, Comm: 912-767-8384, ATVN: 870-8384. All ranks. DV/VIP, EX-2927.

LOGISTICAL SUPPORT: Complete spt fac avail. Key phone numbers follow.

PX-876-2850	**Chap-767-4219**	**Chld C-876-2586**
Commiss-767-3070	**Conv S-876-2785**	**EM C-767-4922**
Fam Svc-767-3434/3700	**Gas S-876-5265**	**Legal-767-2952**
Loc-767-2862	**Med-767-6666**	**NCO C-767-2837**
Off C-767-2837	**Pkg S-368-5080**	**Pol-767-2822**
SDO-767-8666	**Theat-767-3069/7083**	

HEALTH & WELFARE: Winn Army Community Hospital, clinics and depts, Emerg, EX-6161, Appoint, EX-6100, 165 bed. Chapels serving all faiths, EX-4993/4219.

RECREATION: Auto Hobby, EX-2069, Golf, EX-2370, Gym, EX-7090/2143 and EX-7675/8897, Swimming, EX-8575/3045, Skeet, EX-3519, Youth Ctr, EX-2815. Also, Arts & Crafts, EX-3770, Bowling, EX-4127/4866, Library, EX-7149. Holbrook Pond Army Travel Camp, on Post, year round, Comm: 912-767-2717, ATVN: 870-2717, 20 camper spaces w/hookups, 20 camper spaces w/o hookups, complete facilities and all activities.

SPACE-A: Robins AFB, Bldg 110, 0630-1700 hrs dly, Comm: 912-926-3166, ATVN: 468-3166, flts to CONUS and overseas.

ATTRACTIONS: Jekyll Island off coast of GA; fishing, beaches, and boating; historic Savannah and other historic sites; Cumberland Island National Seashore; Okefenokee Swamp National Wildlife Refuge. Hunting on Ft. Stewart.

HUNTER ARMY AIRFIELD (GA10R1)

Hunter Army Airfield, GA 31409-5000

Comm: 912-352-5500
ATVN: 971-1110

LOCATION: From I-95 to GA-204 E for 13 mi to Savannah. Turn left onto Stephenson Ave, proceed straight into Wilson Ave Gate to Installation. RM: p-23, Z/14. NMC: Savannah, in SW part of city.

GENERAL INFORMATION: Aviation and other Army support battalions.

TEMPORARY MILITARY LODGING: Biltg Ofc, Bldg 6010, Duncan at Leonard St,

Hunter AAF, Cont'd

0730-2400 M-F, 0800-1700 Sat, Sun, Hol, Comm: 912-352-5910/5834, ATVN: 971-5910/5834. All ranks. DV/VIP, Comm: 912-767-8610/8779, ATVN: 870-8610/8779.

LOGISTICAL SUPPORT: Complete spt fac avail. Key phone numbers follow.

PX-352-8380	**Chap-352-5515/5111**	**Chld C-352-5261**
Commiss-352-5711/3070	**Conv S-354-0075**	**Fam Svc-352-5259/5301**
Gas S-354-9546	**Legal-352-5115**	**Loc-352-5913**
Med-352-5551/Emerg-5700	**NCO-352-5000**	**NCO C-352-5262**
Off C-352-5995	**Pkg S-354-8752**	**Pol-352-6133/6135**
Theat-352-5556		

HEALTH & WELFARE: Clinic, Emerg, EX-5551/5700. Inpatient, Winn Army Comm Hosp, Ft Stewart, Comm: 912-767-6001, ATVN: 971-1110. Chapels serving all faiths, EX-5515.

RECREATION: Rec Ctr, EX-4363, Bowling, EX-5205/6279, Gym, EX-5078/5918, Auto Crafts, EX-6244, Craft Shops, EX-5622, Racquet Courts, Library, EX-5816/5688, Swimming, EX-5562, Golf (9 hole), EX-5384, Ball Fields, Skeet and Trap Range, EX-6148, Outdoor Equip check out, EX-5274/5722, and Lotts Island, picnic area, fishing pier and docks for boats, EX-5916, Youth Act and Comm Ctr, EX-3076, Swimming Pool, EX-5562.

SPACE-A: Base Ops, HAAF, limited Space-A via executive aircraft, EX-5500. Also, Savannah CGAS at HAAF limited Space-A, Comm: 912-233-8021.

ATTRACTIONS: Historic Savannah, period homes, General Sherman's headquarters, Cotton Exchange, Fort Pulaski and Fort McAllister.

KINGS BAY NAVAL SUBMARINE BASE (GA03R1)

Comm: 912-673-2000
ATVN: 860-2000

Kings Bay Naval Submarine Base, GA 31547-5015

LOCATION: Off I-95 N of GA/FL border. Take Exit 1 which leads right into base, or Exits 2A or 2B, E to Kings Bay and follow road N to base. RM. p-23, R/14. NMC: Jacksonville, FL, 40 mi S.

GENERAL INFORMATION: East Coast Submarine Base, Hq Submarine Squadron 16. Future home of the East Coast Trident submarines and HQ of the Trident Gp and Sqd Commanders.

TEMPORARY MILITARY LODGING: Biltg Ofc, ATTN: Code N53, Bldg 1006, adj to Geo Bancroft, 0800-1600 hrs dly, Comm: 912-673-2056/9, ATVN: 860-2056/9, Off only. DV/VIP, EX-2164. 26 unit Nav Lodge, Comm: 912-673-2132, ATVN: 860-2132.

LOGISTICAL SUPPORT: Most spt fac avail. Key phone numbers follow. Commissary newly opened.

NX-882-3109	**Caf-882-6229**	**CDO-673-2020**
Chap-673-4501/2	**Chf C-673-4566**	**Chld C-673-2042**
Conv S-882-6944	**EM C-673-4567**	**Fam Svc-673-4512**
Gas S-882-6945	**Legal-673-4732**	**Off C-673-4563**
Pkg S-673-3104	**Pol-673-2145**	**SDO-673-4703**
Theat-673-4507		

HEALTH & WELFARE: Medical/Dental Clinics, Emer, EX-4247. Inpatient, Jacksonville Naval Hosp, Comm: 904-452-2353, ATVN: 942-2353. Chapels serving all faiths, EX-4501.

Kings Bay NSB, Cont'd

RECREATION: Bowling, Gym, Swimming, Racquetball Courts, Outdoor Sport Fields, Gear Issue, ITT, Auto Hobby, Ceramics, Library, Picnic Area, Tennis Courts, Boat and Camping Trailer Rental, Fresh and Salt Water Fishing.

SPACE-A: Jacksonville NAS, Bldg 1853, 24 hrs dly, Comm: 904-455-3311, ATVN: 942-3311, flts to CONUS and overseas.

ATTRACTIONS: Beautiful beaches, sport fishing, hunting, Cumberland Island and Jekyll and St Simmon Island, 1 hr north.

MOODY AIR FORCE BASE (GA02R1)

Moody Air Force Base, GA 31699-5000

Comm: 912-333-4211
ATVN: 460-1110

LOCATION: On GA-125, 10 mi N of Valdosta. Also, can be reached from I-75, via GA-122. RM: p-23, R/9. NMC: Valdosta, 10 mi S.

GENERAL INFORMATION: Tactical Air Command Base.

TEMPORARY MILITARY LODGING: Biltg Ofc, Bldg 320, Cooney St, 24 hrs dly, Comm: 912-333-3893, ATVN: 460-3893. All ranks. DV/VIP, EX-3480.

LOGISTICAL SUPPORT: Complete spt fac avail. Key phone numbers follow.

BX-333-3431	Caf-333-3093	Chap-333-3211
Chld C-333-3935	Commiss-333-3365	Conv S-333-3876
Fam Svc-333-3068	Gas S-333-3451	Legal-333-3414
Loc-333-3585/3516	Med-333-3232	NCO-333-3503
NCO C-333-3792	Off C-333-3258	Pkg S-333-4166
Pol-333-3200/3108	Snack B-333-3101	Theat-333-3557

HEALTH & WELFARE: USAF Hospital, clinics and depts, Emerg, EX-3232, Appoint, EX-3816, 30 bed. Chapels serving all faiths, EX-3211/3646.

RECREATION: Bowling, EX-3872, Golf, EX-3297, Auto Hobby, EX-3056, Ceramics & Wood Hobby, EX-3945, Library, EX-3539, Rec Ctr, EX-3280, Swimming, EX-3560/3864, Youth Ctr, EX-3067. Grassy Pond Rec Area, off Base, 15 mi S of Valdosta, off I-75, yr round, Comm: 912-559-5840, ATVN: 460-5840, 7 cabins furnished, full facilities/all activities.

SPACE-A: Reports no Space-A. Try Jacksonville NAS, FL, Comm: 904-772-2537, ATVN: 942-2537, flts to CONUS and overseas.

ATTRACTIONS: Historic homes in Valdosta. Florida and Gulf Coast nearby. Family swimming at Crystal Lake, 40 mi north. Okefenokee Swamp, 60 mi east.

ROBINS AIR FORCE BASE (GA14R1)

Robins Air Force Base, GA 31098-5000

Comm: 912-926-1110
ATVN: 468-1110

LOCATION: Off US-129 on GA-247 at Warner Robins. Access from I-75 S. RM: p-23, K/7. NMC: Macon, 18 mi NW.

GENERAL INFORMATION: Air Force Logistics Command Base. Warner Robins Air Logistics Center and AFRES hqs stationed here.

TEMPORARY MILITARY LODGING: Biltg Ofc, Bldg 557, Club Dr, 24 hrs dly, Comm: 912-926-2100/4354, ATVN: 468-2100/4412. All ranks. DV/VIP, Biltg Ofc.

LOGISTICAL SUPPORT: Complete spt fac avail. Key phone numbers follow.

GEORGIA

Robins AFB, Cont'd

BX-923-5539	Caf-922-8635	Chap-926-2821/2166
Chld C-923-3838	Commiss-926-2126	Conv S-922-3851
EM C-926-2105/6191	Fam Svc-926-6648	Gas S-923-7292
Legal-926-5995	Loc-926-6027/1110	Med-926-3845
NCO C-923-5581	Off C-922-3011	Pkg S-923-5085
Pol-926-2187	Snack B-923-8690	Theat-926-2919

HEALTH & WELFARE: USAF Hospital, clinics and depts, Emerg EX-3845, Appoint, EX-2841, 30 bed. Chapels serving all faiths, EX-2821/2166.

RECREATION: MWR Ctr, EX-3192, Rec Ctr, EX-2105, Bowling, EX-2112, Library, EX-5378, Gym, EX-2129, Hobby Shops, EX-3004/5282, Golf, EX-4103, Swimming, EX-2105, Rec Equip, EX-3192, Picnic Area, EX-4500, Skeet Range, Riding Stables, Nature Ctr, Youth Ctr, EX-4500, Aero Club, EX-4867. Fam-Camp, on Base, Luna Lake, yr round, Comm: 912-926-4500, ATVN: 468-4500, 16 camper spaces w/hookups, tent camping, other fac, all lake activities.

SPACE-A: Pax Term and Lounges, Bldg 111, 0630-1700 hrs dly, Comm: 912-926-3166, ATVN: 468-3166, flts to CONUS and overseas.

ATTRACTIONS: Middle GA area, Macon, 4th largest city in GA, Andersonville Trail, Ocmulgee National Park, Robins AFB Museum of Aviation.

Small Installations Cont'd from page 67

Muskegon Coast Guard Group
Fulton Ave & Bluff St
Muskegon MI 49441-5000

Traverse City Coast Guard
Air Station
Traverse City MI 49684

Calumet Air Force Station
Hwy 41, Bldg 1
Calumet MI 49913-5000

St Louis Coast Guard Base
Foot of Iron Street
St Louis MO 63111-5000

Pascagoula
SSCR, U S Navy
Pascagoula MS 39567-5000

Fort Harrison Site
Bldg P-40
Helena MT 59601-5000

Fort Fisher Air Force Station
Bldg 122
Kure Beach NC 28449-5000

Bismarck Site
Det 1, 1st CEVG
Bismarck ND 58501-5000

Elkhorn Site
Det 2, 1 ACCOM
Elkhorn NE 68022-5000

Hastings RBS BX
Det 10, USAF NAD
Hastings NE 68901-5000

Gloucester City
Coast Guard Base
Gloucester City NJ 08030-5000

Brooklyn Coast Guard Air Sta
Floyd Bennett Field
Brooklyn NY 11234-5000

Mitchel Field NSHA
Bldg 14
Garden City NY 11530-5000

Watervliet Arsenal
Bldg 40
Watervliet NY 12189-5000

Scotia, U S Navy
NAU
Scotia NY 12302-5000

Coast Guard Marine Safety Off
1055 E Ninth St
Cleveland OH 44114-5000

Cleveland 9th Coast Guard Dist
703 Eagle Ave
Cleveland OH 44115-5000

Youngstown Municipal Airport
Arnold Road, Bldg 120
Vienna OH 44473-5000

Cont'd page 78

IDAHO

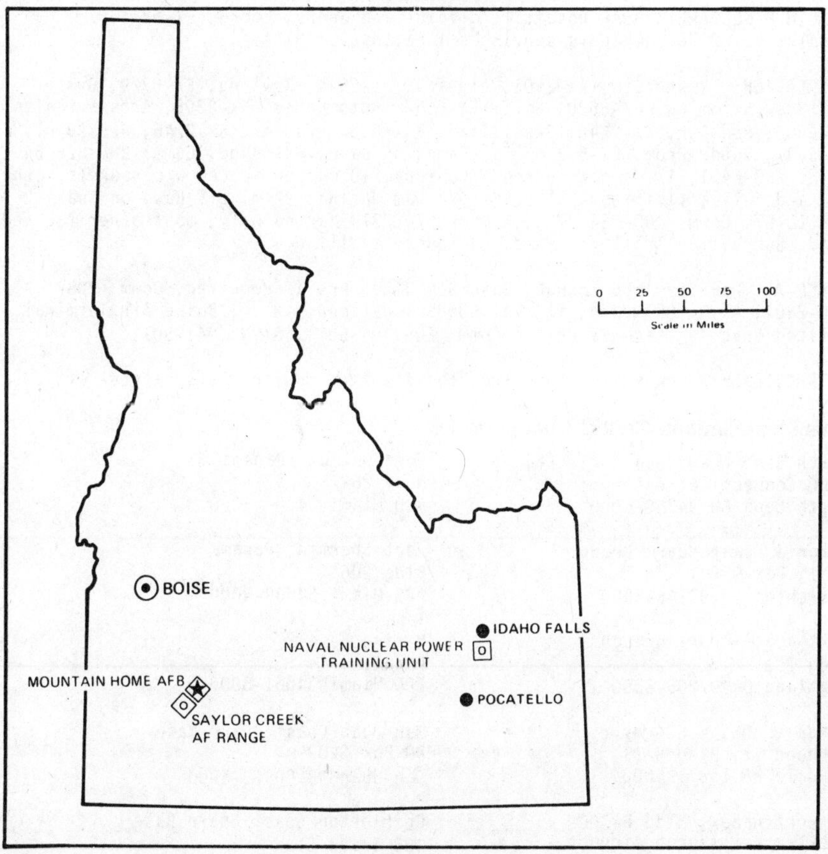

MOUNTAIN HOME AIR FORCE BASE (ID01R4)

Mountain Home Air Force Base, ID 83648-5000

Comm: 208-828-2111
ATVN: 857-1110

LOCATION: Fr Boise, take I-84 SE, 39 mi to Mountain Home exit, follow road through town to Airbase Rd, 10 mi to main gate. RM: p-25, N/4. NMC: Boise, 51 mi NW.

GENERAL INFORMATION: Tactical Air Command Base. F-111A & EF-111A aircraft. Mission is air-to-ground combat and electronic countermeasures.

TEMPORARY MILITARY LODGING: Biltg Ofc, Bldg 2604, F St, 24 hrs dly, Comm: 208-828-6451, ATVN: 857-6451. All ranks.

LOGISTICAL SUPPORT: Complete spt fac avail. Key phone numbers follow.

BX-832-4511	Chap-828-6417	Chld C-828-2443
Commiss-828-2163	Conv S-832-4353	Fam Svc-828-2272
Gas S-832-4459	Legal-828-2238	Med-828-2319
NCO C-828-2105	Off C-828-2597	Pol-828-2256

IDAHO

Mountain Home AFB, Cont'd

Snack B-832-4424 Theat-828-2431

HEALTH & WELFARE: USAF Hospital, clinics and depts, Emerg, EX-6274, Appoint, EX-2276. Chapels serving all faiths, EX-6417.

RECREATION: Youth Ctr, EX-2501, Library, EX-2544, Bowling, EX-6329, Gym, EX-2381, Swimming, EX-6620, Golf, EX-6559, Auto Hobby, EX-2295, Arts & Crafts, EX-6680, Rec Ctr, EX-2246, Skeet/Trap, EX-6093, Stables, EX-2246, Rec Supply, EX-2237, Outdr Prog, EX-6333. Fam-Camp, on Base, yr round, Comm: 208-828-6451, ATVN: 857-6451, 12 camper spaces w/hookups, 10 camper spaces w/o hookups, other fac and full activities. Also, Strike Dam Marina, 22 mi off Base on Snake R via ID-67, Comm: 208-834-2723, ATVN: 857-2237, Marina only, most other fac and act. Sun Valley Trailers, sold, no longer avail.

SPACE-A: Pax Term and Lounge, Base Ops Bldg, hrs as required, Comm: 208-828-2304, ATVN: 857-2304, flts to CONUS locations. Also, Boise Air Terminal, limited Space-A, Reserve Units, Comm: 208-385-5011, ATVN: 941-5011.

ATTRACTIONS: Snow skiing, hunting, and fishing. Boise, ID's capital city.

Small Installations Cont'd from page 76

North Bend Coast Guard Air Sta
2000 Connecticut Ave
North Bend OR 94759-5000

Astoria Coast Guard Group
Rt 1, Box 950
Warrenton OR 97146-5000

Portland Airport Branch
Bldg 203
Portland OR 97205-5000

Oakdale ADS, U S Army
Lobough St, Bldg S-13
Oakdale PA 15071-5000

Airport Annex, Bldg P-300
Greater Pittsburgh Airport
Pittsburgh PA 15231-5000

Warminster, U S Navy
NADC
Warminster PA 18974-5000

Defense Personnel Spt Center
2800 S 30th Street, Bldg M-3
Philadelphia PA 19101-5000

Philadelphia NRMC, U S Navy
17th St & Pattison Ave
Philadelphia PA 19145-5000

Corozal, Panama
Bldg 654
APO Miami 34004-5000

Fort Davis, Panama
Bldg 32
APO Miami 34005-5000

Fort Gulick, Panama
Bldg 202
APO Miami 34008-5000

Fort Sherman, Panama
Bldg 206
APO Miami 34009-5000

Marine Base
Panama
FPO Miami 34061-5000

San Juan Coast Guard Base
PO Box S2029
San Juan PR 00903-5000

Charleston Coast Guard Base
196 Tradd St
Charleston SC 29401-5000

Charleston Naval Regional
Medical Center
Charleston SC 29408-5000

Beaufort Naval
Hospital
Beaufort SC 29902-5000

McGhee Tyson Air National Gd
215 White St, Bldg 221
Knoxville TN 37901-5000

Defense Depot Memphis
2163 Airways Bldg
Memphis TN 38114-5000

Galveston Coast Guard Base
General Delivery
Galveston TX 77550-5000

Cont'd page 94

ILLINOIS

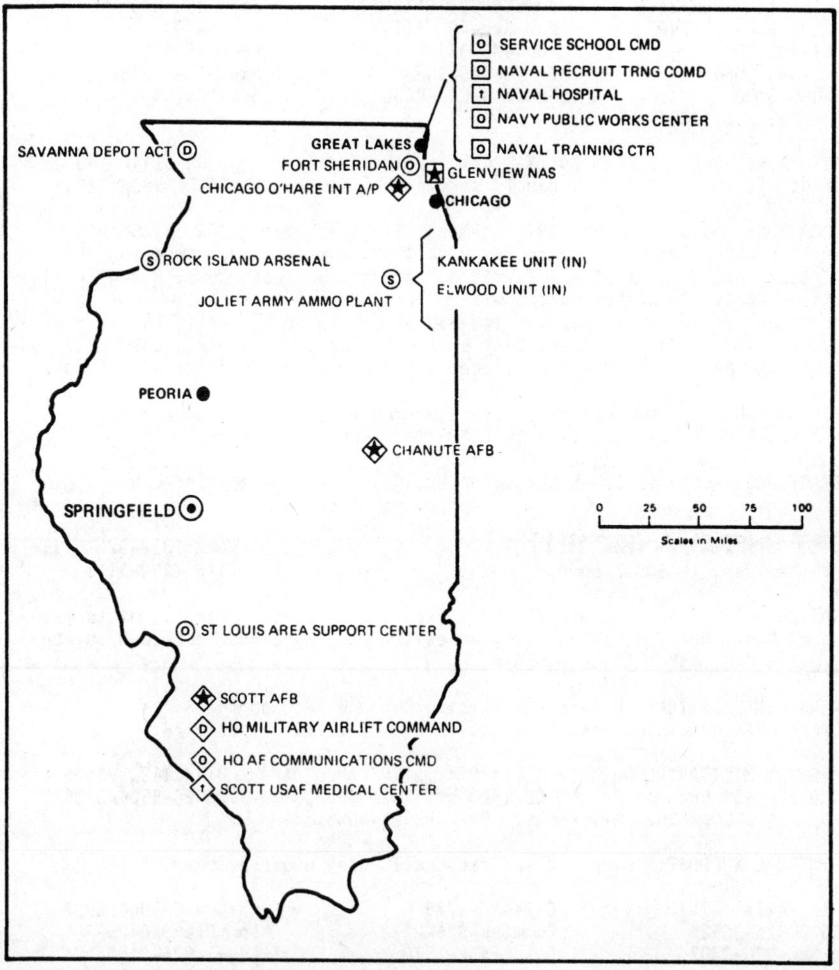

CHANUTE AIR FORCE BASE (IL06R2)

Chanute Air Force Base, IL 61868-5225

Comm: 217-495-1110
ATVN: 862-1110

LOCATION: North of Champaign (14 mi) at Rantoul. Access from I-57 and US-45. RM: p-27, J/14. NMC: Chicago, 120 mi N.

GENERAL INFORMATION: Air Training Command Base. Technical training center for USAF and other services.

TEMPORARY MILITARY LODGING: Biltg Ofc, Bldg 306, 24 hrs dly, Comm: 217-495-2962, ATVN: 862-2962. All ranks.

LOGISTICAL SUPPORT: Complete spt fac avail. Key phone numbers follow.

Chanute AFB, Cont'd

BX-892-9621	Chap-495-2282/3/4122	Chld C-495-3666
Commiss-495-2868/2405	Conv S-893-0113	EM C-893-3138
Fam Sup Ctr-495-3402	Fam Svc-495-3119	Gas S-893-8067
Legal-495-3002	Loc-495-3545	Med-495-3133
NCO C-495-3064	Off C-495-3088	Pkg S-495-2768
Pol-495-3100	Snack B-893-8450	SDO-495-4566
Theat-495-2411/2687		

HEALTH & WELFARE: USAF Hospital, clinics and depts, Emerg, EX-3134/3133, Appoint, EX-2081, 30 bed. Chapels serving all faiths, EX-2282/2283/4122.

RECREATION: Auto Shop, EX-4197, Bowling, EX-2163, Golf, EX-3265, Sports Rental, EX-3241, Racquet Sports, EX-2094/4202, Gym, EX-2094, Swimming, EX-3270/3367, Rec Ctr, EX-3444. Fam-Camp, on Base, Mar-Oct, Comm: 217-495-3444 ATVN: 862-3444, 10 camper spaces w/elec & wa, 20 camper spaces w/elec only, most other fac and activities. Also operates Chanute Rec Area, 15 mi NW of AFB, off I-57 at Paxton/Gibson City exit, W on IL-9 for 2.5 mi, Comm: 217-495-3444, ATVN: 862-3444, 30 camper spaces w/o hookups, other fac and most act.

SPACE-A: None, inactive runway. Try Glenview NAS, Hangar 1, Comm: 312-657-2155, ATVN: 932-2326, flts to CONUS locations.

ATTRACTIONS: Univ of IL at Champaign-Urbana, Chicago, easy drive N.

FORT SHERIDAN (IL01R2)

Fort Sheridan, IL 60037-5000 — **Comm: 312-926-4111** — **ATVN: 459-4111**

LOCATION: Take US-41 to Old Elm Rd (Lake Forest) exit, E for 1.5 mi to main gate of Post. Or I-294 to Half-Day Rd exit, E to US-41 and proceed N as above. RM: p-28, D/8. NMC: Chicago, 28 mi S.

GENERAL INFORMATION: Hq Army Recruiting Command, Hq Fourth US Army. National Historic Landmark District. Centennial Year 11/86-11/87.

TEMPORARY MILITARY LODGING: Biltg Ofc, ATTN: AFZO-DEH-H, Bldg 142, Patten Rd, 0730-1930 hrs Mon-Fri, 0700-1500 hrs Sat-Sun, Comm: 312-926-5506/2735, ATVN: 459-5506/2735. All ranks. DV/VIP, EX-3096.

LOGISTICAL SUPPORT: Complete spt fac avail. Key phone numbers follow.

PX-433-4777	Caf-433-4779	Chap-926-2866/2992
Chld C-926-3652	Commiss-926-2132	CommClb 926-2525
Fam Svc-926-3307	Gas S-432-1510	Legal-926-3184
Loc-926-2274	Med-926-2222/3181	Pol-926-3200
SDO-926-3095/6	Snack B-433-4779	

HEALTH & WELFARE: Clinic, Emer, EX-2222, Appoint, EX-2912. Inpatient, Great Lakes Naval Hosp, 8 mi N, Comm: 312-688-3480, ATVN: 792-3480. Chapels serving all faiths, EX-2123/2866.

RECREATION: Arts & Crafts, EX-3125, Auto Hobby, EX-2477, Bowling, EX-3650, Golf, EX-3554, ITT, EX-3390, Gym, EX-3551, Sports Rental, EX-3653, Library, EX-3188, Swimming, EX-2380, Rec Ctr, EX-3390, Tennis, EX-2995. No Rec Area.

SPACE-A: O'Hare AFRES Facility, Hangar Bldg 30, M-F 0700-2100, Comm: 312-694-6199, ATVN: 930-6199, flts to CONUS and overseas, weekend AFRES flts.

ATTRACTIONS: Adler Planetarium, Art Institute of Chicago, War Memorial Museum of the 1st Division, Field Museum of Natural History, city of Chicago,

Fort Sheridan, Cont'd

major league baseball, football, basketball, ice hockey (in season), theater, symphony, opera (Sep-Dec), city of Milwaukee (60 mi N), Gr Lakes NTC (8 mi N), Six Flags Great America.

GLENVIEW NAVAL AIR STATION (IL03R2)

Glenview Naval Air Station, IL 60026-5000

Comm: 312-657-1000
ATVN: 932-0111

LOCATION: From I-294 N, exit Willow Rd, E two traffic lights, right on Pfingsten Rd to West Lake Ave, left to NAS main gate. RM: p-28, F/8. NMC: Chicago, 10 mi SE.

GENERAL INFORMATION: Navy patrol aircraft units, USCG Air Station, Fort Sheridan Flight Detachment, Army Reserve Center.

TEMPORARY MILITARY LODGING: Biltg Ofc, ATTN: Adm Ofc, Code 04, Bldg 45, Whitehead Dr, Ofc, Comm: 312-651-2275, ATVN: 932-2275, Enl, Bldg 55, Comm: 312-651-2453, ATVN: 932-2453, 24 hrs dly. All ranks. DV/VIP, EX-1000.

LOGISTICAL SUPPORT: Most spt fac avail. Key phone numbers follow.

NX-657-2643	**Chld C-657-2240**	**Conv S-657-2643**
CPO C-657-2396	**EM C-657-2396**	**Fam Svc-657-2429**
Gas S-724-0045	**Legal-657-2112**	**Loc-657-2303**
Med-657-2222/2257	**Off C-724-0202**	**Pkg S-657-2346**
Pol-657-2476	**Theat-657-2471**	

HEALTH & WELFARE: Dispensary/Emerg, EX-2222, Appoint, EX-2257. Inpatient, Great Lakes Nav Hosp, Chicago, Comm: 312-688-4560, ATVN: 792-4560. Chapels serving all faiths, EX-2100.

RECREATION: Bowling, EX-2142, Flying Club, EX-2204, Golf, EX-2141, Sports Rental, EX-2142, Rec Ctr, EX-2120, Library, EX-2440.

SPACE-A: Pax Term and Lounge, Hangar 1, 0700-2300 hrs dly, Comm: 312-657-2155, ATVN: 932-2326, flts to CONUS. Also, CGAS and Flt Det/Fort Sheridan located here and have infrequent flts via exec aircraft to CONUS.

ATTRACTIONS: Museum of Sci & Industry, Museum of Nat History, Chicago.

GREAT LAKES NAVAL TRAINING CENTER (IL07R2)

Great Lakes Naval Training Center, IL 60088-5000

Comm: 312-688-3939
ATVN: 792-3939

LOCATION: From I-94 N or US-41 N of Chicago, exit to IL-137 E to NTC. Clearly marked. RM: p-28, B/7. NMC: Chicago, 30 mi S.

GENERAL INFORMATION: Naval Training Center, Medical Command and Hospital Corps School.

TEMPORARY MILITARY LODGING: Biltg Ofc for Off, Bldg 62, 24 hrs dly, Comm: 312-689-2565, ATVN: 792-3777, Navy Lodge, 578-6196, Enl at Nav Hosp, Bldg 219-H, 24 hrs dly, Comm: 312-688-2374, ATVN: 792-2374. All ranks. DV/VIP, EX-4560.

LOGISTICAL SUPPORT: Complete spt fac avail. Key phone numbers follow.

NX-578-6103	**Caf-578-6120**	**Chap-688-5410/4537**
Chld C-688-5662	**Commiss-689-8200**	**Conv S-578-6247**
CPO C-688-6928	**EM C-688-4641**	**Fam Svc-688-3603**

Great Lakes NTC, Cont'd

Gas S-578-6198	Legal-688-4753	Loc-688-3480/2002
Med-688-6855/4560	Off C-688-6946	Pkg S-688-5352
Pol-688-3430	Theat-688-6763	

HEALTH & WELFARE: US Naval Hospital, clinics and depts, Emerg, EX-6855, Appoint EX-4560. Chapels serving all faiths, EX-5410/4537.

RECREATION: Bowling, EX-5612, Golf, EX-4593, Rec Ctr, EX-4813, Swimming, Tennis, Auto and Craft Shops, Picnic, Boating and Swimming at Nunn Beach on Lake Michigan, EX-4759/5449.

SPACE-A: Glenview NAS, Hangar 1, 0700-2300 hrs dly, Comm: 312-657-2155, ATVN: 932-2326, flts to CONUS. Also, AFRES Fac, O'Hare, IAP, Hangar 30, M-F 0700-2100, Comm: 312-694-6199, ATVN: 930-6199, flts to CONUS and overseas.

ATTRACTIONS: Chicago nearby, great restaurants, entertainment and sports. Milwaukee one hour north. Excellent seasonal recreation available on Lake Michigan and in the Chain o' Lakes area (one-half hour west).

ROCK ISLAND ARSENAL (IL08R2)

Rock Island Arsenal, IL 61299-5000

Comm: 309-782-1110
ATVN: 793-1110

LOCATION: From I-74 N in Moline exit to 3rd Ave W and follow signs to Arsenal Island located in middle of Mississippi River. RM: p-27, Y/3. NMC: Quad Cities of Rock Island and Moline in IL and Davenport and Bettendorf in IA.

GENERAL INFORMATION: Hq US Army Armament, Munitions and Chemical Command, and Rock Island Arsenal.

TEMPORARY MILITARY LODGING: None.

LOGISTICAL SUPPORT: Most spt fac avail. Key phone numbers follow.

PX-788-4940	Caf-782-4337	Chld C-782-6291
Commiss-782-4614/4798	Fam Svc-782-3828/3908	Legal-782-4051
Loc-782-1110	Med-782-5722/6018	NCO-782-6001
NCO C-788-7342	Off C-788-4868	Pol-782-6116/5921

HEALTH & WELFARE: Dispensary/Emerg, EX-25722, Appoint, EX-24791. Inpatient, Great Lakes Naval Hosp, Chicago, 170 mi NE, Comm: Emerg, 312-688-6855, Appoint, 312-688-4560. Chapels sergving all faiths, EX-24736/5000.

RECREATION: Arts & Crafts, EX-24950, Auto Hobby, EX-25365, Rec Svc, EX-24950, Wrkout Rm, EX-26787, Rqtbl Ct, EX-25592, Outdoor Rec, EX-24950, Phys Fit, EX-24950, Skeet, EX-22737/22779.

SPACE-A: None.

ATTRACTIONS: Rock Island Arsenal Museum, National Cemetery, Cemetery for Confederate Prisoners, COL Davenport House.

SCOTT AIR FORCE BASE (IL02R2)

Scott Air Force Base, IL 62225-5000

Comm: 618-256-1110
ATVN: 576-1110

LOCATION: Off I-64 E or W, exit 19E or 19A W to IL-158 S, 2 mi and watch for signs to AFB entry. RM: p-27, S/8. NMC: St Louis, 25 mi W.

GENERAL INFORMATION: Military Airlift Command Base. Hq MAC, Hq Air Force Communications Command, Hq Aerospace Rescue and Recovery Service, Hq Air

Scott AFB, Cont'd

Weather Service, Aeromedical Airlift Wing, 23rd Air Force.

TEMPORARY MILITARY LODGING: Biltg Ofc, Bldg 1510, Scott Dr, 24 hrs dly, Comm: 618-744-1200, ATVN: 576-3224/2045. All ranks. DV/VIP, EX6-6555.

LOGISTICAL SUPPORT: Complete spt fac avail. Key phone numbers follow.

BX-744-0888	**Caf-746-4199**	**Chap-256-3303**
Chld C-256-2612	**Commiss-256-2783**	**Conv S-744-0237**
EM C-744-4328	**Fam Svc-256-3936**	**Gas S-746-2184**
Legal-256-2358	**Loc-256-4108**	**Med-256-1847**
NCO C-744-0300	**Off C-744-0444**	**Pkg S-744-0611**
Pol-256-2223	**Theat-256-5177**	

HEALTH & WELFARE: USAF Hospital, clinics and depts, Emer, EX6-6795, Appoint, EX6-6323/6324, 185 bed. Chapels serving all faiths, EX6-3303/5144.

RECREATION: Arts & Crafts, EX6-4230, Auto Hobby, EX6-4566, Bowling, EX6-4054, Golf, 744-1400, Fitness Center, EX6-4020, Gym, EX6-4524, Sports Rental, EX6-2067, Riding Stables, 233-9362, Library, EX6-3028, Swimming, EX6 2579, Youth Ctr, EX6-2115. Fam-Camp, on Base, year round, Comm: 618-256-5919, ATVN: 576-5919, 12 camper spaces w/hookups, other fac and activities.

SPACE-A: Pax Term and Lounge, Bldg P-8, 0500-2400 hrs dly, Comm: 618-256-3051/4042, ATVN: 576-3051/4042, flts to CONUS and overseas.

ATTRACTIONS: St Louis nearby, Mississippi River.

ST. LOUIS AREA SUPPORT CENTER (IL04R2)

Granite City, IL 62040-1801

Comm: 618-452-4212
ATVN: 892-4212

LOCATION: From I-70 take McKinley Bridge exit, cross Mississippi River, follow signs to Center. From I-270, cross river bridges and take first Granite City exit (Route 3, south) to Center. RM: p-27, R/7. NMC: St Louis, 7 mi W.

GENERAL INFORMATION: Army Area Support Center, Army Material Command Aviation and other material procurement and logistical activities. Supports 53,000 individuals in St Louis area.

TEMPORARY MILITARY LODGING: Biltg Ofc, Bldg 102, Niedringhaus St, 0730-1615 hrs dly, Comm: 618-452-4288, ATVN: 892-4288. All ranks.

LOGISTICAL SUPPORT: Complete spt fac avail. Key phone numbers follow.

PX-452-5230	**ACS-452-4260**	**Aud-452-4245**
Chap-452-4277	**Chld C-452-4480**	**Commiss-452-4232**
CU-451-7596	**Educ Ctr-452-4359**	**EM C-452-5595**
Gas S-877-4275	**ID Card-452-4226**	**Mil Pers-452-4422**
NCO C-452-5595	**Off C-452-5595**	**Pkg S-452-6112**
Pol-452-4224	**SDO-452-4247**	

HEALTH & WELFARE: Clinic, 425-4851. Inpatient, USAF Med Ctr, Scott AFB, 20 mi E, Emerg-256-6795, Appoint-256-6323/6324. Chapels serving all faiths, 452-4277.

RECREATION: Auto Hobby, EX-4279, Bowling, EX-4319, Golf, EX-4444, Library, EX-4332, Rec Ctr, EX-4398, Swimming Pool, EX-4221, Gym, EX-4209.

SPACE-A: Scott AFB, Pax Term and Lounge, Bldg P-8, 0500-2400 hrs dly, Comm: 618-256-3051/4042, ATVN: 576-3051/4042, flts to CONUS and overseas.

ILLINOIS

St Louis Area Spt Ctr, Cont'd

ATTRACTIONS: Gateway Arch, Mississippi River Tours, St Louis, Lacledes Landing, Hockey, Grants Farm, Busch Stadium, Six Flags, Kiel Auditorium, VP Fair (July), Fox Theater, Soulard Market area, museums.

INDIANA

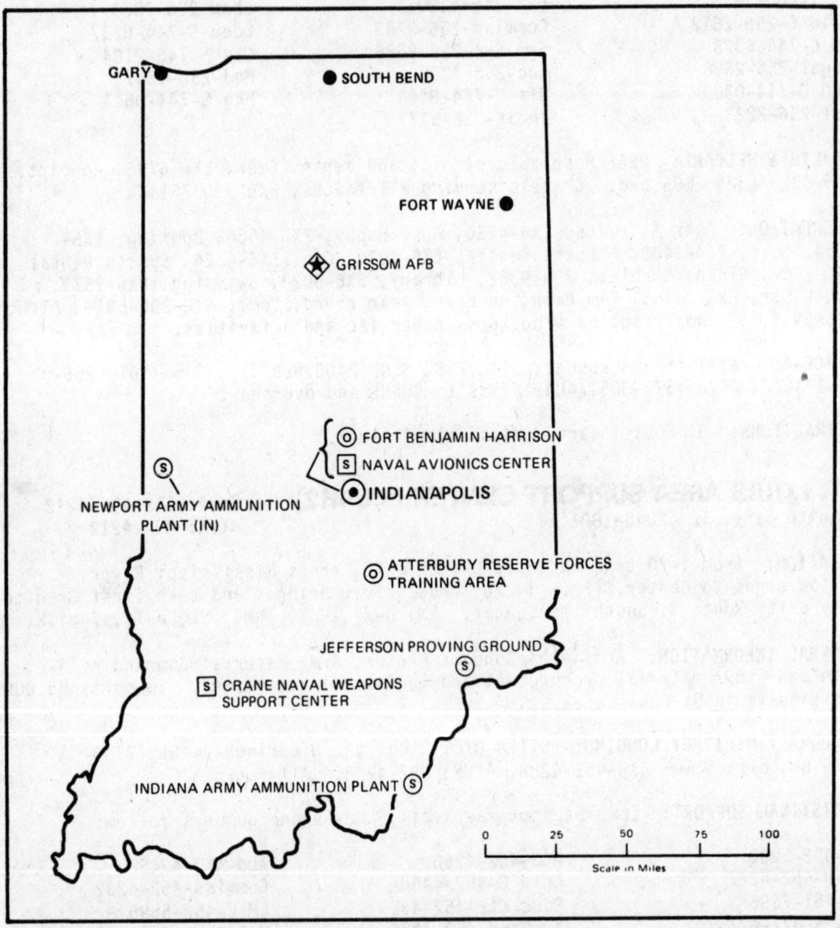

CRANE NAVAL WEAPONS SUPPORT CENTER (IN03R2)

Comm: 812-854-2511
ATVN: 482-1110

Crane Naval Weapons Support Center, IN 47522-5000

LOCATION: From US-231 N or S exit to IN-45 or IN-645 to enter the Center from the W. RM: p-31, S/7. NMC: Bloomington, 34 mi NE.

GENERAL INFORMATION: A major Naval Weapons and Ammunition Support Center.

TEMPORARY MILITARY LODGING: Limited lodging, Comm: 812-854-1176, ATVN:

Crane NWSC, Cont'd

482-1176, 0800-1600 hrs. All ranks. DV/VIP.

LOGISTICAL SUPPORT: Limited spt fac avail. Key phone numbers follow.

NX-854-1392	**Caf-854-1381**	**CM-854-1501**
Commiss-854-1297	**Gas S-854-1864**	**Loc-854-2511**
Med-854-1220	**Pkg S-854-1392**	**Pol-854-3300**

HEALTH & WELFARE: Dispensary/Emerg, EX-1220. Inpatient, Hawley Army Comm Hosp, Fort Benjamin Harrison, Indianapolis, 80 mi NE, Comm: 317-546-5153, ATVN: 699-5153.

RECREATION: Rec Ctr, EX-1586, Tickets, EX-1586, Library, EX-1586, Golf, EX-1242, Bowling, EX-1586.

SPACE-A: None.

ATTRACTIONS: Indiana University, Bloomington, easy drive 34 mi NE.

FORT BENJAMIN HARRISON (IN02R2)

Fort Benjamin Harrison, IN 46216-5000

Comm: 317-546-9211
ATVN: 699-1110

LOCATION: Take I-465 E to Ft Harrison exit 40, E on 56th St, or take Pendleton Pike (IN-67/US-36) exit 42 to Post Rd and N to Fort Harrison. RM: p-29, K/26. NMC: Indianapolis, 8 mi SW.

GENERAL INFORMATION: Army Soldier Support Ctr, Army Finance & Accounting Ctr, Enlisted Records & Evaluation, Defense Info School, Soldier Support Inst.

TEMPORARY MILITARY LODGING: Biltg Ofc, Bldg T-609, Green Ave (Post Rd), 24 hrs, Comm: 317-549-5455, ATVN: 699-5455. All ranks. DV/VIP, EX-3491.

LOGISTICAL SUPPORT: Complete spt fac avail. Key phone numbers follow.

PX-549-5698	**Chap-549-5251**	**Chld C-542-4375**
Commiss-549-5293	**Conv S-546-1027**	**EM C-542-4485**
Fam Svc-542-4357	**Gas S-546-2224**	**Legal-549-5260**
Loc-542-4537	**Med-549-5195/4**	**NCO C-542-4425**
Off C-542-2582	**Pkg S-542-4487**	**Snack B-549-5699**
Theat-549-5702		

HEALTH & WELFARE: Hawley Army Community Hospital, clinics and depts, Emerg, 549-5195/4, 25 bed. Chapels serving all faiths, 549-5251.

RECREATION: Auto Hobby, 542-4478, Bowling, 542-4352, Youth Ctr, 542-4480, Craft Shops, 542-4356, Golf, 542-4655, Library, 542-4476 (Post), Outdoor Rec, 542-4862, Gym, 542-4760, Swimming, Rec Ctr, 542-4509, Wood Shp, 542-4479. Outdoor Rec Travel Camp, on Post, yr round, Comm: 317-542-4862, ATVN: 699-4862, 20 camper spces w/hookups, other spt fac & act.

SPACE-A: Grissom AFB, Peru, Comm: 317-689-3620, ATVN: 928-3620, flts via KC-135/10 aircraft to CONUS and overseas.

ATTRACTIONS: Indianapolis Motor Speedway and Museum, state capital, park and zoo, universities, Union Station with shops and restaurants and the Sports Dome.

GRISSOM AIR FORCE BASE (IN01R2)

Grissom Air Force Base, IN 46971-5000

Comm: 317-689-5211
ATVN: 928-1110

Grissom AFB, Cont'd

LOCATION: On US-31, 15 mi N of Kokomo and 7 mi S of Peru. RM: p-31, G/10. NMC: Indianapolis, 72 mi S.

GENERAL INFORMATION: Strategic Air Command Base. Air Refueling Wing, AFRES units.

TEMPORARY MILITARY LODGING: Biltg Ofc, Bldg 549, Lancer St, 24 hrs dly, Comm: 317-689-2618, ATVN: 928-2618. All ranks. DV/VIP, EX-2240.

LOGISTICAL SUPPORT: Complete spt fac avail. Key phone numbers follow.

BX-689-9157	Chap-689-2191	Chld C-689-7177
Commiss-689-2111	Conv S-689-8347	EM C-689-9804
Fam Svc-689-2476	Gas S-689-7033	Legal-689-2231
Loc-689-3032	Med-689-3302	NCO C-689-9151
Off C-689-9135	Pkg S-682-7218	Snack B-689-7095
Theat-689-3065		

HEALTH & WELFARE: USAF Hospital, clinics and depts, Emer, EX-3302, Appoint, EX-3331, 10 bed. Chapels serving all faiths, EX-2191.

RECREATION: Aero Club, EX-3157, Auto Hobby, EX-2530, Bowl, EX-2504, Ceramics, EX-3170, Golf, EX-3166, Gym, EX-2000, Lib, EX-2056, MWR Ofc, EX-2002, Rec Ctr, EX-3030, Rec Supp, EX-3293, Swimming, EX-2070, Youth Ctr, EX-2511.

SPACE-A: Pax Term and Lounge, Bldg S-14, 24 hrs dly, Comm: 317-689-2256, ATVN: 928-2256, flts to CONUS and overseas (Panama) via KC-135.

ATTRACTIONS: City of Peru, Mississinewa Reservoir developed by Army Corps of Engineers at Peoria. Indianapolis Motor Speedway, 72 mi S.

IOWA

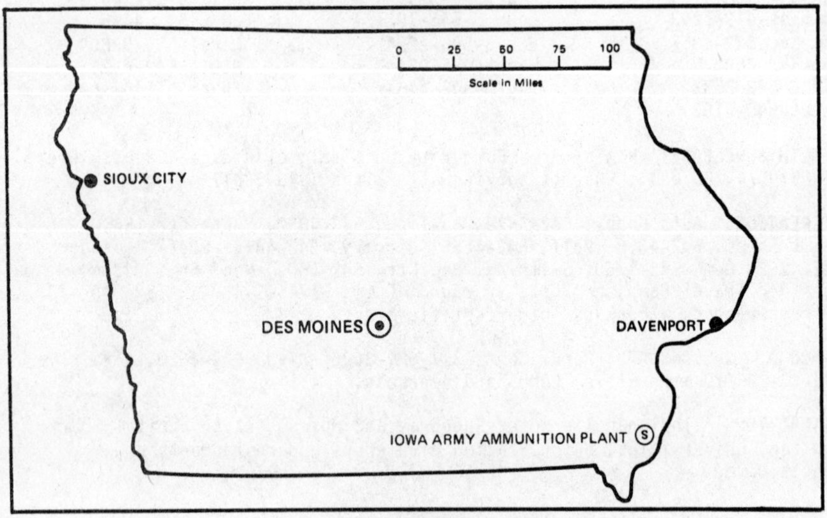

- None -

KANSAS

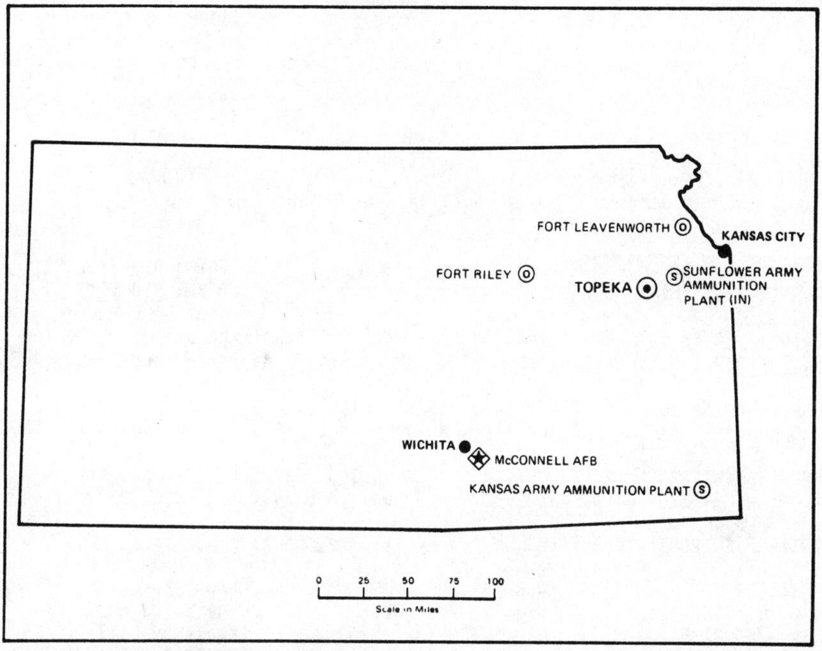

FORT LEAVENWORTH (KS04R3)

Fort Leavenworth, KS 66027-5000

Comm: 913-684-4021
ATVN: 552-4021
FTS: 753-4021

LOCATION: From I-70 W exit US-73 N to Leavenworth. Fort is adjacent to city of Leavenworth. RM: p-35, D/23. NMC: Kansas City, 30 mi SE.

GENERAL INFORMATION: Combined Arms Center, US Disciplinary Barracks, HQ 35th Infantry Division (mech).

TEMPORARY MILITARY LODGING: Biltg Ofc, Bldg 225, Truesdell Hall, Stimson Av, 24 hrs dly, Comm: 913-684-4931, ATVN: 552-4931. All ranks. DV/VIP, EX-4064.

LOGISTICAL SUPPORT: Complete spt fac avail. Key phone numbers follow.

PX-651-7270	Caf-651-7272	Chap-684-2210
Chld C-651-6353	Commiss-684-4903	Conv S-651-7278
EM C-651-6852	Fam Svc-684-4357	Gas S-651-6541
Legal-684-4941	Loc-684-3651/4021	Med-684-6100
NCO C-651-6852	Off C-651-7013	Pkg S-651-7175
Pol-684-2111	SDO-684-4154	Snack B-651-6573/7272
Theat-651-6920		

HEALTH & WELFARE: Munson Army Community Hospital, clinics and depts, Emerg EX-3570/2171, Appoint, EX-4041, 55 bed. Chapels serving all faiths, EX-2017.

RECREATION: Bowling, EX-2755, Flying Club, 651-6808, Golf, 651-7176, Gym, EX-3588/3895/4396/2037, Pre-Teen Ctr, EX-3261, Teen Ctr, EX-3261, Health &

Fort Leavenworth, Cont'd

Exercise Rooms, EX-3895, Library (Post), EX-3569, CGSC, EX-4035, ITT, EX-2580, Morale Spt Act, EX-2468, Multi-Crafts, EX-3386, Rec Ctr, EX-2580, Squash & Handball, EX-2037, Stables, 651-7307, Swimming, 651-4437/3998/3088/2187, Tennis, EX-3392, Youth Act, EX-3850. Rec Vehicle Park, on Base, year round, Comm: 913-684-4931/3045, ATVN: 552-4931/3045, 24 camper spaces w/hookups, other fac & activities.

SPACE-A: None

ATTRACTIONS: Ft Leavenworth Museum, Agriculture Hall of Fame, Bonner Springs, KS, and Harry S. Truman Lib & Museum, Independence, MO.

FORT RILEY (KS02R3)

Fort Riley, KS 66442-5000 **Comm: 913-239-3911** **ATVN: 856-1110**

LOCATION: On KS-18 and off I-70 in the central part of the state. Junction City, 5 mi SW and Manhattan, 10 mi NE. RM: p-35, E/18. NMC: Topeka, 50 mi E.

GENERAL INFORMATION: Hq First Infantry Division (mech), Fort Riley Readiness Gp, Army Correctional Activity, HQ Army Third ROTC Region and Fort Riley.

TEMPORARY MILITARY LODGING: Biltg Ofc, Bldg 145, Barry Ave, 24 hrs dly, Comm: 913-239-2830, ATVN: 856-2830. All ranks. DV/VIP, EX-3926/3927.

LOGISTICAL SUPPORT: Complete spt fac avail. Key phone numbers follow.

PX-784-4439	**ACS-239-3091**	**Chap-239-HELP**
Chld C-784-5909	**Commiss-239-6621**	**Conv S-784-6037**
Fam Svc-239-3235	**Gas S-784-5081/5969**	**Legal-239-3117**
Loc-239-9067	**Med-239-7777**	**NCO/EM C-784-4075**
Off C-784-5999	**Pkg S-784-5182**	**Pol-239-3053/MPMP**
Theat-239-3760/239-6668		

HEALTH & WELFARE: Irwin Army Community Hospital, clinics and depts, Emerg EX-7777, 250 bed. Chapels serving all faiths, EX-3730.

RECREATION: Multi-Crafts, 239-2536, Auto Hobby, 239-4028, Bowling, 239-4366, Skating Rink, 239-2243, Library, 239-2392, Gyms, 239-3868/5562/5771/4683, Tennis, 239-3288, Comm Ctrs, 239-2813/5517, Youth Ctr, 239-2840, Wood Shop, 239-2210, ITT, 239-5614, Golf, 239-5412, Flying Club, 239-2205, Swimming, 239-3945, Riding Club, 239-6651, Outdoor Rec, 239-6189, Rec Ch Out, 239-6189. Moon Lake Rec Area, on post, 236-6189, no camping fac. Operates Kiernan/Smith Marina, Travel Camp and Beach on Milford Lake, near Junction City, year round, Comm: 913-239-4478, ATVN: 856-4478, 10 camper spaces w/hookups, 5 mobile homes, comp fac and all activities.

SPACE-A: Marshall Army Airfield, flts via exec aircraft, 239-2575. Also try Forbes Field, 3 mi S of Topeka, Comm: 913-862-4663, ATVN: 720-4663, flts to CONUS and overseas via KC-135/10 aircraft.

ATTRACTIONS: Custer House, US Cavalry Museum, & First Territorial Capitol.

McCONNELL AIR FORCE BASE (KS03R3)

McConnell Air Force Base, KS 67221-5000 **Comm: 316-652-6100** **ATVN: 743-1110**

LOCATION: Take I-35 to Wichita, exit at Kellogg St (US-54) west to Rock Rd S and McConnell AFB. RM: p-35, Q/4. NMC: Wichita, 6 mi NW.

GENERAL INFORMATION: Strategic Air Command Base. Air Refueling Wing,

McConnell AFB, Cont'd

Tactical Fighter Group and future B-1B Bombardment Wing (1988).

TEMPORARY MILITARY LODGING: Biltg Ofc, Bldg 193, Manhattan St, 24 hrs dly, Comm: 316-681-5361/2, ATVN: 743-5361/2. All ranks. DV/VIP, EX-6100.

LOGISTICAL SUPPORT: Complete spt fac avail. Key phone numbers follow.

BX-685-0231	**Chap-681-5341/5218**	**Chld C-681-5496**
Commiss-681-5266/5164	**Conv S-685-0291**	**EM C-681-5004**
Fam Svc-681-5188	**Gas S-681-5673**	**Legal-681-6361**
Loc-681-6322	**Med-681-5743/4**	**NCO C-681-5004/685-5377**
Off C-681-5070	**PAO-681-5781/2**	**Pkg S-685-5214**
Theat-681-5618		

HEALTH & WELFARE: USAF Hospital, clinics and depts, Emerg EX-5743/4 & 6758, Appoint EX-5843. Chapels serving all faiths, EX-5341/5218.

RECREATION: Arts & Crafts, EX-6750, Auto Hobby, EX-5250, Bowling, EX-5783, ITT, EX-5353, Golf, EX-6272, Gym, EX-5310, Sports Rental, EX-5563, Library, EX-5414, Rec Ctr, EX-6447/5003, Tickets, EX-5353, Youth Ctr, EX-5932.

SPACE-A: Pax Term and Lounge, Bldg 1112, 24 hrs dly, Comm: 316-681-5271/2, ATVN: 743-5271/2, flts to CONUS and overseas via KC-135R aircraft.

ATTRACTIONS: Rebuilt downtown core area has resulted in outstanding civic and cultural complex, great zoo, museums of art and history.

KENTUCKY

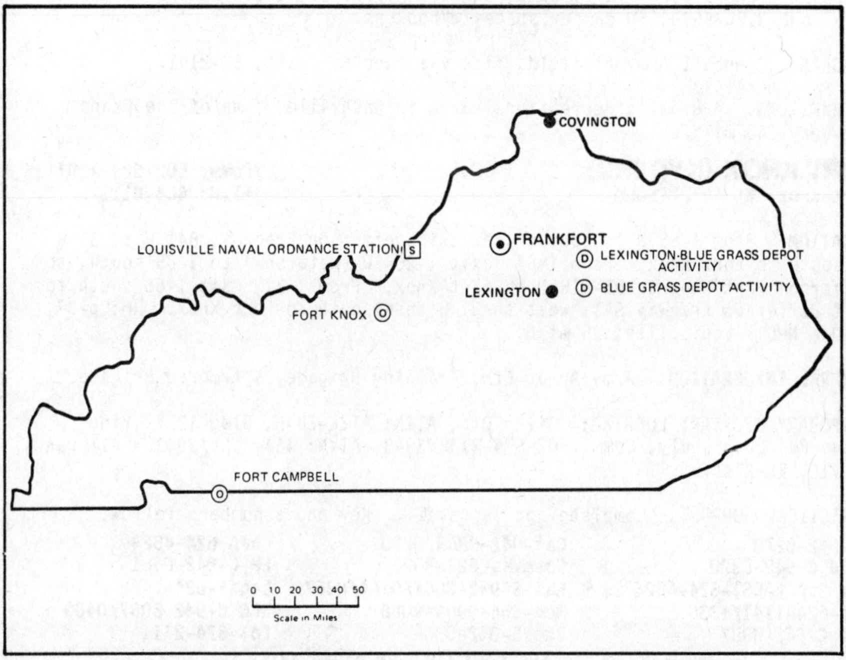

FORT CAMPBELL (KY02R2)

Fort Campbell, KY 42223-5000

Comm: 502-798-2151
ATVN: 635-1110

LOCATION: In the SW part of KY, 4 mi S of intersection of US-41A and I-24. Ten mi NW of Clarksville, TN. RM: p-37, P/4. NMC: Hopkinsville, 15 mi N.

GENERAL INFORMATION: 101st Airborne Division (Air Assault), Air Assault School, Support Brigade.

TEMPORARY MILITARY LODGING: Biltg Ofc, Bldg B-2601, 24 hrs dly, Comm: 502-798-5012, ATVN: 635-5012. All ranks. DV/VIP, EX-3213.

LOGISTICAL SUPPORT: Complete spt fac avail. Key phone numbers follow.

PX-439-1846	**Caf-631-3779**	**Chap-798-6123/2707**
Chld C-798-6217	**Commiss-798-3663**	**Conv S-439-1914**
EM C-431-3610	**Fam Svc-798-5080**	**Gas S-439-1689**
Legal-798-5890	**Loc-798-7196**	**Med-798-8041/8045**
NCO C-431-3610	**Off C-431-5603**	**Pkg S-431-3621**
Pol-798-7111	**Theat-798-3833**	

HEALTH & WELFARE: Blanchfield Army Community Hospital, clinics and depts, Emerg, EX-8001/8100 and 2101, Appoint EX-7201 and 439-3114. Chapels serving all faiths, EX-3181/2707.

RECREATION: Arts/Crafts, EX-3625, Auto Hobby, EX-6317, Bowling, EX-5887/3780, ITT, EX-7436, Golf, EX-4906, Gym, EX-3320/7504, Sports Rental, EX-3919/6806, Riding Stables, EX-2629/2487, Outdr Rec, EX-3126/5590, Library, EX-5729, Swimming, EX-6290, Rec Ctr, EX-6383/4616/7446/5818, Tennis, EX-6973/5225, Show Ctr, EX-6086. Eagles' Rest Travel Camp, on Post, yr round, Outdoor Rec, Comm: 502-798-3126/5590, ATVN: 635-3126/5590, 25 camper spaces w/hookups, full fac & acts. Fletchers Fork Travel Camp, on Post, yr round, Comm: 502-798-3126/5590 ATVN: 635-3126/5590, 50 camper spaces w/hookups.

SPACE-A: Campbell Army Airfield, flts via exec aircraft, EX-2151.

ATTRACTIONS: A growing recreational area in Nashville, home of the Grand Ole Opry, 45 mi SE.

FORT KNOX (KY01R2)

Fort Knox, KY 40121-5000

Comm: 502-624-1181
ATVN: 464-0111

LOCATION: From I-65 N in Louisville, exit Jefferson Freeway, 841 W to 31 W go south to Fort Knox. From I-64, exit I-264 W (Waterson) to I-65 south, to Jefferson Fort Knox to US-31W N to Fort Knox. From I-71, exit I-65 south to exit Jefferson Freeway 841, west to 31 W then south to Fort Knox. RM: p-37, J/10. NMC: Louisville, 25 mi N.

GENERAL INFORMATION: Army Armor Ctr, Training Brigade, & Armored Brigade.

TEMPORARY MILITARY LODGING: Biltg Ofc, ATTN: ATZK-EH-H, Bldg 1224, Vine Grove Rd, 24 hrs dly, Comm: 502-624-3138/3943, ATVN: 464-3138/3943. All ranks. DV/VIP, EX-2744.

LOGISTICAL SUPPORT: Complete spt fac avail. Key phone numbers follow.

PX-942-0278	**Caf-942-0805/0230**	**Chap-624-4524**
Chld C-942-8320	**Commiss-624-8525**	**EM C-942-0581**
Fam Svc (ACS)-624-4826	**Gas S-942-0841/0164/0367**	**Legal-624-7448**
Loc-624-1141/1336	**Med-624-9999/9000**	**NCO C-942-8067/0409**
Off C-942-8383	**Pkg S-942-0166**	**Pol-624-2111**

THIS LISTING CONTINUED ON PAGE 244

LOUISIANA

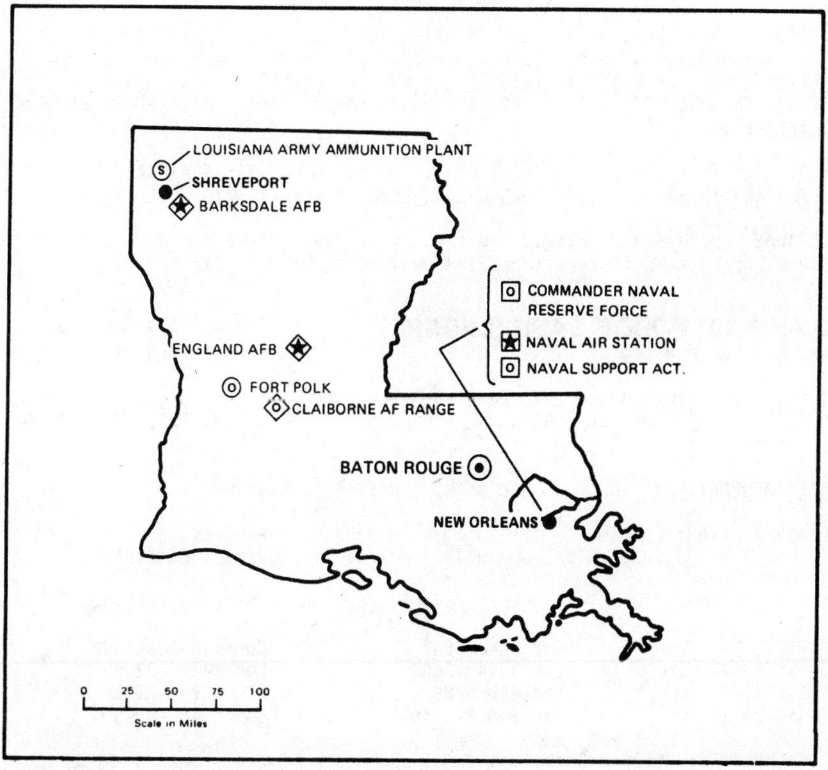

BARKSDALE AIR FORCE BASE (LA01R2)

Barksdale Air Force Base, LA 71110-5000

Comm: 318-456-2252
ATVN: 781-1110

LOCATION: Exit I-20 at Airline Dr, go S to Old Minden Rd (.25 mi), left on Old Minden Rd (1 block), then right on North Gate Dr (1 mi) to North Gate of AFB. RM: p-38, A/13. NMC: Shreveport, 1 mi W. Co-located with Bossier City and Shreveport.

GENERAL INFORMATION: Strategic Air Command Base, Hq 8th Air Force, 2nd Bomb Wing, Aerial Refueling Aircraft, AFRES units stationed here.

TEMPORARY MILITARY LODGING: Biltg Offc, Bldg 5155, Hangarline Rd, 24 hrs dly. Comm: 318-456-3138, ATVN: 781-3138. All ranks. DV/VIP, EX-2161.

LOGISTICAL SUPPORT: Complete spt fac avail. Key phone numbers follow.

BX-746-2554	Chap-456-2111	Chld C-456-4139
Command Post-456-2151	Commiss-456-3377	Conv S-746-7582
EM C-456-3190	Fam Svc-456-3331	Gas S-746-0953
Legal-456-2561	Loc-456-2252	NCO C-746-3366
Off C-746-2203	Pkg S-742-2284	Pol-456-2551
Snack B-746-5441	Theat-746-3666	

Barksdale AFB, Cont'd

HEALTH & WELFARE: USAF Hospital, clinics and depts, Emer, EX-4051, Appoint, EX-2511 and 749-1915. Chapels serving all faiths, EX-2111 and 797-6401.

RECREATION: Aero Club, Auto Hobby, Bowling, EX-4162, Golf, EX-3832, MWR Supply, Wood Hobby, Swimming, Tennis, Youth Ctr, Riding, Rod & Gun Club, Rec Ctr, EX-3190 and for other activities listed. Fam-Camp, on Base, year round, Comm: 318-456-3496, ATVN: 781-3496, 12 camper spaces w/hookups, all facilities and complete activities.

SPACE-A: Pax Term and Lounge, Bldg 6404, 24 hrs, Comm: 318-456-4641/3738, ATVN: 781-4641/3738, flts to CONUS and overseas via KC-135/10 aircraft.

ATTRACTIONS: Moderate Climate. Great water sports. Horse racing (Louisiana Downs) Apr-Oct, NCAA sanctioned Independence Bowl late Dec.

ENGLAND AIR FORCE BASE (LA05R2)

England Air Force Base, LA 71311-5000

Comm: 318-488-2100
ATVN: 683-2100

LOCATION: From Alexandria, take LA-1 to Airbase Rd, enter main gate, or LA-28 to Vandenberg Dr (LA-3054), enter back gate. RM: p-38, F/5. NMC: Alexandria, 5 mi E.

GENERAL INFORMATION: Tactical Air Base, Tactical Fighter Wing.

TEMPORARY MILITARY LODGING: Biltg Ofc, Bldg 1102, Lighting St, 24 hrs dly, Comm: 318-448-5313, ATVN: 683-5313. All ranks. DV/VIP, EX-5181.

LOGISTICAL SUPPORT: Complete spt fac avail. Key phone numbers follow.

BX-487-8768	**Chap-448-5661**	**Commiss-448-5392**
Fam Svc-448-5425	**Gas S-443-3431**	**Legal-448-2285**
Loc-448-2311	**Med-448-5809**	**NCO C-448-2525**
Off C-448-2421	**Pkg S-448-5746**	**Theat-448-5295**

HEALTH & WELFARE: USAF Hospital, clinics and depts, Emerg, EX-5528, Appoint EX-5416, 40 bed. Chapels serving all faiths, EX-5661.

RECREATION: MWR, EX-5421, Arts/Crafts, EX-5807, Auto Hobby, EX-5821, Bowling, EX-2540, Golf, EX-5737, Gym, EX-5521, Sports Rental, EX-2358, Boats, EX-2358, Handball/Squash, EX-5521, Riding, EX-5421, Rod/Gun, EX-5848, Library, EX-5760, Swimming, EX-5787, Rec Ctr, EX-2503, Tickets, EX-5421, Youth Ctr, EX-2479. Cotile Lake Rec Area, off Base 12 mi, take LA-496, Comm: 318-448-2503, ATVN: 683-2503, 12 mobile homes, A/C and furn, 2 camper spaces w/o hookups, all facilities and activities.

SPACE-A: Pax Term and Lounge, Bldg 107, M-F 0700-2200, Sat-Sun 0900-1100, Comm: 318-448-5721/2460, ATVN: 683-5721/2460.

ATTRACTIONS: Cajun Country. Good food.

FORT POLK (LA07R2)

Fort Polk, LA 71459-5000

Comm: 318-535-2911
ATVN: 863-1110
FTS: 528-1110

LOCATION: Off US-171, at LA-10, 9 mi S of Leesville. RM: p-38, G/4. NMC: Alexandria, 45 mi NE.

GENERAL INFORMATION: Hq 5th Infantry Division (Mech) and support units.

Fort Polk, Cont'd

TEMPORARY MILITARY LODGING: Biltg Ofc, ATTN: Trans Fac, Bldg 1560, 0800-1630, 5 days/wk, Comm: 318-535-2013/4025, ATVN: 863-2013/4025. All ranks. DV/VIP, (04/above) EX-4227/2104-4327.

LOGISTICAL SUPPORT: Complete spt fac avail. Key phone numbers follow.

PX-537-1001	**ACS 535-2205**	**Amb-535-3361/2**
Chap-535-7338/6179	**Commiss-535-6573**	**Conv S-537-1050**
EM C-535-6795	**Gas S-537-1354**	**Legal-535-2580**
Loc-535-6622/6523	**NCO C-535-7943**	**Off C-535-4440**
Pkg S-537-0356	**Pol-535-2677**	**Snack B-537-3540**
SNCO-535-2228	**Theat-535-2087**	

HEALTH & WELFARE: Bayne-Jones Army Community Hospital, clinics and depts, Emerg, EX-3368/9, Appoint, 535-2921. Chapels serving all faiths, EX-2232.

RECREATION: Bowling, EX-6273, Golf, EX-4661, Gyms, Ball fields, Rod/Gun Club, Skeet/Trap Range Club, Flying Club, Library, 578-2527, Youth Ctr, Arts/Crafts, Auto Hobby, EX-2911. Fitness course, Theater presentations, Rec Ctr, Tour/Travel info. Toledo Bend Rec Site, off Post, at Toledo Bend Reservoir, yr round, Comm: 318-535-5350, ATVN: 863-5350, FTS: 528-5350, 16 camper spaces w/hookups, 12 mobile homes full equip, all fac and activities.

SPACE-A: Polk Army Airfield, flts via exec fixed wing aircraft, EX-4831/7328, Rotary Wing, EX-2722/2044.

ATTRACTIONS: Fort Polk Military Museum, Baton Rouge, state cap, 155 mi SE. New Orleans, 232 mi SE.

NEW ORLEANS NAVAL AIR STATION (LA11R2)

Comm: 504-393-3011
New Orleans Naval Air Station, LA 70143-5000
ATVN: 363-3011

LOCATION: Off LA-23 in Belle Chasse. Clearly marked. RM: p-38, J/11. NMC: New Orleans, 10 mi N.

GENERAL INFORMATION: Navy Patrol Aircraft Squadrons, Coast Guard Air Station, Reserve and ANG units, and US Customs Service.

TEMPORARY MILITARY LODGING: Biltg Ofc, Bldg 40, Comm: 504-393-3842/3841, ATVN: 363-3842/3841. Officers only. DV/VIP, EX-3201.

LOGISTICAL SUPPORT: Most spt fac avail. Key phone numbers follow.

NX-393-3580	**Chap-393-3525**	**Conv S-393-3510**
CPO C-393-3520	**EM C-393-3508**	**Gas S-393-3506**
Legal-393-3266	**Loc-393-3253**	**Med-393-3660**
Off C-393-3840	**Theat-393-3230**	

HEALTH & WELFARE: Dispensary/Emerg, EX-3610, Appoint, EX-2670. Inpatient, USAF Med Ctr, Keesler AFB, Biloxi, MS, 90 mi E, 601-377-4129. Chapels serving all faiths, EX-3525.

RECREATION: Arts/Crafts, EX-3516, Auto Hobby, EX-3448, Bowling, EX-3514, Golf, EX-3453, Gym, EX-3230, Sports Rental, EX-3230, Ceramics, EX-3516, Gear Issue, EX-3230, Library, EX-3525, Rec Ctr, EX-3231. Campgrounds, on Base, year round, Comm: 504-393-3232, ATVN: 363-3232, 17 camper spaces w/hookups, 4 camping trailers, other facilities and activities.

SPACE-A: Pax Term and Lounge, Bldg 1, ATC, 0700-2300 hrs dly, Comm: 504-393-3103, ATVN: 363-3100, flts to CONUS.

New Orleans NAS, Cont'd

ATTRACTIONS: New Orleans French Quarter, Mississippi River.

NEW ORLEANS NAVAL SUPPORT ACTIVITY (LA06R2)

Comm: 504-361-2011
ATVN: 485-2011

New Orleans Naval Support Activity, LA 70142-5000

LOCATION: On the west bank of Mississippi River. From I-10 E take Canal St-Mississippi River Bridge exit, right on Clairborne Ave, right at Gretna exit. Take Gen DeGaulle E exit after passing over bridge and turn left at Shirley Dr which leads to NSA. RM: p-38, F/17. NMC: New Orleans LA, 5 mi E.

GENERAL INFORMATION: Naval Reserve Center for Personnel, Hq 4th Marine Corps Division and Air Wing.

TEMPORARY MILITARY LODGING: Navy Lodge, Bldg 702, 0700-1800, Comm: 504-366-3266, ATVN: 485-2700. All ranks. DV/VIP, EX-2540.

LOGISTICAL SUPPORT: Complete spt fac avail. Key phone numbers follow.

NX-361-2702	**Chap-361-2244**	**Chld C-361-2508**
Commiss-361-2182	**CPO C-361-2218**	**EM C-361-2219**
Fam Svc-361-2647	**Gas S-361-2747**	**Off C-361-2702**
Pkg S-362-0661	**Pol-361-2570**	**Snack B-361-2739**
SNCO-361-655 (aft hrs)	**Theat-361-2206**	

HEALTH & WELFARE: Dispensary/Emerg, EX-2692, Appoint, EX-2443. Inpatient, USAF Med Ctr, Keesler AFB, Biloxi, MS, 90 mi E, 601-377-4129. Chapels serving all faiths, EX-2244.

RECREATION: Spec Svc, EX-2268, Gear Issue, EX-2205, Swimming, EX-2654, Boats & Campers, EX-2205, Courts, EX-2527, Hobby Shops, EX-2207, Bowling, EX-2204. Recreation Area, on Base, year round, Spec Svc, Comm: 504-361-2205, ATVN: 485-2205, 16 camper spaces w/hookups, other fac and activities.

SPACE-A: New Orleans NAS, Bldg 1, 0700-2300 hrs dly, Comm: 504-393-3103, ATVN: 363-3100, flts to CONUS locations.

ATTRACTIONS: New Orleans French Quarter, Bourbon Street, Sports, Zoo, River Boat Cruises.

Small Installations Cont'd from page 78

Camp Bullis, Fort Sam Houston
Training Site, Bldg 3700
Camp Bullis TX 78234-5000

Utah Air National
Guard
Salt Lake City UT 84100-5000

Defense Depot
Bldg 46
Ogden UT 84401-5000

Chesapeake Naval Security
Group Activity
Chesapeake VA 23322-5000

Camp Elmore
Marine Corps
Camp Elmore VA 23511-5003

Portsmouth Coast Guard Spt Ctr
4000 Coast Guard Blvd
Portsmouth VA 23703-5000

Ethan Allen
Firing Range
Jericho VT 05465-9706

Seattle Coast Guard Spt Center
1519 Alaskan Way St
Seattle WA 98134-5000

Cont'd page 98

MAINE

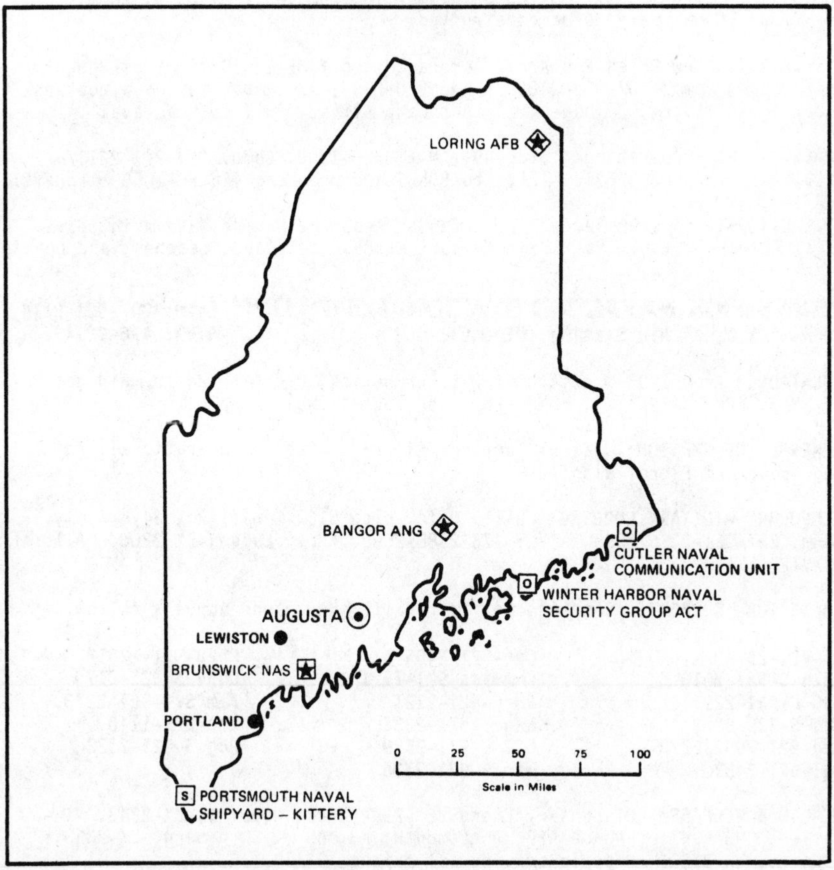

BANGOR AIR NATIONAL GUARD BASE (ME10R1)

Comm: 207-941-2211/2215
ATVN: 698-6700

Bangor Air National Guard Base, ME 04401-3099

LOCATION: Located in Bangor city limits. North from I-95 exit on Ohio Street, drive two blocks, turn left for one block, turn right on Union Street (State Route 222), pass Bangor IAP, turn left on Griffin Road, entrance to is 300 yards on right. RM: p-39, L/7. NMC: Bangor, in city limits.

GENERAL INFORMATION: Air National Guard Air Refueling Wing with KC-135E aircraft, Tactical Air Command Over-the-Horizon Backscatter Radar Operations Center.

TEMPORARY MILITARY LODGING: The Army National Guard operates Transient Quarters - "Pine Tree Inn" for military members and retirees and families. Facility is located at 22 Cleveland Avenue on Bangor International Airport adjacent to Bangor ANG Base. Open 24 hours per day, Comm: 207-942-2081, ATVN: 698-6700, Off-Net patch to 942-2081. All ranks.

Bangor ANG Base, Cont'd

LOGISTICAL SUPPORT: BX located on Bangor International Airport, Comm: 941-2233. Commissary to be constructed.

RECREATION: Dow Pines Rec Area, Aurora, ME, off ME-9, 37 mi NE of Bangor, year round, Comm: 207-999-2309, ATVN: 920-2309, 18 camper spaces w/hookups, 5 cottages, 1 dorm, 2 lodges, 24 tent camp sites. Full spt fac avail.

SPACE-A: Base Operations, Bldg 491, M-F 0730-1600, Comm: 207-941-2211/2215, ATVN: 698-6211/6215. Flts to CONUS and overseas via KC-135E aircraft.

ATTRACTIONS: Lobster and other seafood, Penobscot Marine Museum in Searsport, Acadia National Park, Bar Harbor, Mt Desert Island, beaches, and boating.

BRUNSWICK NAVAL AIR STATION (ME07R1)

Brunswick Naval Air Station, ME 04011-5000

Comm: 207-921-2214
ATVN: 476-2214

LOCATION: From I-95 N exit US-1 N to Brunswick, Old Bath Rd to main gate of BNAS. RM: p-39, I/4. NMC: Portland, 30 mi SW.

GENERAL INFORMATION: Patrol Tactical Aircraft, Special Operational Training Group and Patrol Wing.

TEMPORARY MILITARY LODGING: Biltg Offc, bldg 512, Sewall Rd, 24 hrs dly, Comm: 207-921-2386/2568, ATVN: 476-2386/2568. Navy Lodge, EX-2206. All ranks. DV/VIP, EX-2323.

LOGISTICAL SUPPORT: Complete spt fac avail. Key phone numbers follow.

NX-921-2378	**Caf-921-2455**	**Chap-921-2231/2**
Chld C-921-2610	**Commiss-921-7121**	**Conv S-921-2179**
CPO C-921-2291	**EM C-921-2121**	**Fam Svc-921-2273**
Gas S-725-8863	**Legal-921-2331**	**Loc-921-1110**
Med-921-2992/2222	**Off C-921-2589**	**Pkg S-921-2128**
Pol-921-2457	**Theat-921-2736**	

HEALTH & WELFARE: Dispensary/Emerg, EX-2992/2222, Appoint, EX-2243. Inpatient, USAF Hosp, Pease AFB, Portsmouth, NH, 85 mi SW, Emerg, EX-2333, Appoint, EX-3122. Chapels serving all faiths, EX-2231/2.

RECREATION: Flying Club, EX-729-1468, Gear Locker, EX-2932, Bowling, EX-2145, Auto/Wood Hobby, EX-2488, Crafts, EX-2168, Golf, EX-2155, Library, EX-2639, Ski Shop, EX-2155, Snow Central, EX-2632, Swimming, EX-2181.

SPACE-A: Pax Term and Lounge, ATC bldg, 24 hrs dly, Comm: 207-921-2689, ATVN: 476-2689/2682, flts to CONUS and overseas.

ATTRACTIONS: Great snow skiing area. Lobsters and other seafood. Maine Maritime Museum in Bath, and Penobscot Marine Museum in Searsport.

CUTLER NAVAL COMMUNICATIONS UNIT (ME08R1)

East Machias, ME 04630-5000

Comm: 207-259-8218
ATVN: 476-7218

LOCATION: From I-95 take Alt US-1 to Ellsworth ME, US-1 E to East Machias, right on ME-191 to Base at Cutler. RM: p-39, F/12. NMC: Bangor, 100 mi W.

GENERAL INFORMATION: The world's most powerful transmitter. Provides fleet communications.

Cutler NCU, Cont'd

TEMPORARY MILITARY LODGING: BOQ. Biltg Ofc, Bldg 500, 0800-1600 hrs dly, Comm: 207-259-8204, ATVN: 476-7204. All ranks.

LOGISTICAL SUPPORT: Many spt fac avail. Key phone numbers follow.

NX-EX-244	**Caf-EX-225**	**Commiss-EX-270**
Cons C-EX-277	**Gas S-EX-232**	**Med-EX-209/219**
Pkg S-EX-277	**Pol-EX-261**	**Rec Ser-EX-285**

HEALTH & WELFARE: Dispensary, Appoint, EX-209/219, Emerg after hrs, EX-229. Inpatient, USAF Hosp, Loring AFB, Limestone, 150 mi N, 207-328-9175. No chapel facilities on Base.

RECREATION: Bowling, EX-276, Hobby Shop, EX-285. Sprague's Neck, on Base, year round, road conditions permitting, Comm: 207-259-8218, 1 log cabin, 4 camper spaces w/o hookups, some fac and activities. Rocky Lake, on ME-191, 20 mi NW of unit, year round, road conditions permitting, Comm: 207-259-8218, 1 very rustic cottage (no elec or water), limited fac, snow skiing.

SPACE-A: Bangor ANG Base, Bldg 491, M-F 0730-1600, Comm: 207-947-0572 EX-211/215, ATVN: 476-6211/6215, flts to CONUS and overseas via KC-135 aircraft.

ATTRACTIONS: Beaches, seafood, boating, outdoor sports, and snow skiing.

LORING AIR FORCE BASE (ME06R1)

Loring Air Force Base, ME 04751-5000

Comm: 207-999-1110
ATVN: 920-1110

LOCATION: Take I-95 to Houlton, exit to US-1 N to Caribou (50 mi). The AFB is 1 mi N of ME-89 & ME-223 intersection. Clearly marked. RM: p-39, B/17. NMC: Caribou, 3 mi SW.

GENERAL INFORMATION: Strategic Air Command Base. Bomb Wing and support units here.

TEMPORARY MILITARY LODGING: Biltg Ofc, Bldg 2501, Virginia Place, 24 hrs dly, Comm: 207-999-2227, ATVN: 920-2227. DV/VIP, 2227.

LOGISTICAL SUPPORT: Complete spt fac avail. Key phone numbers follow.

BX-328-7031	**Chap-999-6132**	**Chld C-999-4227/7239**
Commiss-999-6248	**Conv S-764-7031**	**Fam Svc-999-6100**
Gas S-328-6249	**Legal-999-7216**	**Loc-999-2100**
Med-999-3200/4250	**NCO C-328-7335**	**Off C-328-7358**
Pkg S-328-7871	**Pol-999-2115**	**SDO-207-999-2170**
Snack B-328-7337/7729	**Theat-999-4242**	

HEALTH & WELFARE: USAF Hospital, clinics and depts, Emerg, EX-3200/4250, Appoint, EX-2163, 20 bed. Chapels serving all faiths, EX-6132/7571.

RECREATION: Arts/Crafts, EX-2640, Auto Hobby, EX-2615, Bowling, EX-2617, Golf, EX-2501, Gym, EX-4245, Rec Ctr, EX-2646/2382, Rec Sup, EX-2432, Swimming, EX-6116, Youth Ctr, EX-2442/2469. Dow Pines Rec Area, Aurora, off ME-9, 38 mi NE of Bangor, year round, Comm: 207-999-2309/2646, ATVN: 920-2309/2646, 18 camper spaces w/hookups, 5 cottages, 1 dormitory, 2 lodges, 24 tent camp sites, full fac/support avail, comp acts. Also operates Malbeam Lake and Fam-Camp, on Base, 9 hardstands, 10 camper spaces w/o hookups, 8 tent spaces, Comm: 207-999-2506, ATVN: 920-2506.

SPACE-A: Base Ops, 24 hrs, Comm: 207-999-2294/2170, ATVN: 920-2294/2170,

Loring AFB, Cont'd

flts to CONUS and overseas.

ATTRACTIONS: Snow skiing, hunting, fishing, snow mobiling, Bar Harbor resort area, and Canada.

WINTER HARBOR NAVAL SECURITY GROUP ACTIVITY (ME09R1)

Comm: 207-963-5534
ATVN: 476-9011

Winter Harbor, ME 04693-0900

LOCATION: From Ellsworth, take US-1 N to ME-186 E to Acadia Natl Park. Nav Security Stn on Schoodic Point in the park. RM: p-39, G/9. NMC: Bangor, 45 mi NW.

GENERAL INFORMATION: Provides communications/other security for Nav units.

TEMPORARY MILITARY LODGING: Biltg Ofc, Bldg 84, dty hrs, Comm: 207-963-5534 EX-223, ATVN: 476-9223. All ranks. DV/VIP, no fac.

LOGISTICAL SUPPORT: Most spt fac avail. Key phone numbers follow.

NX-963-2223	**Commiss-963-5827**	**Conv S-963-7105**
CPO C-EX-283	**Gas S-963-7391**	**Legal-EX-210**
Loc-963-5534/5535	**Med-EX-297**	**Off C-EX-283**
Pkg S-963-7105	**Pol-EX-202**	**Snack B-963-2223**

HEALTH & WELFARE: Dispens/Emer, EX-297. Inpatient, USAF Hosp, Loring AFB, Limestone, 180 mi N, Comm: 207-328-9175. Chapel serving all faiths, EX-279.

RECREATION: Bowling, Hobby Shop, Gear Lock, Rec Svc, EX-297. Winter Harbor Rec Area, on Base, May-Sep, Comm: 207-963-5534/5315, ATVN: 476-9304/9315, 6 camper spaces w/hookups, 4 camper spaces w/o hookups, 5 mobile homes (avail year round), other fac and all activities.

SPACE-A: See Cutler Naval Communications Unit, ME listing.

ATTRACTIONS: Ocean sports, seafood, winter snow skiing.

Small Installations Cont'd from page 94

Port Angeles Coast Guard
Air Station
Port Angeles WA 98362

Vancouver Barracks
Bldg 752
Vancouver WA 98661-5000

Yakima Firing Range
Bldg T-157
Yakima WA 98901-5000

Billy Mitchell BX, Bldg 102
300 E College Avenue
Milwaukee WI 53207-5000

Milwaukee BX, Bldg 304
4828 W Silver Spring Dr
Milwaukee WI 53218-5000

Martinsburg Air National
Guard Station
Martinsburg WV 25401-5000

End

-NOTES-

MARYLAND

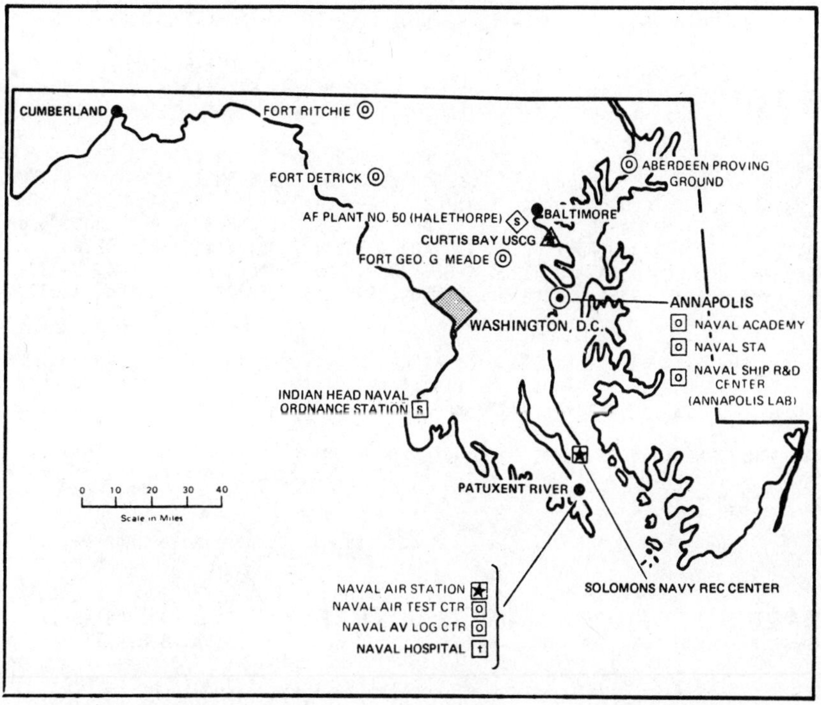

ABERDEEN PROVING GROUND (MD11R1)

Aberdeen, MD 21005-5055

(Aberdeen Area) Comm: 301-278-5201
(Edgewood Area) Comm: 301-278-5201
(Aberdeen Area) ATVN: 298-1110
(Edgewood Area) ATVN: 584-1110

LOCATION: Aberdeen Area, take exit 5 E from I-95 N on MD-22 E for 2 mi to main gate. Also, from US-40 N to right on Maryland Blvd, entrance to main gate. NMC: Baltimore, 23 mi SW. Edgewood Area, take exit 4 from I-95 N on MD-24 E for 2 mi to main gate. Also, from US-40 right on MD-24 to main gate. RM: p-41, D/20. NMC: Baltimore, 13 mi SW.

GENERAL INFORMATION: Hq US Army Test & Evaluation Command, Ordnance Center & School, Ballistics R&D Lab, Human Engineering Lab, Chemical R&D Center.

TEMPORARY MILITARY LODGING: Biltg Ofc, Bldg 2207, 24 hrs dly, Comm: 301-278-4373/5148, ATVN: 298-4373/5148. All ranks. DV/VIP, EX-4373.

LOGISTICAL SUPPORT: Complete spt fac avail. Key phone numbers follow.

PX-A-272-6828	Caf-278-2621	Chap-278-4333
E-676-2882	Chld C-272-6621	Commiss-A-278-3926
Conv S-272-1681	EM C-272-8873	E-278-2114
Fam Svc-A-278-4530	Gas S-272-7407	Legal-278-3036/3096
E-671-3362	Loc-278-5138/1110	Med-278-2086

Aberdeen Prov Gd, Cont'd

NCO-278-4500	NCO C-A-272-6269	Off C-A-272-9107
Pkg S-A-272-6269	E-671-4297	E-676-5400
E-671-4297	Pol-278-5291/92	Theat-278-3351

HEALTH & WELFARE: Clinics only, Emerg, EX-5251, Appoint, EX-5253. Inpatient, Kimbrough Army Comm Hosp, Fort Meade, 40 mi SW, 677-4124. Chapels serving all faiths, A-278-2886, E-278-4109.

RECREATION: Flying Club, A-272-8699, E-278-4109; Rod & Gun Club, A-272-6648, E-671-2579; Riding & Boat Clubs, A-272-6273, E-676-9831; Library, A-278-3417, E-671-3650; Teen Club, EX-2794; Golf, A-278-4794, A-278-2541, E-278-2213; Arts & Crafts, A-278-4438, E-671-2153; Auto, A-278-5178, E-671-3889; Boat Dock, A-278-3344, E-671-3439; Bowling, A-278-4041, E-671-2294; Gym, A-278-4216, E-671-3375; Rec Equip, A-278-4124, E-671-3939; Rec Center, A-278-5684, E-671-2713; Tennis, A-278-3351; Youth Activities, A-278-4995/2322; Swimming, A-278-4217, E-671-2113.

SPACE-A: Phillips AAF, flts via exec aircraft to CONUS locations, EX-3483. Also, Warfield ANG Base, Glenn L Martin State Arprt, Bldg 3198, 0600-2400 hrs dly, Comm: 301-981-2740/2744, ATVN: 858-2740/2744, flts to CONUS locations.

ATTRACTIONS: Ordnance Museum, Baltimore, easy drive. One of the world's largest ports, and Fort McHenry, where Francis Scott Key composed the "Star Spangled Banner."

NOTE: A - Aberdeen Area, E - Edgewood Area. Single number is for Aberdeen.

ANDREWS AIR FORCE BASE (MD02R1)

Andrews Air Force Base, MD 20331-5000

Comm: 301-981-9111
ATVN: 858-9111

LOCATION: From I-95 (east portion of Capital Beltway, I-495) N or S, exit 9, first traffic light after leaving exit ramp turn right into main gate of AFB. Also, from I-395 N, exit So Capitol St, cross Anacostia River on So Capitol St, bear left to Suitland Parkway E, exit Parkway at Morningside on Suitland Rd E to main gate of AFB. Clearly marked. RM: p-42, J/11. NMC: Washington, DC, 6 mi NW.

GENERAL INFORMATION: Military Airlift Command Base. Hq Air Force Systems Command, 1776 Air Base Wing, 89th Military Airlift Wing, and support units.

TEMPORARY MILITARY LODGING: Biltg Ofc, ATTN: 1776 SVSSQN/SVH, Bldg 1375, Arkansas St, 24 hrs dly, Comm: 301-981-4624, ATVN: 858-2606. All ranks. DV/VIP, EX-4525.

LOGISTICAL SUPPORT: Complete spt fac avail. Key phone numbers follow.

BX-568-1500	Caf-568-2381	Chap-981-2111
Chld C-981-3035	Commiss-981-7106	Fam Svc-981-7322/23
Gas S-735-0868/568-2320	Legal-981-3622	Loc-981-6161
Med-981-2333	NCO C-568-3100	Off C-420-4744
Pol-981-2000/2001	Shoppette-568-2364	Snack B-568-0180/2357
Theat-981-6641		

HEALTH & WELFARE: USAF Hospital, clinics and depts, Emerg, EX-2333, Appoint, EX-7511, 320 bed. Chapels serving all faiths, EX-2111.

RECREATION: Aero Club, EX-6900, Auto Hobby, EX-3917/6643, Bowling, EX-6452, Golf, EX-4404, Gym & Racquet, EX-7101, Library, EX-6454, Multi-Craft, EX-2697, Rec Ctr, EX-6560, Skeet/Trap, EX-5985, Swimming, EX-4260/4877/6738,

Andrews AFB, Cont'd

Tennis, EX-5710, Youth Ctr, EX-5636. Fam-Camp, on Base, year round, Comm: 301-981-6560, ATVN: 858-6560, 12 camper spaces w/hookups, all fac and act.

SPACE-A: Pax Term/Lounge, Bldg 1245, 0600-2200 hrs dly, Comm: 301-981-3528/3527, ATVN: 858-3528/3527, flts to CONUS/overseas. Also, Washington NAF, E side of AFB, Bldg 3198, 24 hrs dly, Comm: 301-981-2740/2744, flts to CONUS.

ATTRACTIONS: Aerial gateway to Washington, DC, home of "Air Force One," the President's aircraft. Monuments, White House, parks, and museums.

BETHESDA NAVAL MEDICAL COMMAND NATIONAL CAPITAL REGION (MD06R1)

Comm: 301-295-5960/2075
ATVN: 295-5960/2075

Bethesda, MD 20814-5000

LOCATION: From I-495 (Capital Beltway) take exit 25, Wisconsin Ave (MD-335, Rockville Pike) S 1 mi to Naval Medical Center on left. Enter first gate, Wilson Dr for spt fac. Also, can enter the center from Jones Bridge Rd off Wisconsin Ave. RM: p-42, F/7. NMC: Washington, DC, 1 mi SE.

GENERAL INFORMATION: The Navy's largest and most advanced medical center, Navy Health Science Education and Training Command, Uniformed Services University of the Health Sciences, Naval Hospital.

TEMPORARY MILITARY LODGING: Biltg Ofc, Bldg 1, Wisconsin Ave, 0800-1630 hrs dly, Navy Lodge, 24 hrs dly, Comm: 301-295-0798, ATVN: 295-1138. Navy Lodge, Comm: 301-654-1795. All ranks. DV/VIP, EX-1111/1110.

LOGISTICAL SUPPORT: Complete spt fac avail except Commiss. Key phone numbers follow.

NX-652-5361	**Caf-652-2282**	**Chap-295-1510**
Chld C-295-0014	**Conv S-652-5362**	**CPO C-295-0256**
EM C-652-0256	**Gas S-530-1380**	**Legal-295-2298**
Loc-295-5387	**Med-295-4810/1400**	**Off C-652-6318**
Pkg S-652-6318	**Pol-295-0501**	**Snack B-652-2282**
SNCO-295-5960	**Theat-295-4611**	

HEALTH & WELFARE: USN Hospital, clinics and depts, Emerg, EX-4810, Appoint, EX-1400, 500 bed. Chapels serving all faiths, EX-1510.

RECREATION: Bowling, EX-2060, Gym, EX-0031, Library, EX-0993, Recreation Service, EX-0031, Swimming, EX-0031, Tennis, EX-0031, Picnic area on South Palmer Rd.

SPACE-A: See Andrews AFB listing.

ATTRACTIONS: National Zoo in nearby Rock Creek Park. Wisconsin Ave S ends in the Georgetown section of Washington, DC. Washington Cathedral is also off Wisconsin Ave at Massachusetts Ave.

CURTIS BAY COAST GUARD YARD (MD01R1)

Comm: 301-789-1600
ATVN: None
FTS: 922-6229

Curtis Bay, MD 21226-1797

LOCATION: Take I-695 to exit 1, bear to your right, right on Hawkins Point Rd, left into Coast Guard Yard. RM: p-41, P/5. NMC: Baltimore, 2 mi N.

Curtis Bay CG Yard, Cont'd

GENERAL INFORMATION: The Coast Guard's only shipbuilding and repair facility, Curtis Bay Station, CG Group Baltimore, 2 home-port ships.

TEMPORARY MILITARY LODGING: Biltg Ofc, Bldg 33, NAFA Ofc, 0730-1530 hrs M-F, Comm: 301-789-1600 EX-530, FTS: 922-6487. All ranks. DV/VIP, EX-1600.

LOGISTICAL SUPPORT: Most spt fac avail. Key phone numbers follow.

CGX-EX-294	**Chap-EX-218/219**	**Conv S-EX-526**
CPO C-EX-387	**EM C-EX-386**	**Fam Svc-EX-227**
Gas S-EX-484	**Legal-EX-391/393**	**Loc-1600/6229**
Med-EX-481	**Off C-EX-387**	**Pkg S-EX-294**
Pol-EX-485	**Snack B-EX-294**	**SNCO-EX-370**

HEALTH & WELFARE: Dispensary/Emerg, EX-481. Inpatient, Kimbrough Army Comm Hosp, Fort Meade, 30 mi S, Emerg, Comm: 301-677-4288, ATVN: 923-4288. Chapels serving all faiths, EX-218/219.

RECREATION: Auto Hobby, EX-492, Bowling, 789-9796 or EX-360, Marina, EX-219, Picnic Area, EX-219, Rec Ctr, 374/375, Spec Svc Ofc, EX-219, Swimming, EX-374/375.

SPACE-A: Warfield ANG Base, Glenn L Martin State Airport, Bldg 3198, 0600-2400 hrs dly, Comm: 301-981-2740/2744, ATVN: 858-2740/2744, flts to CONUS and overseas locations.

ATTRACTIONS: Aquarium, maritime and streetcar museums, Civic Center, Fort McHenry, World Trade Center, new Inner Harbor area.

FORT DETRICK (MD07R1)

Fort Detrick, MD 21701-5000

Comm: 301-663-8000
ATVN: 343-1110
FTS: 935-7101

LOCATION: From Washington, DC, take I-270 N to US-15 N. From Baltimore, take I-70 W to US-15 N. From US-15 N, in Frederick, exit Seventh Street. Clearly marked to Post. RM: p-41, D/13. NMC: Baltimore, 50 mi E and Washington, DC, 50 mi SE.

GENERAL INFORMATION: Army Medical R&D Command, Research in Infectious Diseases and Bioengineering R&D Lab, Telecommunications Center, Quad-Service Medical Logistics.

TEMPORARY MILITARY LODGING: Biltg Ofc, Bldg 901, Sulton St & Doughten Dr, 0745-1630 hrs dly, Comm: 301-663-2154, ATVN: 343-2154. All ranks. DV/VIP, EX-7114.

LOGISTICAL SUPPORT: Most spt fac avail. Key phone numbers follow.

PX-662-7755	**Chap-663-7371**	**Chld C-663-2620**
Comm C-663-2823	**EM C-663-2823**	**Fam Svc-663-2452**
Gas S-662-7777	**Legal-663-2643**	**Loc-663-8000/1110**
Med-663-7175	**Pol-663-7114**	**SNCO-663-7114**

HEALTH & WELFARE: Clinic, Appoint, EX-7175. Inpatient, Walter Reed Army Med Ctr, Washington, DC, 50 mi SE, 202-576-3501. Chapel serving Protestants and Catholics, EX-7371.

RECREATION: Arts & Crafts, EX-2964, Auto Hobby, EX-2849, Bowling, EX-2816, Driving Rnge, EX-2964, Field Hse & Gym, EX-2498, Fishing Pnd, EX-2498, Lib, EX

Fort Detrick, Cont'd

2806, Outdr Rec, EX-2920, Swim, EX-2498, Tennis, EX-2498, Youth Ctr, EX-2403.

SPACE-A: See Andrews AFB listing.

ATTRACTIONS: Baltimore and Washington, easy drives. Near snow ski resorts. Quaint shops and excellent restaurants in Historic Frederick.

FORT GEORGE G. MEADE (MD08R1)

Fort George G. Meade, MD 20755-5000

Comm: 301-677-6261
ATVN: 923-6261

LOCATION: Off Baltimore-Washington Parkway, I-295, exit MD-198 E which is Fort Meade Rd. Clearly marked. RM: p-42, C/12. NMC: Baltimore and Washington, DC, 30 mi from each city.

GENERAL INFORMATION: Hq First US Army, National Security Agency, Army Intelligence, and Army Field Band.

TEMPORARY MILITARY LODGING: Biltg Ofc, ATTN: AFZI-EH-HB, bldg 4707, Brett Hall, Ruffner Rd, 24 hrs dly, Comm: 677-5660/5884, ATVN: 923-5884/6529. All ranks. DV/VIP, EX-4550.

LOGISTICAL SUPPORT: Complete spt fac avail. Key phone numbers follow.

PX-674-7170	**Chap-677-3703/04**	**Chld C-677-5201/7568**
Commiss-677-3250	**Conv S-672-3287**	**EM C-677-4546**
Fam Svc-677-3418/3586	**Gas S-672-3238**	**Legal-677-7535**
Loc-677 AR-4547, AF-3295, NA-6354, MC-3233		**Med-677-4392**
NCO C-677-3983/4546	**Off C-677-5358/5298**	**Pol-677-6222**
Snack B-674-5588	**Theat-677-5324**	

HEALTH & WELFARE: Kimbrough Army Community Hospital, clinics and depts, Emerg, EX-4392, Appoint, EX-2151. Chapels serving all faiths, EX-6703/04.

RECREATION: Amusement Ctr, 674-8585, Auto Hobby, EX-5542, Bowling, EX-5541, Equip Rental, EX-1242, Flying Club, 674-7030, Sports Arena, EX-3716, Golf, EX-4308, Gym, EX-3042, Library, EX-4509, Multi Craft, EX-4310, Outdoor Rec, EX-3825, Picnic Area, EX-3810, Rec Ctr, EX-4666, Rod & Gun, 672-3673, Stables, EX-7101, Swimming, EX-4121, Tennis, EX-3042, Youth Ctr, EX-5437. Fort Miles Rec Area, DE, on Atlantic Ocean at Cape Henlopen State Park, Lewes, DE, May-Sep, Outdoor Rec, bldg 940, Comm: 301-677-4945, ATVN: 923-4945, 20 mobile homes, 19 2-3 bdrm apts, 1 fam duplex, 1 dorm, full rec ctr fac, all act.

SPACE-A: Tipton Army Airfield, flts via exec aircraft to CONUS locations, EX-6261.

ATTRACTIONS: Museum on Post, horse races at Laurel and Bowie.

FORT RITCHIE (MD13R1)

Fort Ritchie, MD 21719-5010

Comm: 301-241-1300
(Wash, DC, Area Dial
Comm: 301-878-1300)
ATVN: 988-1300

LOCATION: From US-15 N exit at Thurmont, to MD-550 N for 7 mi to Cascade, and main gate. Also, from Hagerstown, take MD-64 E to MD-491 N to MD-550 N to Cascade and main gate. RM: p-41, B/12. NMC: Hagerstown, 16 mi SW, Baltimore, 50 mi SE, Washington, DC, 55 mi SSE.

GENERAL INFORMATION: Hq, 7th Signal Command, US Army Information Systems Engineering Command - Continental United States.

Fort Ritchie, Cont'd

TEMPORARY MILITARY LODGING: Biltg Ofc, Bldg 520, West B Ave and I Ave, 0800-1845 M-F, other times SDO, Comm: 301-241-4445, or 878-5171, DC Dial, 878-4700, ATVN: 988-5171. All ranks. DV/VIP, EX-1300.

LOGISTICAL SUPPORT: Complete spt fac avail. Key phone numbers follow.

PX-241-4600/878-4103	**Chap-878-5070**	**Chld C-878-5380**
Commiss-878-5820	**EM C-878-5588**	**Fam Svc-878-5100**
Gas S-241-4334	**Legal-878-5771**	**Loc-878-5685**
Med-878-4132/5439	**NCO C-878-5588**	**Off C-241-4700/878-4361**
Pkg S-878-4290	**Pol-878-4228**	**Snack B-878-4192**
Theat-878-5867		

HEALTH & WELFARE: Health Clinic, Emerg, EX-4132/4455. Inpatient, Walter Reed Army Med Ctr, Washington, DC, Comm: 202-576-1199, ATVN: 291-1199. Chapels servng all faiths, EX-5070.

RECREATION: Auto Craft, EX-5891, Bowling, EX-4192, Comm Ctr, EX-5881, Golf, EX-5990, Gym, EX-4485, ITT, EX-5881, Lake Royer, EX-4186, Library, EX-5060, Skill Devl Ctr, EX-5340, Outdoor Rec, EX-4186, Ski Club, 878-4463, Swimming, EX-4186, Tennis, EX-4485, Wood Craft, EX-5984, Youth Act, EX-5500. Outdoor Rec Ctr, on Post, year round, Comm: 301-878-4186, ATVN: 988-4186, lake, picnic area, boat rental, swimming, fishing, ice skating.

SPACE-A: Warfield ANG Base, Glenn L Martin State Arprt, Baltimore, Comm: 301-981-2740/2744, ATVN: 858-2740/2744, flts to CONUS locations.

ATTRACTIONS: Several snow ski areas nearby, historic Gettysburg, PA, 35 mi NE, historic Civil War sites surround the area.

INDIAN HEAD NAVAL ORDNANCE STATION (MD04R1)

Comm: 301-743-4000
ATVN: 364-4011

Indian Head, MD 20640-5000

LOCATION: Take I-495 (Capital Beltway) E, exit to MD-210 S for 25 mi to station. RM: p-41, J/14. NMC: Washington, DC, 25 mi N.

GENERAL INFORMATION: Naval Explosive Ordnance Disposal Technology Center, Naval School Explosive Ordnance Disposal.

TEMPORARY MILITARY LODGING: Biltg Ofc, Bldg 1659, Jackson St, 0730-1600 hrs dly, Comm: 301-743-4609, ATVN: 364-4609. Off only. DV/VIP, EX-4000.

LOGISTICAL SUPPORT: Limited spt fac avail. Key phone numbers follow.

NX-743-5414	**Chap-743-3183**	**Chld C-743-4822**
CPO C-743-4648	**EM C-743-4638**	**Gas S-743-3183**
Loc-743-4000	**Med-743-4601/2/3**	**Off C-743-4557**
Pol-743-4000	**Theat-743-4622**	

HEALTH & WELFARE: Dispensary/Emer, EX-4601/2/3. Inpatient, USAF Hosp, Andrews AFB, 25 mi N, Comm: 301-981-6867/2195, ATVN: 858-6867/2195. Chapel serving all faiths, EX-3183.

RECREATION: Rec Svc Ofc, EX-4850/4761, Gym, EX-4850, Bowling, EX-4761, Auto Hobby, EX-1396, Ceramics, EX-4661, Racquetball, EX-1239, Lib, EX-4747, Weight Rm, 743-1396.

Indian Head NOS, Cont'd

SPACE-A: Andrews AFB, Bldg 1245, 0600-2200 hrs dly, Comm: 301-981-3528, ATVN: 858-3528, flts to CONUS and overseas locations.

ATTRACTIONS: The Potomac River and Washington, DC, 28 mi N.

PATUXENT RIVER NAVAL AIR STATION (MD09R1)

Comm: 301-863-3000
ATVN: 356-3000

Patuxent River Naval Air Station, MD 20670-5409

LOCATION: From I-95 (east portion of Capital Beltway, I-495) exit 7A to Branch Ave (MD-5) S. Follow MD-5 until it turns into MD-235 near Oraville, on to Lexington Park, and the NAS. Main gate is on MD-235 and MD-246 (Cedar Point Rd). RM: p-41, L/18. NMC: Washington, DC, 65 mi NW.

GENERAL INFORMATION: Naval Air Test Center and Naval Aviation Log Ctr.

TEMPORARY MILITARY LODGING: Housing Ofc, Bldg 423, 0730-1600 hrs dly, Comm: 301-863-3244, ATVN: 356-3848. All ranks. DV/VIP.

LOGISTICAL SUPPORT: Complete spt fac avail. Key phone numbers follow.

NX-863-8814	**Caf-863-8855**	**Chap-863-3812**
Chld C-863-3636	**Commiss-863-3789**	**Conv S-863-8715**
CPO C-863-3685	**Fam Svc-863-4911**	**Gas S-863-5828**
Legal-863-1044	**Loc-863-1079**	**Med-863-1418**
Off C-863-3656	**Pkg S-863-9315**	**Pol-863-3208**
SNCO-863-1097	**Theat-863-3572**	

HEALTH & WELFARE: Naval Hospital, clinics, Emerg, 863-1418, Appoint, 863-3904. Chapels serving all faiths, 863-3812.

RECREATION: Auto Hobby, EX-3507, Bowling, EX-3994, Ceramics, EX-3160, Gear Issue, EX-3519, Golf, EX-3597, Gym, EX-3519/3508, Library, EX-1927, Marina, EX-3573, Stables, EX-3296, Swimming, EX-4225/5960, Wood Hobby, EX-3569, Youth Ctr, EX-3329. Rec Area, on Base, closed winters, Comm: 301-863-3244, ATVN: 356-3244, 14 camper spaces w/hookups, 14 mobile homes, 60 camper spaces w/o hookups, full facilities & all activities.

SPACE-A: Pax Term and Lounge, Bldg 103, 24 hrs, Comm: 301-863-3836, ATVN: 356-3836, flts to CONUS locations.

ATTRACTIONS: Calvert Cliffs Nuclear Power Plant Museum, Calvert Marine Museum, Naval Air Test & Evaluation Museum, Cecil's Mill, Tobacco Auctions and Oyster Festivals.

SOLOMONS NAVY RECREATION CENTER (MD05R1)

Comm: 301-326-4216
301-863-9074
DC Metro: 261-2816

PO Box 147
Solomons, MD 20688-0147

LOCATION: On Patuxent River in Solomons. From I-95 (east portion of Capital Beltway, I-95) exit 11A, Pennsylvania Ave and MD-4/2 for 65 mi SE. Clearly marked. RM: p-41, L/18. NMC: Washington, DC, 65 mi NW.

GENERAL INFORMATION: A Navy operated/managed rec center of 262 acres on delta where Patuxent River merges with Chesapeake Bay. Open year round.

TEMPORARY MILITARY LODGING: Reservation Ofc, outside main gate, 0800-1800

Solomons Navy Rec Ctr, Cont'd

hrs dly (later hrs in Summer), reservations numbers above. Call for peak season reservation priority information. 50 units include: furnished cottages, 1-5 brm; equipped kitchens. Most have 1 bath, some have 2. TVs, telephones not provided. Linens not provided but may be rented. All units recently renovated and upgraded.

LOGISTICAL SUPPORT: Limited. The following are available: ice vending, campers' store (summer season), and laundry. NEX, commissary, located at NAS Patuxent River (10 miles).

HEALTH & WELFARE: Navy Hospital at Patuxent River NAS, Comm: 301-863-3904, ATVN: 356-3904. Chapels at NAS, 301-863-3812. Security: 301-326-2436.

RECREATION: Swim Pools, Beach, Bowling Ctr, Fitness Ctr, MiniGolf, Basketball, Volleyball, Tennis, Marina, Golf Dr Range, Picnic Areas, Multi-Purpose Rm (Roller Skating, Table Tennis, Movies), Recr Gear Issue, Amusement Machines Arcade, 278 camper spaces w/hookups, 80 camper spaces w/o hookups, and all rec facilities including fishing pier & complete activities. Recreation Info: 301-326-3217 or 301-863-3566.

SPACE-A: See Patuxent River NAS listing.

ATTRACTIONS: A super camping, crabbing, fishing, and recreation area. Washington, DC, and Annapolis are 1.5 hrs drive NNW. Solomons Island area, historic St Marys City, Cliffs of Calvert, Chesapeake Bay.

UNITED STATES NAVAL ACADEMY/ ANNAPOLIS NAVAL STATION (MD10R1)

Annapolis, MD 21402-5054

Comm: 301-267-6100
ATVN: 281-0111
FTS: 930-0111

LOCATION: Two miles off US-50/301, two exits to Naval Academy. Clearly marked. Main gate is on King George St in Annapolis. Naval Station is across the Severn River off US-50/301 E, first exit. Clearly marked. RM: p-41, I/6. NMC: Annapolis, in city limits near city dock.

GENERAL INFORMATION: Navy education facility for officers and their support, Brigade of Midshipmen.

TEMPORARY MILITARY LODGING: Biltg Ofc, Off Club, 0800-1600 hrs dly, Comm: 301-267-3906, ATVN: 281-3906/2803. Officers only. DV/VIP, EX-2803.

LOGISTICAL SUPPORT: Complete spt fac avail. Key phone numbers follow.

NX-757-0005	Caf-757-0005	Chld C-267-3947
Commiss-267-2697/2596	Conv S-757-2120	CPO C-267-3660
EM C-267-3660	Fam Svc-267-2602/2607	Gas S-757-2120
Legal-267-2166/2782	Loc-267-2385	Med-267-2535/2061/2/3
Off C-263-8280/261-1563	Pkg S-267-3232	Pol-267-2886/7
Snack B-757-0005	SNCO-267-2485/3823	Theat-267-3580/2888

HEALTH & WELFARE: Naval Regional Clinic, Emer, EX-2535, Appoint, EX-3617. Inpatient, Bethesda Naval Hosp, Washington, DC, Comm: 202-295-5388, ATVN: 295-5388. Chapels serving all faiths, EX-2891/267-6100.

RECREATION: Auto Hobby, EX-3859, Bowling, EX-3034, Check Out Hut, EX-2077, Golf, EX-2022, Hobby Shops, EX-3410, Gym, EX-3850/2518, Library, EX-2311, Marina, EX-2058, Rec Ctr, EX-3472/3691, Swim, EX-3691/4092, Youth Act, EX-3580/3691. Annapolis Fam-Camp, on Base, Mar-Nov, Comm: 301-267-2518, ATVN: 281-2518, 10 camper spaces w/hookups, comp fac and all act.

US Naval Academy & NS, Cont'd

SPACE-A: See Andrews AFB listing.

ATTRACTIONS: Walking tour of the Naval Academy, museum, tour of historic Annapolis, city dock area, and the MD Capitol building.

MASSACHUSETTS

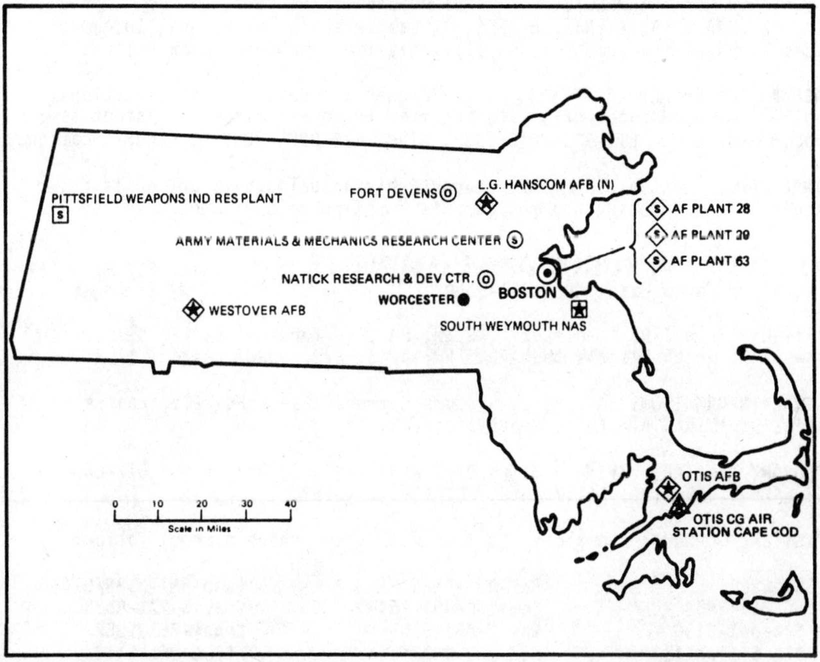

FORT DEVENS (MA09R1)

Fort Devens, MA 01433-5000

Comm: 617-796-3911
ATVN: 256-3911

LOCATION: From Boston follow MA-2 W and exit at Fort Devens, about 5 mi W of I-495. In Ayer. RM: p-45, D/14. NMC: Boston, 35 mi SE.

GENERAL INFORMATION: Army Intelligence School, 10th Special Forces Group (Airborne), 39th Engineer Battalion (Combat), 36th Medical Battalion, Combat Support Battalion, other units.

TEMPORARY MILITARY LODGING: Biltg Ofc, ATTN: AFZD-FEG-B, Sherman Ave, 24 hrs, Comm: 617-796-3354, ATVN: 256-3354. All ranks. DV/VIP, EX-2500.

LOGISTICAL SUPPORT: Complete spt fac avail. Key phone numbers follow.

PX-772-4825/6425	Caf-772-4828	Chap-796-2936
Chld C-796-3604/3751	Commiss-796-2140	Conv S-772-5944
Fam Svc-796-3528/3023	Gas S-772-4820	Legal-796-3108/2541
Loc-796-7954/2748	Med-796-6816	NCO/EM C-796-2535
Off C-772-5968/796-2149	Pkg S-772-4803/796-3848	Pol-796-3333
Snack B-772-5955	Theat-796-3470	

Fort Devens, Cont'd

HEALTH & WELFARE: Cutler Army Community Hospital, clinics except maternity, depts, Emer, EX-6822/2560, Appoints, EX-6714, 116 bed. Chapels serving all faiths, EX-2619/2936.

RECREATION: Auto Crafts, EX-2319, Bowling, 772-4817/796-2591, Crafts Ctr, EX-2177, Rec Ctr, EX-2342, Golf, 772-3273/796-3164, Gym, EX-2472/3965, Library, EX-2431, Mirror Lake, EX-3506, Outdoor Rec, EX-3255, Swimming, EX-2588/2589, Stables, EX-2077, Sports Arena, EX-2848/2587, Youth Act, EX-2472/3965, Youth Ctr, EX-3387/2920. Robbins Pond Travel Camp, on Post, Apr-Oct, Comm: 617-796-3255, ATVN: 256-3255, 14 camper spaces w/hookups, 10 camper spaces w/o hookups, comp fac and all activities including snow skiing.

SPACE-A: Moore Army Airfield, flts via exec aircraft to CONUS locations, EX-3130. Also, Hanscom AFB, flts are from the civil airport adjacent to Hanscom AFB, Comm: 617-861-2285/3333, ATVN: 478-2285/3333 to CONUS locations.

ATTRACTIONS: Boston, 35 mi SE, seaport, historical sites, universities, museums, great shopping and restaurants. Lexington and Concord nearby.

HANSCOM AIR FORCE BASE (MA06R1)

Hanscom Air Force Base, MA 01731-5000

Comm: 617-377-4441
ATVN: 478-4441

LOCATION: From I-95 N take exit 45 NW, MA-2A W for 2 mi to right on Hartwell Rd which bisects the AFB. RM: p-43, E/2. NMC: Boston, 17 mi SE.

GENERAL INFORMATION: Air Force Systems Command Base. Hq, Electronics Systems Division, Air Force Geophysics Lab.

TEMPORARY MILITARY LIVING: Biltg Ofc, Bldg 1427, 24 hrs, Comm: 617-861-2112, ATVN: 478-2112. All ranks. DV/VIP, EX-5151.

LOGISTICAL SUPPORT: Complete spt fac avail. Key phone numbers follow.

BX-861-5258	**Caf-861-2189**	**Chap-861-3538/5144**
Chld C-861-2858	**Commiss-861-2544**	**Conv S-274-8588**
Fam Svc-861-3436	**Gas S-861-5155**	**Legal-861-2362**
Loc-861-5111/4441	**Med-861-2333**	**NCO C-861-2123**
Off C-861-5740	**Pkg S-861-3675**	**Pol-861-4357/2314**
Rec Ctr-861-2610	**Snack B-861-2395**	**Theat-861-5200**

HEALTH & WELFARE: Clin/Emer EX-2333, Appoint, EX-4706. Inpatient, Cutler Army Comm Hosp, Ft Devens, Comm: 617-796-6816, ATVN: 256-6816. Chapels serving all faiths, EX-3538/5144.

RECREATION: Activity Schedule, EX-2687, Aero Club, EX-5160, Auto Hobby, EX-2612, Bowling, EX-2237, Gym, EX-3435/3639, Hobby Shops, EX-2612, Rec Svc, EX-3901, Rec Sup, EX-3348, Svc Club, EX-2610, Swimming, EX-2455, Library, EX-2177. Fam-Camp, on Base, yr round, Comm: 617-861-4670, ATVN: 478-4670, 21 camper spaces w/hookups, other fac and activities. Fourth Cliff Family Rec Area, 30 mi S of Boston, Comm: 617-861-3262, ATVN: 478-3262, 11 camper spaces w/hookups, 20 camper spaces w/o hookups, 18 cabins, comp fac/all activities.

SPACE-A: At Hanscom Civil Airport, adjacent to AFB, Comm: 617-861-3333, ATVN: 478-3333, flts to CONUS locations.

ATTRACTIONS: The Bedford flag, oldest flag in America, Concord and North Bridge, and Lexington. Historic sites and taverns.

OTIS AIR NATIONAL GUARD BASE (MA07R1)

Otis Air National Guard Base, MA 02542-5000

Comm: 508-968-5300
ATVN: 557-5300
FTS: 829-5300

LOCATION: Off Route 28 at the base of Cape Cod, on former Otis AFB. RM: p-45, L/21. NMC: New Bedford, 30 mi SW.

GENERAL INFORMATION: Air Force ANG, Army NG, CG Air Stn, Cape Cod.

TEMPORARY MILITARY LODGING: Very ltd, call EX-5461. Also, try North Truro AFS in Provincetown, on tip of Cape Cod, Comm: 508-487-1248 EX-322. Transient apts, 968-5461, rms, apts, ltd vac apts, call for reservations.

LOGISTICAL SUPPORT: Limited spt fac avail. Key phone numbers follow.

BX, Cons C, Pkg S, Commiss, Gas S, EX-968-5639, CGAS. Child Care Center, 968-5450. (Nonappropriated-funds grocery.)

HEALTH & WELFARE: Dispensary, 968-5570. Inpatient, Newport Naval Hosp, RI, 80 mi SW, 401-841-3915. Chapels serving all faiths, EX-4901.

RECREATION: Golf Course (9 hole), 968-5454, Movie Theater, 968-5452, Tennis, Stables, Teen Center, 968-5451, Auto Hobby Shop, 968-5449.

SPACE-A: Base Ops, CGAS, Comm: 617-968-3330, ATVN: 881-3330, CONUS flts.

ATTRACTIONS: Beautiful Cape Cod, beaches and water sports.

SOUTH WEYMOUTH NAVAL AIR STATION (MA05R1)

South Weymouth Naval Air Station, MA 02190-5000

Comm: 617-786-2500
ATVN: 478-5980
955-XXXX

LOCATION: From MA-3 (Pilgrims Highway) exit 16 to MA-18 (Main St) S to 3rd traffic light, turn left, follow Union St to White St, turn right, NAS at end of White St. Located in residential area. RM: p-45, H/19. NMC: Boston, 15 mi NW.

GENERAL INFORMATION: Training of Navy and Marine Corps Combat Air Units.

TEMPORARY MILITARY LODGING: Biltg Ofc, Adm Bldg 16, Shea Memorial Dr, 0800-1600 hrs dty dys, Comm: 617-786-2738/2692, ATVN: 955-2738/2692. All ranks. DV/VIP, EX-786-2963. Space very limited on week ends.

LOGISTICAL SUPPORT: Most spt fac avail. Key phone numbers follow.

NX-786-2952	**Chap-786-2604**	**CDO/JOOD-786-2933**
EM C-786-2766/2898	**Fam Svc-786-2581**	**Gas S-786-2814**
Legal-786-2540	**Med-786-2674**	**Off C-786-2938/2932**
Pkg S-331-1620	**Pol-786-2610/2605**	**Snack B-786-2980**

HEALTH & WELFARE: Clinic/Emerg, EX-2674. Inpatient, Cutler Army Comm Hosp, Fort Devens, 50 mi NW, Comm: 617-796-6816, ATVN: 256-6816. Chaplain, EX-2604.

RECREATION: Auto Hobby, EX-2886, Spec Svc, EX-2909, Swimming, EX-2824, Bowling, EX-2921.

SPACE-A: Pax Term, Lounge, Hangar 1, 0700-2300 hrs dly (closed Mondays), Comm: 617-786-2712/3, ATVN: 955-2712/2713, flts to CONUS and overseas.

South Weymouth NAS, Cont'd

ATTRACTIONS: Historic Quincy is nearby. Massachusetts Bay and Boston are easy drives N and NW.

WESTOVER AIR FORCE BASE (MA03R1)

Westover Air Force Base, MA 01022-5000

Comm: 413-557-1110
ATVN: 589-1110

LOCATION: Take exit 5 off I-90 (MA Turnpike) in Chicopee. Westover is on Rte 33, N of I-90; signs mark way to Base. RM: p-45, H/7. NMC: Springfield, 8 mi S.

GENERAL INFORMATION: Air Force Reserve Base. 439th Tactical Airlift Wing, Army Reserve, Navy Reserve, Marine Corps Reserve, and Air Force Reserve units.

TEMPORARY MILITARY LODGING: Biltg Ofc, Bldg 2200, Outer Dr, 0745-2245 hrs dly, Comm: 413-557-3032, ATVN: 589-3032. All ranks.

LOGISTICAL SUPPORT: Complete spt fac avail. Key phone numbers follow.

BX-593-5941	**Commiss-593-3288**	**Gym-557-3958**
Cons C-593-5531	**Gas S-593-5102**	**Pol-557-3557**
Snack B-557-3990	**Theat-557-3152**	

HEALTH & WELFARE: Inpatient, Cutler Army Comm Hosp, Fort Devens, 80 mi NE, Comm: 617-796-6816, ATVN: 256-6816. No Chaplains.

RECREATION: Bowling, EX-3990, Swimming Pool, EX-2980.

SPACE-A: Bldg 7091, Base Ops, 0700-2300 hrs, Comm: 413-557-2951, ATVN: 589-2951, flts to CONUS and infrequently overseas.

ATTRACTIONS: Springfield, Mt Tom snow ski resort, Holyoke. Basketball Hall Of Fame, Springfield.

MICHIGAN

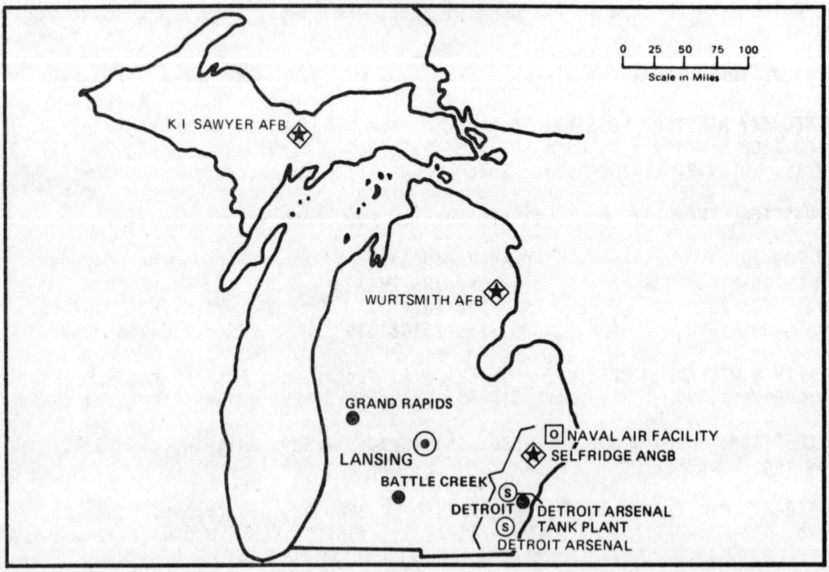

K.I. SAWYER AIR FORCE BASE (MI02R2)

K. I. Sawyer Air Force Base, MI 49843-5000

Comm: 906-346-6511
ATVN: 472-1110

LOCATION: In the MI upper peninsula. From US-41 S of Marquette, take MI-460 W to gate 2. RM: p-47, F/2. NMC: Marquette, 23 mi NW.

GENERAL INFORMATION: Strategic Air Command Base. Bomb Wing (Bomb Squadron, 2 Tanker Squadrons).

TEMPORARY MILITARY LODGING: Biltg Ofc, Bldg 801, A St, 24 hrs dly, Comm: 906-346-2145, ATVN: 472-2145. All ranks. DV/VIP, EX-6511.

LOGISTICAL SUPPORT: Complete spt fac avail. Key phone numbers follow.

BX-346-9216	**Caf-346-3518**	**Chap-346-2922/2924**
Chld C-346-2921	**Commiss-346-2374**	**Conv S-346-3714**
Fam Svc-346-2366	**Gas S-346-9242**	**Legal-346-2216**
Loc-346-6511/1110	**Med-346-2233**	**NCO C-346-2842**
Off C-346-2480	**Pkg S-472-2053/346-2660**	**Pol-346-2131**
Theat-346-2378		

HEALTH & WELFARE: USAF Hospital, clinics, depts, Emerg, EX-2233, Appoint, EX-2881, 25 bed. Chapels serving all faiths, EX-2922/2924.

RECREATION: Par course, nature trails, cross country skiing, snowmobile trails and rentals; Base Lake (swimming, fishing, boat rentals); ski hill and rentals; 18-hole golf course; rec ctr; arts and crafts ctr; auto/wood hobby shops; library; athletic fields/gym; bowling ctr; youth act; Wonder Hut (health spa, indoor tennis, archery, roller skating; disco dancing); Officers' and NCO clubs; amusement ctr. EX 6511 or 1110 for all.

SPACE-A: Pax Term and Lounges, Bldg 427, 24 hrs dly, Comm: 906-346-2345, ATVN: 472-2345, flts to CONUS and overseas via MEDEVAC and KC-135 aircraft.

ATTRACTIONS: Isle Royale Natl Park in Lake Superior; historic Fort Wilkins on the Keweenaw Peninsula; Lake of the Clouds in the Porcupine Mountains; Mount Marquette; National Ski Hall of Fame in Ishpeming, MI; cruises on the Tahquamenon R; 3 national parks and 18 state parks (more than 165,000 acres of rec land); iron ore loading docks/giant ore boats at Marquette and Escanaba; numerous rec areas and sightseeing tours along Lakes Superior and Michigan.

SELFRIDGE AIR NATIONAL GUARD BASE (MI01R2)

Selfridge Air National Guard Base, MI 48045-5004

Comm: 313-466-4011
ATVN: 273-0111

LOCATION: Take I-94 N from Detroit, to Selfridge exit, then E on MI-59 to main gate of Base. RM: p-48, D/12. NMC: Detroit, 25 mi SW.

GENERAL INFORMATION: Air National Guard units, Reserve units of Air Force, Navy, Marine Corps, and Army. Coast Guard Air Station, Detroit.

TEMPORARY MILITARY LODGING: Biltg Ofc, Bldg 410, George Ave, 24 hrs Wed-Sun, 0700-2245 hrs Mon-Tue, Comm: 313-466-4435, ATVN: 273-4435. All ranks. DV/VIP, EX-5444.

LOGISTICAL SUPPORT: Complete spt fac avail. Some spt also available at Detroit Arsenal. Key phone numbers follow.

BX-466-5788	**Chap-466-4763**	**Chld C-466-4711**
Commiss-466-5570	**Comm Svc-466-5903**	**Gas S-466-4256**

Selfridge ANG Base, Cont'd

Legal-466-4809	Loc-466-4021	Med-466-5000
NCO C-466-4786	Off C-466-4785	Pkg S-466-5072
Pol-466-4673	Snack B-469-3660	Theat-466-4754

HEALTH & WELFARE: Dispensary/Emerg, EX-5000, Appoint, EX-4857. Inpatient, USAF Hosp, Wurtsmith AFB, 200 mi N, Emerg 517-739-6333, Appoint 517-739-6821. Chapels serving all faiths, EX-4763.

RECREATION: Auto, EX-4535, Wood, EX-4535, Multi-Crafts, EX-5155, Rec Ctr, EX-4524/4464, Youth Act, EX-4524/4564, Bowling, EX-5941, Swimming, Tennis, EX-5765, Picnic Area, 463-3070, Golf Course and Outdoor Sports Ctr, EX-4344. Selfridge Outdoor Rec Area, on Base, Apr-Sep, Comm: 313-463-3070, ATVN: 273-3070, marina only, no camping, other activities on lake, cross country skiing.

SPACE-A: Base Ops, and CGAS Ops, Comm: 313-466-4011, and ATVN: 273-4011, flts to CONUS via C-130 aircraft.

ATTRACTIONS: Air Museum. City of Detroit, Lake St Clair, Canadian cities of Windsor and Sarnia.

WURTSMITH AIR FORCE BASE (MI03R2)

Wurtsmith Air Force Base, MI 48753-5000

Comm: 517-739-2011
ATVN: 623-1110

LOCATION: Off US-23, 2 mi NW of Oscoda. On F-41. Clearly marked. RM: p-47, N/14. NMC: Bay City, 65 mi SW.

GENERAL INFORMATION: Strategic Air Command Base. Hq 40th Air Division, and 379th Bomb Wing.

TEMPORARY MILITARY LODGING: Biltg Ofc, Bldg 1600, 24 hrs dly, Comm: 517-747-6692/6033, ATVN: 623-6692/6033. All ranks. DV/VIP, Biltg switchboard, 517-739-8252/9007 EX-49/50.

LOGISTICAL SUPPORT: Complete spt fac avail. Key phone numbers follow.

BX-739-7466	Chap-747-6266	Chld C-747-6356
Commiss-747-6401	Conv S-739-8992	Fam Svc-747-6054
Gas S-739-8120	Legal-747-6436	Med-747-6333/6568
NCO C-747-6493	Off C-747-6300	Pkg S-747-6455
Pol-747-6023	Theat-747-6368	

HEALTH & WELFARE: USAF Hospital clinics and depts, Emerg EX-6333, Appoint, EX-6821, 20 bed. Chapels serving all faiths, EX-6266.

RECREATION: Rec Ctr, EX-6532, Boats and Trailers, EX-6467/6468, Gym, EX-8063, Hobby Shops, EX-6063, Bowling, EX-6100. Wurtsmith Air Force Beach and Fam-Camp, on Van Etton Lake, May-Oct, Comm: 517-747-6083, ATVN: 623-6083, 7 camper spaces w/hookups, beach rec fac & activities. Point Betsie Rec Cottage, operated by the Frankfort CG Station, 5 mi N of Frankfort, year round, Comm: 616-352-4242, 2 bdrm cottage, boat, 2 camper spaces w/o hookups, limited facilities and activities on Lake Michigan.

SPACE-A: Pax Term and Lounges, Bldg 47, 24 hrs, Comm: 517-747-6263, ATVN: 623-6263, flts to CONUS and overseas.

ATTRACTIONS: Lake Huron, outdoor & water sports, cross country ski trails.

MINNESOTA

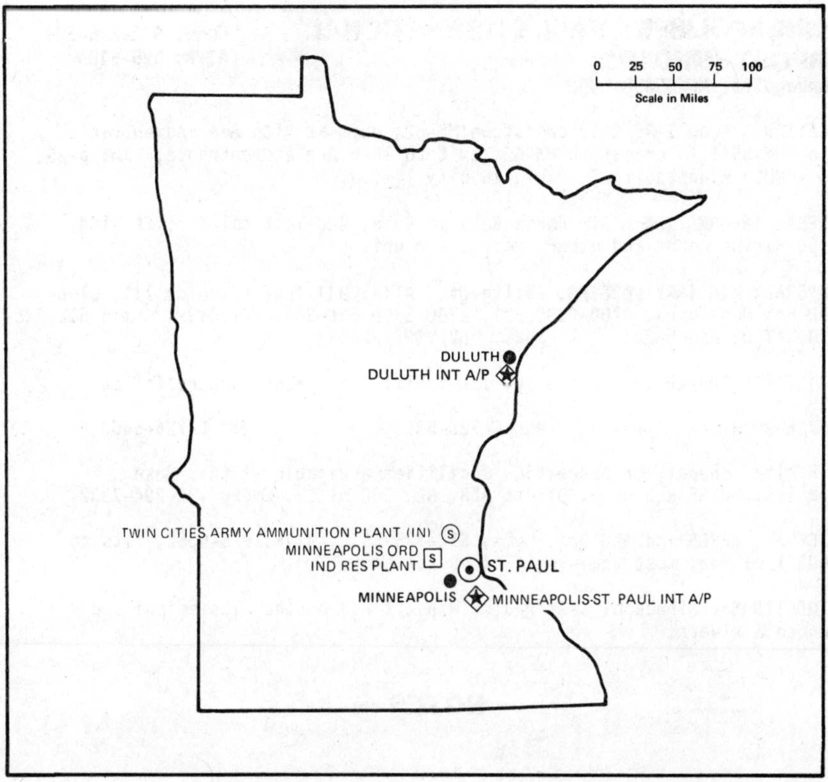

DULUTH INTERNATIONAL AIRPORT (MN02R2)

Duluth, MN 55811-5000

Comm: 218-727-6886
ATVN: 825-7210

LOCATION: From Duluth, take US-53, right on Haines Rd which leads to main gate. RM: p-51, A/9. NMC: Duluth, 8 mi SE.

GENERAL INFORMATION: Air National Guard Fighter Interceptor Group.

TEMPORARY MILITARY LODGING: No military lodging.

LOGISTICAL SUPPORT: Very limited spt fac avail. Key phone numbers follow.

BX-723-7450	ID Cards-723-7212	Loc-727-6886
NCO C-723-7309	Pol-723-7442	

HEALTH & WELFARE: None. Inpatient, USAF Hosp, K. I. Sawyer AFB, Marquette, 250 mi E, 906-346-2233.

RECREATION: None.

SPACE-A: Base Ops, 148th Ftr Intcp Group, very limited Space-A opportunities, 727-6886.

ATTRACTIONS: Port of Duluth/Superior, Lake Superior, outdoor sports.

MINNEAPOLIS-ST. PAUL INTERNATIONAL AIRPORT (MN01R2)

Minneapolis, MN 55450-5000

Comm: 612-725-5011
ATVN: 825-5100

LOCATION: From I-35 E to crosstown MN-62, exit at 34th Ave and entrance. Also, MN-55 S to crosstown MN-62, exit to 34th Ave and entrance. RM: p-49, J/6. NMC: Minneapolis-St Paul, in city limits.

GENERAL INFORMATION: Air Force Reserve Base, ANG Tactical Airlift Wing, Navy, Marine Corps and other reserve air units.

TEMPORARY MILITARY LODGING: Biltg Ofc, ATTN: 934 TAG/SVH, Bldg 711, 0700-1800 hrs Mon-Thurs, 0700-2100 Fri, 0700-1300 Sat-Sun, Holidays, Comm: 612-725-5320, ATVN: 825-5320. All ranks. DV/VIP, EX-5320.

LOGISTICAL SUPPORT: Limited spt fac avail. Key phone numbers follow.

BX-726-9023	**NCO C-726-5390**	**Off C-726-5403**

No health, chapel, or recreation facilities available at this Base. Inpatient, USAF Reg Hosp, Offutt AFB, NE, 380 mi SW, Emerg 402-294-7332.

SPACE-A: AFRES and ANG Ops, Comm: 612-725-5552, ATVN: 825-5552, flts to CONUS location, most training flts on weekends.

ATTRACTIONS: Minneapolis-St Paul (Twin Cities) on the Mississippi and Minnesota rivers.

- NOTES -

MISSISSIPPI

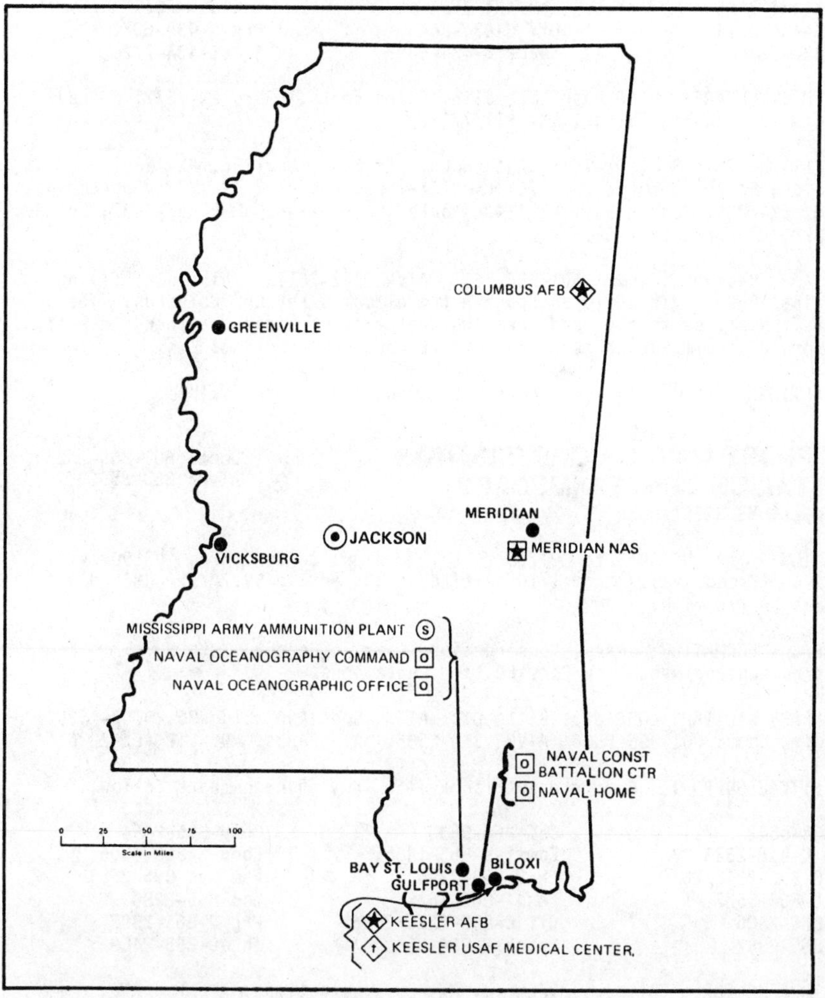

COLUMBUS AIR FORCE BASE (MS01R2)

Columbus Air Force Base, MS 39701-5000 — **Comm: 601-434-7322** **ATVN: 742-7322**

LOCATION: Off US-45 N, 60 mi W of Tuscaloosa, AL via US-82. RM: p-52, E/11. NMC: Columbus, 10 mi S.

GENERAL INFORMATION: Air Training Command Base. Undergraduate pilot tng.

TEMPORARY MILITARY LODGING: Biltg Ofc, ATTN: 14th ABG/SVH, Bldg 956, B St, 24 hrs dly, Comm: 601-434-7477, ATVN: 742-7477. All ranks. DV/VIP, EX-7974.

LOGISTICAL SUPPORT: Complete spt fac avail. Key phone numbers follow.

Columbus AFB, Cont'd

BX-434-6013/8484	Chap-434-7511/7734	Commiss-434-7463
Conv S-434-6864	Fam Svc-434-7527	Gas S-434-6161
Legal-434-7406	Loc-434-7666	Med-434-7577
NCO C-434-6714	Off C-434-8843	Pkg S-434-6026
Pol-434-7718	Snack B-434-6498	Theat-434-7728

HEALTH & WELFARE: USAF Hospital, clinics and depts, Emerg EX-7577, 20 bed. Chapels serving all faiths, EX-7511/7734.

RECREATION: Rec Svc, EX-7437, Auto Hobby, EX-7633, Bowling, EX-7288, Ceramics, EX-7578, Golf, EX-7879, Gym, EX-7501, Rec Ctr, EX-7772, Sportsman's Lodge, EX-7225, Wood Hobby, EX-7142, Youth Ctr, EX-7841, MWR, EX-7437, Rec Sup, EX-7229, Library, EX-7762.

SPACE-A: Base Ops, Comm: 601-434-7611, ATVN: 742-7611. This is a "flying training" Base. Virtually no Space-A travel comes through Columbus. The only aircraft assigned to this Base are two-seater jet trainers. Other aircraft that could accommodate passengers comes through very seldom.

ATTRACTIONS: Period historic homes, Columbus and NE Mississippi.

GULFPORT NAVAL CONSTRUCTION BATTALION CENTER (MS03R2)

Comm: 601-865-2555
ATVN: 363-2555

Gulfport, MS 39501-5000

LOCATION: Take US-49 S to Gulfport. Follow signs to Ctr. From US-90 exit to US-49 (Broad Ave). From I-10 exit to US-49. RM: p-52, P/1. NMC: New Orleans LA, 70 mi W.

GENERAL INFORMATION: Naval Construction Training Center, 20th Naval Construction Regiment, and Construction Battalions.

TEMPORARY MILITARY LODGING: Biltg Ofc, ATTN: Code 450, Bldg 85, 0700-1530 hrs dly, Comm: 601-865-2586, ATVN: 363-2586. All ranks. DV/VIP, EX-2226.

LOGISTICAL SUPPORT: Complete spt fac avail. Key phone numbers follow.

NX-864-5542	Caf-864-5530	Chap-865-2454
Chld C-865-2323	Commiss-863-1885	Conv S-863-4506
CPO C-865-2245/2619	EM C-865-2396	Fam Svc-865-2581
Gas S-864-5527	Legal-865-2626	Loc-865-2286
Med-865-2806	Off C-865-2226/2616	Pkg S-865-2392
Pol-865-2361	Snack B-864-5530	Theat-865-2489

HEALTH & WELFARE: Dispensary/Emerg, EX-2809. Inpatient, Keesler AFB, Comm: 601-377-6655, ATVN: 868-6655. Chapels serving all faiths, EX-2454.

RECREATION: Auto Hobby, EX-2804, Bowling, EX-2739, Craft Shop, EX-2560, Wood Hobby, EX-2518, Golf, EX-2494, Gym, EX-2353/2668, Library, EX-2409, Marina, EX-2439, Racquetball, EX-2668, Rec Lake/Park, EX-2231, Rec Rental, EX-2519, Swimming, EX-2768, ITT, EX-2231, Youth Ctr, EX-2251.

SPACE-A: Pax Term and Lounge, Bldg 4205, Hangar 5, Keesler AFB, 0600-2300 hrs dly, Comm: 601-377-2120/2629, ATVN: 868-2120/2629, flts to CONUS.

ATTRACTIONS: Gulf Coast, beaches, and water sports. Cities of Gulfport and New Orleans, short drives.

KEESLER AIR FORCE BASE (MS02R2)

Keesler Air Force Base, MS 39534-5000

Comm: 601-377-1110
ATVN: 868-1110

LOCATION: From I-10 exit 46, follow signs to Base. From US-90, N on White Ave to main gate. RM: p-52, P/4. NMC: Biloxi, in the city.

GENERAL INFORMATION: Air Training Command Base, Technical Training Center, USAF Medical Center, MAC, TAC, and AFRES units.

TEMPORARY MILITARY LODGING: Biltg Ofc, Bldg 2101, Muse Manor, Larcher Blvd near Meadows, 24 hrs dly, Comm: 601-377-2631/3224, ATVN: 868-2631. All enlisted ranks. DV/VIP, Officers, Show House, Comm: 601-377-3663/3774.

LOGISTICAL SUPPORT: Complete spt fac avail. Key phone numbers follow.

BX-435-2216	Chap-377-2110	Chld C-377-3324
Commiss-377-2830/3205	EM C-377-2424	Fam Svc-377-4293
Gas S-432-2404	Legal-377-3510	Loc-377-2755
Med-377-6550	NCO C-377-4146	Off C-377-2719
Pkg S-377-2094	Pol-377-3040	Sub Shop-436-3124
Theat-377-4280		

HEALTH & WELFARE: USAF Regional Medical Center, clinics and depts, Emer EX-6655/56, Appoint, EX-4430, 325 bed. Chapels serving all faiths, EX-6337.

RECREATION: Bowling, EX-2817, Golf, EX-3832, Gyms, swimming and outdoor sports, EX-2800, Youth Ctr, EX-3349, Library, Ex-2181, Arnold Rec Ctr, EX-4109, Vandeberg Rec Ctr, EX-3646, Aero Club, EX-4671. Marina, on Base, year round, Comm: 601-377-3160, ATVN: 868-3160, 20 camper spaces w/hookups, all marina facilities, boat rental and full rec activities.

SPACE-A: Pax Term and Lounge, Bldg 4205, Hangar 5, 0600-2200 hrs dly, Comm: 601-377-2120/2123, ATVN: 868-2120/2163, flts to CONUS locations.

ATTRACTIONS: Gulf Coast beaches, seafood, Pensacola, FL, New Orleans, LA, and Mobile, AL, all easy drives.

MERIDIAN NAVAL AIR STATION (MS04R2)

Meridian Naval Air Station, MS 39309-5000

Comm: 601-679-2211
ATVN: 446-2211

LOCATION: Take MS-39 N from Meridian, for 12 mi to 4-lane access road. Clearly marked. Right for 3 mi to NAS main gate. RM: p-52, I/11. NMC: Meridian, 15 mi SW.

GENERAL INFORMATION: Naval Technical Training Center, Naval Air Training Wing, and Marine Corps Aviation Training Support Group.

TEMPORARY MILITARY LODGING: Biltg Ofc, Bldg 208, Highley Rd, 24 hrs dly, Comm: 601-679-2414/2616. ATVN: 446-2414/2616. All ranks. DV/VIP, EX-2211.

LOGISTICAL SUPPORT: Complete spt fac avail. Key phone numbers follow.

NX-679-2725	Chap-679-2319	Chld C-679-2652
Commiss-679-2554	CPO C-679-2650	EM C-679-2636
Fam Svc-679-2358/2360	Gas S-679-2664	Legal-679-2340
Loc-679-2211	Med-679-2444	Off C-679-2667
Pkg S-679-2653	Pol-679-2361	Snack B-679-2531
Theat-679-2306		

HEALTH & WELFARE: Clinics, Emerg, EX-2444, Appoint, EX-2683. Inpatient, USAF Hosp, Columbus AFB, 90 mi NE, 601-434-7577. Chapels serving all faiths, EX-2319/2475.

Meridian NAS, Cont'd

RECREATION: Auto/Woodworking Hobby, EX-2609, Bowling, EX-2651, Golf, EX-2615/2526, Ceramics, EX-2608, Gym, EX-2379, Stables, EX-2507, Library, EX-2623, Swimming, EX-2667/2650/2684, Racquetball, EX-2379, Rec Gear, EX-2562, Sportsmen's Assoc, EX-2666, ITT, EX-2687, Youth Ctr, EX-2687.

SPACE-A: Pax Term and Lounge, Ops Bldg, 24 hrs dly, Comm: 601-679-2454, ATVN: 446-2454, flts to CONUS. Also, Jackson Mun Aprt, 7 mi E of Jackson, ANG, Comm: 601-939-3633, ATVN: 731-9310, flts to CONUS locations.

ATTRACTIONS: Lake Okatibbee, 10 mi W, Jimmie Rodgers Museum and Monument in Meridian.

MISSOURI

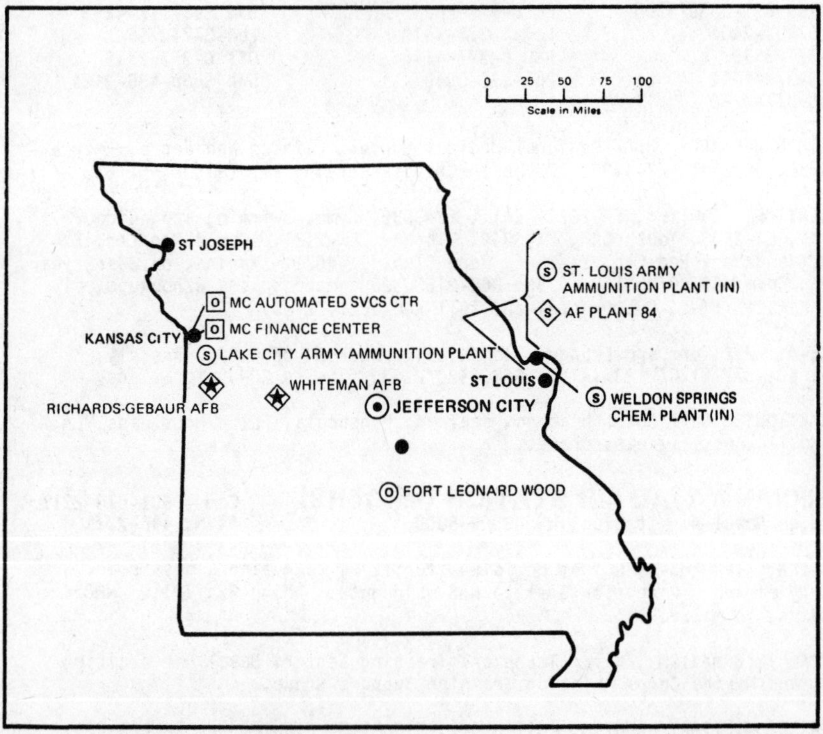

FORT LEONARD WOOD (MO03R2)

Fort Leonard Wood, MO 65473-5000

Comm: 314-368-0113
ATVN: 581-0110
FTS: 270-0110

LOCATION: Two mi S of I-44, adjacent to St Robert & Waynesville, at Ft Leonard Wood exit. RM: p-55, L/16. NMC: Springfield, 85 mi SW.

GENERAL INFORMATION: Army Training Center for Engineer Corps personnel, Army Engineer units. Site of Army's Engineer School beginning 1989.

TEMPORARY MILITARY LODGING: Biltg Ofc, ATTN: DEH-HB, Bldg 315, Welcome

Fort Leonard Wood, Cont'd

Ctr, MO Ave at Hq Ave, 24 hrs dly, Comm: 314-368-6169, ATVN: 581-6169. All ranks. DV/VIP, EX-5161.

LOGISTICAL SUPPORT: Complete spt fac avail. Key phone numbers follow.

PX-329-2200	**Caf-329-3611**	**Chap-368-9559**
Chld C-368-1243	**Commiss-368-1749**	**Conv S-329-2777/6200**
EM C-329-3477	**Fam Svc-368-1126**	**Gas S-329-3373**
Legal-368-8171	**Loc-368-2151**	**Med-368-2157**
NCO C-329-6533	**Off C-329-6565**	**Pkg S-329-6333**
Snack B-329-3266	**Theat-368-1267/2531**	

HEALTH & WELFARE: General Leonard Wood Army Community Hospital, clinics and depts, Emerg, EX-2157, Appoint, EX-9701. Chapels serving all faiths, EX-2125.

RECREATION: Rec Ctr, ITT, Library, Gym, Field House, Outdoor Trails, Swimming, Stadium, Riding Stables, Golf, Tennis, Bowling, Auto Crafts, Youth Act, Multi-Crafts, all EX-4033. Operates Lake of the Ozarks Rec Area, located on Grand Glaize Arm of the Lake of the Ozarks, 50 mi NW of Fort, Comm: 314-346-5640, Apr-Oct, 16 camper spaces w/hookups, 20 camper spaces w/o hookups, 40 mobile homes, 2-3 bdrms, comp support facilities, all water & other activities.

SPACE-A: Forney Army Airfield, 329-2400.

ATTRACTIONS: Near Lake of the Ozarks recreational areas.

RICHARDS-GEBAUR AIR FORCE BASE (MO02R2)

Richards-Gebaur Air Force Base, MO 64030-5000

Comm: 816-348-2132
ATVN: 463-1110
FTS: 926-2132

LOCATION: From US-71 S take Belton-Air Base exit, W on County Line Rd (or 155th St) to Base. Between Granview & Belton. RM: p-55, H/9. NMC: Kansas City, 17 mi N.

GENERAL INFORMATION: Air Force Res Base. Also, Navy and Army Res units.

TEMPORARY MILITARY LODGING: Biltg Ofc, Bldg 250, across street from Con Op Mess, 0730-1700 M-Thurs, 0730-2400 F-Sat, 1000-1700 Sun, Comm: 816-348-2125/2055, ATVN: 463-2125/2055. All ranks. DV/VIP, 2125/2055.

LOGISTICAL SUPPORT: Limited spt fac avail. Key phone numbers follow.

BX-331-2019	**Med-348-2114**	**Loc-348-2000/1110**
Cons C-331-1601	**Pol-348-2118**	

HEALTH & WELFARE: Clinic/Emerg, EX-2114. Inpatient, USAF Hosp, Whiteman AFB, 50 mi S, 816-687-3733.

RECREATION: None.

SPACE-A: Pax Term and Lounge, Bldg 901, 0700-2230 dly except holidays, Comm: 816-348-3132/2022, ATVN: 463-3132/2022, flts via reserve units for training to CONUS locations. Also, Rosecrans Memorial Aprt, St Joseph, ANG, Bldg 17, 0730-1600 M-F, Comm: 816-271-1260, ATVN: 720-9260, flts via C-130 aircraft to CONUS locations.

ATTRACTIONS: Kansas City, the city of fountains, stockyards, and great restaurants. Harry S. Truman Library and home in Independence, short drive.

WHITEMAN AIR FORCE BASE (MO04R2)

Whiteman Air Force Base, MO 65305-5000

Comm: 816-687-1110
ATVN: 975-1110

LOCATION: From I-70 E exit to US-13 S to US-50 E for 10 mi, then left on Route J which leads to AFB. RM: p-55, H/12. NMC: Kansas City, 60 mi NW.

GENERAL INFORMATION: SAC Base. 351st Strategic Missile Wing.

TEMPORARY MILITARY LODGING: Biltg Ofc, ATTN: 351/SS/SVH, Bldg 3006, Mitchell Ave, 24 hrs dly, Comm: 816-687-3937, ATVN: 975-3937. All ranks. DV/VIP, EX-3738.

LOGISTICAL SUPPORT: Complete spt fac avail. Key phone numbers follow.

BX-563-3001	Caf-687-3355	Chap-687-3652
Chld C-687-3588	Commiss-687-3617	Conv S-563-5445
Fam Svc-687-3498	Gas S-563-3113	Legal-687-3422
Loc-687-3201	Med-687-3733	NCO C-687-2241
Off C-563-2273	Pkg S-563-2110	Pol-687-3700
Theat-687-3016		

HEALTH & WELFARE: USAF Hospital, clinics and depts, Emerg, EX-3733/2186, Appoint, EX-2242/2245, 25 bed. Chapels serving all faiths, EX-3652.

RECREATION: Rec Ctr, EX-3272, Youth Ctr, EX-3315/3319, Golf, EX-3048, Rec Supply, EX-3370, Bowling, EX-3270, Swimming, 687-3463, Gym, EX-3295/3386, Library, EX-3089, ITT, EX-3424, Auto Hobby, EX-3340, Ceramics Hobby, EX-3218.

SPACE-A: Base Ops, flts to CONUS locs, EX-816-687-3261, ATVN: 975-3261.

ATTRACTIONS: Outdoor sports, Lake of the Ozarks region. Kansas City: prof sports, shopping, Zoo, Worlds of Fun.

MONTANA

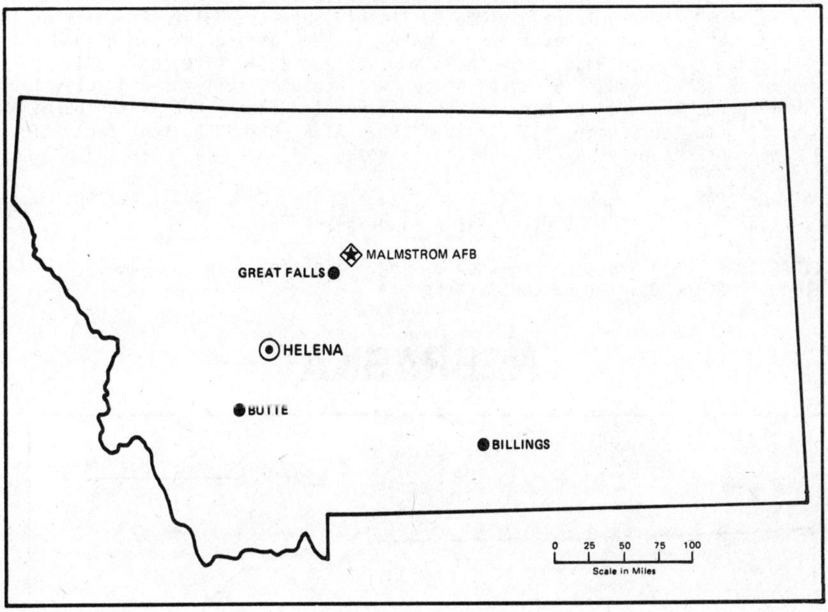

MALMSTROM AIR FORCE BASE (MT03R3)

Malmstrom Air Force Base, MT 59402-5000

Comm: 406-731-9990
ATVN: 632-1110

LOCATION: From I-15 N or S take 10th Ave S exit to AFB. From E take Malmstrom exit off US-87/89 to AFB. Clearly marked. RM: p-57, P/22. NMC: Great Falls, 1 mi W.

GENERAL INFORMATION: Strategic Air Command Base. 341st Strategic Missle Wing; Det 5, 37th Air Rescue and Recovery Squadron.

TEMPORARY MILITARY LODGING: Housing Referral, Bldg 1085, Mon-Fri, 0745-1630, 731-3273. Biltg Off in Malmstrom Hotel, Bldg 1680, 24 hrs, 731-3394. DV/VIP, EX-9990

LOGISTICAL SUPPORT: Complete spt fac avail. Key phone numbers follow.

BX-761-8004	**Caf-727-0671**	**Chap-731-3721/2/3**
Chld C-731-2417	**Commiss-452-6473**	**EM C-452-6421**
Fam Svc-731-3686	**Fire-731-1117**	**Legal-731-2878**
Loc-731-2214	**Med-731-3483**	**NCO C-452-6421**
Off C-761-6430	**Pkg S-454-1772**	**Pol-731-3894**
Serv S-452-4903	**Shoppette-761-8004**	**Sm Animal Cl-731-3178**
Snack B-731-2494	**Theat-731-3236**	

HEALTH & WELFARE: USAF Hospital, Bldg 2055, clinics and depts, Emerg, EX-3483, Appoint, EX-3425, 29 bed. Ambulance, 731-3300. Chapels serving all faiths, EX-3721/2/3.

Malmstrom AFB, Cont'd

RECREATION: Bowling, 731-2494, Library, 731-2748, Rec Ctr, 731-3964, Arts & Crafts, 731-3691, Auto Craft Shop, 731-3777, Wood Craft Shop, 731-3232, Yth Ctr, 761-4079, Sports Arena, 731-3621, Ticket/Tour Ctr, 731-2248. Swimming is available at the Morony Natatorium located at 12th St and 1st Ave N in Great Falls. Fam-Camp, on Base, Apr-Oct, Comm: 406-731-3394, ATVN: 632-3394, 10 camper spaces w/hookups, 5 camper spaces w/o hookups, full fac and activities. St Mary's Rec Camp, on St Mary's Lake adjacent to Glacier Nat Park, 165 mi NW of Base, June-Sept, Comm: 406-731-3263/2903, ATVN: 632-3263, tent facilities, 10 camper spaces w/wat & elec, 15 mobile homes, 2/3 bdrm, all fac and act.

SPACE-A: Pax Term, Base Ops, Bldg 360, 24 hrs dly, Comm: 406-731-2290, ATVN: 632-2290, flts to CONUS locations, mostly MEDEVAC.

ATTRACTIONS: Big Sky Country, snow skiing, wildlife, fish and game. Yellowstone and Glacier National Parks.

NEBRASKA

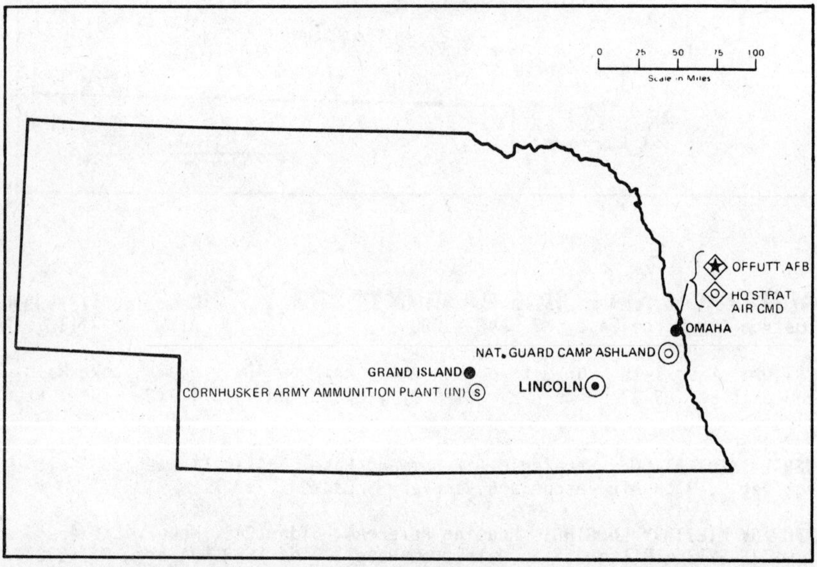

OFFUTT AIR FORCE BASE (NE02R3)

Offutt Air Force Base, NE 68113-5000

Comm: 402-294-1110
ATVN: 271-1110

LOCATION: From I-80 exit to US-73/75 S to AFB exit, 6.5 mi S of I-80/US-73/75 interchange. RM: p-59, L/24. NMC: Omaha, 8 mi N.

GENERAL INFORMATION: SAC Base. Reconnaissance Wing, Intelligence Wing, Weather Central, Weather Wing, Communications and other support units.

TEMPORARY MILITARY LODGING: Biltg Ofc, Bldg 44, Grant Pass St, 24 hrs dly, Comm: 402-294-3671/9000, ATVN: 271-3671/9000. All ranks. DV/VIP, EX-4461.

LOGISTICAL SUPPORT: Complete spt fac avail. Key phone numbers follow.

Offutt AFB, Cont'd

BX-291-9100	Airmens C-291-6785	Caf-291-9596
Chap-294-6244	Chld C-294-2203	Commiss-294-5920
Conv S-292-0218	Fam Svc-294-3111	Gas S-291-8743
Legal-294-3732	Loc-294-5125	Med-294-7332
NCO C-292-1600	Off C-292-1560	Pkg S-292-1400
Pol-294-6110	Theat-294-5189	

HEALTH & WELFARE: USAF Hospital, clinics and depts, Emerg EX-7332, Appoint-Active Duty/EX-6477, Dep/292-5510. Chapels serving all faiths, EX-6244/6051.

RECREATION: Aero Club, EX-3503, Auto Hobby, EX-5564, Arts/Crafts, EX-3872, Bowling, EX-2514, Golf, EX-3362 or 292-1680, Gym, EX-5904, Library, EX-2533, Photo Hobby, EX-3872, Rec Ctr, EX-6247, Swimming, EX-2274/3593/6466, Wood Hobby, EX-3318, Youth Ctr, EX-5152. Fam-Camp, on Base, Apr-Nov, Comm: 402-294-3671, ATVN: 271-3671, 10 camper spaces w/hookups, full facilities and all activities on the Base lake.

SPACE-A: Pax Term and Lounge, Bldg T-47, 0500-2200 M-F, other times based on flt schedule, Comm: 402-294-6235, ATVN: 271-6235, flts to CONUS locations including MEDEVAC.

ATTRACTIONS: City of Omaha, steaks, zoo, art and music.

NEVADA

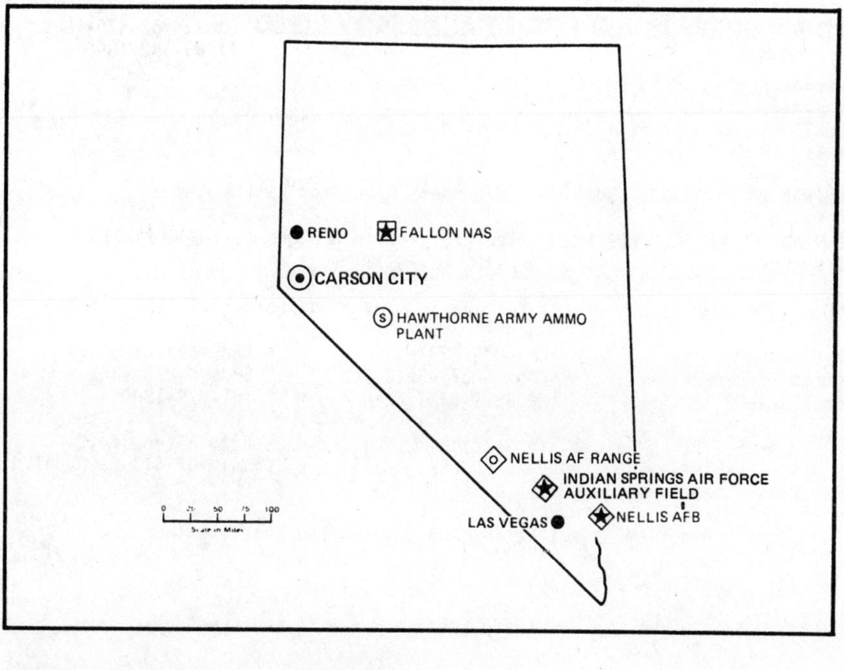

FALLON NAVAL AIR STATION (NV02R4)

Fallon Naval Air Station, NV 89406-5000

Comm: 702-426-5161
ATVN: 830-2110

LOCATION: From US-50 exit to US-95 S at Fallon, for 5 mi to left on Union St to NAS. RM: p-60, G/4. NMC: Reno, 72 mi W.

Fallon NAS, Cont'd

GENERAL INFORMATION: Fleet training squadrons.

TEMPORARY MILITARY LODGING: Biltg Ofc, Navy Lodge, 0800-1600 hrs, dty dys, Comm: 702-426-2522, ATVN: 830-2522. All ranks. DV/VIP, EX-5161/2522.

LOGISTICAL SUPPORT: Complete spt fac avail. Key phone numbers follow.

NX-EX-2818	Caf-EX-2501	Chap-EX-2742
Commiss-EX-2465	Conv S-EX-2489	EM C-EX-2449
Fam Svc-EX-2708	Gas S-EX-2583	Legal-EX-2711
Loc-EX-2801	Med-EX-2410	CPO C-EX-2483
Off C-EX-2464	Pkg S-EX-2464	Pol-426-2471
Theat-EX-2580		

HEALTH & WELFARE: Dispensary/Emerg, EX-2410, Appoint, EX-3110. Inpatient, USAF Hosp, Mather AFB, Sacramento, CA, 200 mi SW, Comm: 916-364-3212, ATVN: 828-3212. Chapels serving all faiths, EX-2742.

RECREATION: Bowling, Gear Issue, Gym, Library, Swimming, Boating, Auto & Ceramics Hobby, Picnic Area, Car Rental, Spc Svc Ofc, EX-2819.

SPACE-A: Pax Term, Hangar 4, 1st deck, N side, 0730-1600 hrs dly, Comm: 702-426-3462, ATVN: 830-3462, flts to CONUS locations.

ATTRACTIONS: Reno and Lake Tahoe nearby.

INDIAN SPRINGS AIR FORCE AUXILIARY FIELD (NV03R4)

Comm: 702-879-1800
ATVN: 682-1800

Indian Springs AFAF, NV 89018-5000

LOCATION: Off US-95, 47 miles NW of Nellis AFB. RM: p-60, M/9. NMC: Las Vegas, 45 mi SE.

GENERAL INFORMATION: Bombing and Gunnery Range for Nellis AFB.

TEMPORARY MILITARY LODGING: Biltg Ofc, Nellis AFB, Comm: 702-643-2710, ATVN: 682-2710. All ranks. DV/VIP, EX-2987.

LOGISTICAL SUPPORT: Most spt fac avail. Key phone numbers follow.

BX-879-3449	Caf-879-6237	Chap-643-2446
Commiss-879-6224	Conv S-879-3449	Cons C-879-6229
Loc-879-1800	Med-879-6214/6246	Pol-879-1800

HEALTH & WELFARE: Medical Aid Station, Emerg, EX-6286/6214 after dty, Appoint, EX-6258. Inpatient, USAF Hosp, Nellis AFB, Comm: 702-643-2343, ATVN: 682-2252. Chapels serving all faiths, EX-6254.

RECREATION: Complete fac at Nellis AFB, please see that listing.

SPACE-A: See Nellis AFB listing.

ATTRACTIONS: Desert area, Tolyabe National Forest, and Lee Canyon ski area, S of US-95, nearby.

NELLIS AIR FORCE BASE (NV01R4)

Comm: 702-643-1800
ATVN: 682-1800

Nellis Air Force Base, NV 89191-5000

Nellis AFB, Cont'd

LOCATION: Off I-15 N or S. Also, accessible from US-91/93. Clearly marked. RM: p-60, N/10. NMC: Las Vegas, 8 mi SW.

GENERAL INFORMATION: Tactical Air Command Base. Tactical Fighter Weapons Center. Tactical Fighter units. USAF Thunderbirds Air Demonstration Squadron.

TEMPORARY MILITARY LODGING: Biltg Ofc, ATTN: 554 CSG/SBH, Bldg 502, Fitzgerald St, 24 hrs dly, Comm: 702-643-2710, ATVN: 682-2710. All ranks. DV/VIP, EX-2987.

LOGISTICAL SUPPORT: Complete spt fac avail. Key phone numbers follow.

BX-644-2062	Chap-643-2950	Chld C-643-5720
Commiss-643-6950	Conv S-644-6375	EM C-643-5272
Fam Svc-643-5271	Gas S-643-1686	Legal-643-4213
Loc-643-2317	Med-643-2343	NCO C-643-5272
Off C-644-2582	Pkg S-644-1433	Pol-643-2311
Theat-643-5020		

HEALTH & WELFARE: USAF Hospital, clinics and depts, Emerg, EX-2343, Appoints, EX-2252, 40 bed. Chapels serving all faiths, EX-2950.

RECREATION: Arts & Crafts, EX-2794, Auto Hobby, EX-2284, Bowling, EX-2160, Golf, EX-2602, Gym, EX-5108/2626, Library, EX-2280, Rec Ctr, EX-2192/2600, Rec Supply, EX-2514, Youth Ctr, EX-4706. Fam-Camp, on Base, Comm: 702-643-6768, ATVN: 682-6768, 10 camper space w/hookups.

SPACE-A: Pax Term and Lounge, Bldg 805, 0600-2200 hrs dly, Comm: 702-643-4600, ATVN: 682-4600, flts to CONUS and overseas.

ATTRACTIONS: Las Vegas, Lake Mead, beautiful deserts.

NEW HAMPSHIRE

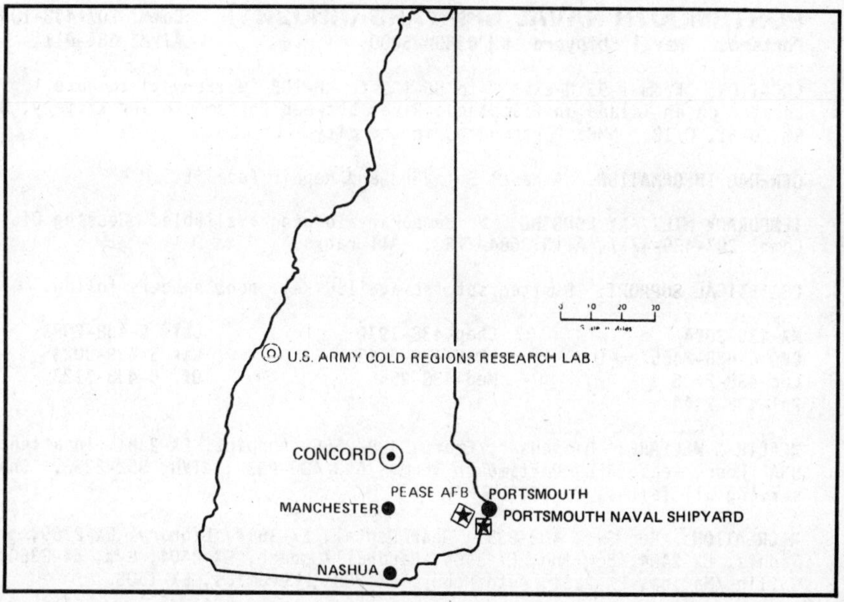

PEASE AIR FORCE BASE (NH01R1)

Pease Air Force Base, NH 03803-5000

Comm: 603-430-0100
ATVN: 852-1110

LOCATION: From I-95 N to Spaulding Turnpike, then follow signs. Base is at intersection of Spaulding Turnpke & Newington Rd. RM: p-61, O/12. NMC: Portsmouth, 3 mi NE.

GENERAL INFORMATION: Strategic Air Command Base. Air Division, Bomb Wing and Air Refueling Group.

TEMPORARY MILITARY LODGING: Biltg Ofc, Bldg 96, 24 hrs dly, Comm: 603-430-2264/2265, ATVN: 852-3385/3426. All ranks. DV/VIP, EX-3385.

LOGISTICAL SUPPORT: Complete spt fac avail. Key phone numbers follow.

BX-436-0302	**Caf-430-3748**	**Chap-430-2343/3413**
Chld C-436-7342	**Commiss-430-2618**	**Conv S-436-2006**
Fam Svc-430-3330	**Gas S-436-6707**	**Legal-430-4124/2124**
Loc-430-1841	**Med-430-2333**	**NCO C-430-3559**
Off C-430-2558	**Pkg S-431-4692**	**Pol-430-2747/8**
Snack B-436-6806	**Theat-430-3758**	

HEALTH & WELFARE: USAF Hospital, clinics and depts, Emerg, EX-2333, Appoint, EX-3122/3713/4, 70 bed. Chapels serving all faiths, EX-2343

RECREATION: Bowling, EX-3748, Gym, EX-2616, Auto Hobby, EX-3796, Photo Hobby, EX-3675, Arts & Crafts, EX-3675, Library, EX-3734, Youth Ctr, EX-2431, Rec Ctr, EX-2665, Rec Supply, EX-2509, Golf, EX-2218.

SPACE-A: Pax Term and Lounge, Bldg 237, 0800-1700 hrs dly, Comm: 603-430-2103/3643, ATVN: 852-2103/3643, flts to CONUS and overseas.

ATTRACTIONS: City of Portsmouth, harbor, beaches, skiing in White Mountains 1 hr drive.

PORTSMOUTH NAVAL SHIPYARD (NH02R1)

Portsmouth Naval Shipyard, NH 03804-5000

Comm: 207-438-1000
ATVN: 684-0111

LOCATION: From I-95 N exit 2 to US-1 S to NH-103 (Walker St) to gate 1. Located on an island on Piscataqua River between Portsmouth and Kittery, ME. RM: p-61, O/12. NMC: Portsmouth, in the city.

GENERAL INFORMATION: A Naval Shipyard and Repair Facility.

TEMPORARY MILITARY LODGING: No temporary lodging available. Housing Ofc, Comm: 207-439-4777, ATVN: 684-2732. All ranks.

LOGISTICAL SUPPORT: Limited spt fac avail. Key phone numbers follow.

NX-439-3024	**Chap-438-1970**	**Chld C-438-2083**
CPO C-438-2455	**EM C-438-2565**	**Gas S-439-3024**
Loc-438-2436	**Med-438-2555**	**Off C-438-3123**
Pol-438-2444		

HEALTH & WELFARE: Dispensary/Emerg, 438-2555, Appoint, EX-2381. Inpatient, USAF Hosp, Pease AFB, Portsmouth, Comm: 603-430-2333, ATVN: 852-2333. Chapels serving all faiths, EX-1970.

RECREATION: Rec Svc, 438-2351, Gear Rental, EX-1514, Library, EX-2769, Tennis, EX-2404, Bowling, EX-2404, Handball/Squash, EX-2404, Gym, EX-2360, Sailing/Marina, EX-2351, Auto Hobby, EX-2981, Ceramics, EX-1503.

Portsmouth Naval Shipyard, Cont'd

SPACE-A: See Pease AFB listing. **ATTRACTIONS:** Beaches, snow skiing nearby.

NEW JERSEY

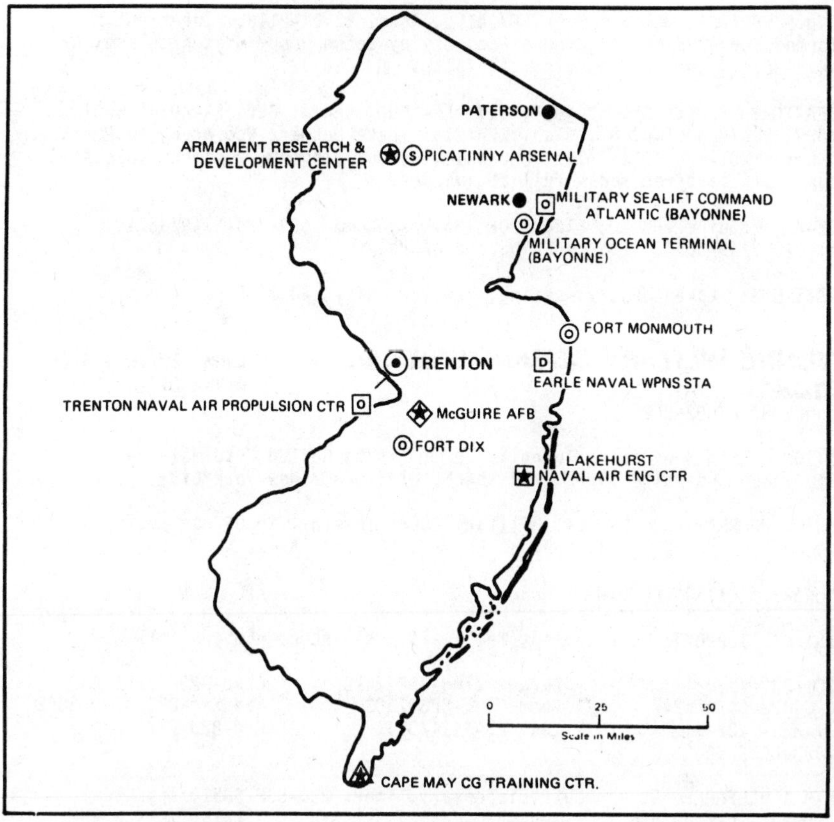

ARMAMENT RESEARCH & DEVELOPMENT CENTER (NJ01R1)

Dover, NJ 07801-5001

Comm: 201-724-4021
ATVN: 880-4021

LOCATION: Take I-80 W, exit 34B, follow signs to Ctr, 1 mi N. From I-80 E, exit 33 follow signs to CTR. RM: p-63, E/10. NMC: Newark, 30 mi E.

GENERAL INFORMATION: An R&D element of the Army Armament Munitions and Chemical Command located at old Picatinny Arsenal.

TEMPORARY MILITARY LODGING: Biltg Ofc, ATTN: SMCAR-ISE-H, Bldg 3124, Main Rd, 0800-1630 hrs dly, Comm: 201-724-2633/3506, ATVN: 880-2633/3506. All ranks. DV/VIP, EX-5024/5025.

LOGISTICAL SUPPORT: Complete spt fac avail. Key phone numbers follow.

PX-989-2410	Caf-989-2420	Chap-724-4139
Chld C-724-4337	Commiss-724-2918	EM C-724-2639

Armament R & D Ctr, Cont'd

Fam Svc-724-2145	Gas S-724-3902	Legal-724-2121
Loc-724-4021	Med-724-2113	NCO C-724-2639
Off C-724-7119	Pol-724-6666	Theat-724-2343

HEALTH & WELFARE: Clinic/Emer, EX-4190, Appoint, EX-2113. Inpatient, Uniformed Services Med Treatment Fac, Bayley-Seton Hosp, Staten Island, NY, 212-447-3010. Chaplain serving all faiths, EX-4139.

RECREATION: Crafts Sh, EX-4014, ITT, EX-3500, Rental Ctr, EX-4014, Bowl, EX-4629, Rod/Gun Club, Ski Club, EX-4014. Lake Denmark Rec Area, on Post, May-Oct, Comm: 201-724-4014, ATVN: 880-4014, 9 mobile homes, 2 camper spaces w/ hookups, all lake rec fac & full activities.

SPACE-A: McGuire AFB NJ, Bldg 1706, 24 hrs, Comm: 609-724-2749/3671, ATVN: 440-2749/3671, flts to CONUS and overseas.

ATTRACTIONS: Lake Country, skiing, New York City, 40 mi E.

BAYONNE MILITARY OCEAN TERMINAL (NJ10R1)

Bayonne, NJ 07002-302

Comm: 201-823-5111
ATVN: 247-0111

LOCATION: From New Jersey Turnpike, exit 14A to NJ-169 E to main gate. Follow green and white signs. RM: p-65, Q/7. NMC: New York City, 10 mi NE.

GENERAL INFORMATION: Largest Military Ocean Terminal in US, an activity of the Military Traffic Management Command, Eastern Area.

TEMPORARY MILITARY LODGING: None.

LOGISTICAL SUPPORT: Limited spt fac avail. Key phone numbers follow.

PX-201-339-4520	Burger King-858-2490	Chap-823-7265/7833
Chld C-823-6670/6247	Comm Club-823-7603	Fam Svc-823-5690/6062
Legal-823-7122	Loc-823-5111/0111	Med-823-7371
Pol-823-6666/6000		

HEALTH & WELFARE: Ft Monmouth Dispensary, Emer, 201-532-3851. Inpatient, Uniformed Services Med Treatment Fac, Baley-Seton Hosp, Staten Island, NY, 212-447-3010. Chaplain serving all faiths, 823-7265.

RECREATION: Gym, Rec Equip, 823-7767, Morale Support Activities, 823-5650.

SPACE-A: See McGuire AFB, listing.

ATTRACTIONS: New York City, short drive. Atlantic City, 3 hr drive.

CAPE MAY COAST GUARD TRAINING CENTER (NJ13R1)

Cape May, NJ 08204-5000

Comm: 609-884-6900
FTS: 483-6900

LOCATION: Take Garden State Parkway or US-9 to Cape May. In Cape May, take Pittsburgh Ave to Pennsylvania Ave to main gate of Center. RM: p-63, Z/7. NMC: Atlantic City, 45 mi NE.

GENERAL INFORMATION: Coast Guard Training Center for Recruits, Electronic Engineering Ctr, Cape May Air Station, several Coast Guard ships.

Cape May CGTC, Cont'd

TEMPORARY MILITARY LODGING: Housing Ofc, 0845-1630 hrs, Comm: 609-884-6917, FTS:483-6917. All ranks.

LOGISTICAL SUPPORT: Limited spt fac avail. Key phone numbers follow.

CGX-884-6940	**Chap-884-6900 EX-6223**	**Commiss-884-6940**
CPO C-884-6937	**EM C-884-6937**	**Fam Svc-884-6925**
Loc-884-6900	**Med-884-6963**	**Off C-884-6935**
OOD-884-6950	**Pkg S-884-6900 EX-6355**	**Pol-884-8451**

HEALTH & WELFARE: USCG Hospital, clinics and depts, Emerg, EX-5661/8451, Appoint, EX-5661, 25 bed. Chapel serving all faiths, EX-6223.

RECREATION: Gym, EX-6937, Swimming, EX-6937. Wildwood Rec Area, at Wildwood Crest, 5 mi N of CGTRACE, May-Sep, Comm: 609-729-8913, FTS: 346-7213, 8 camper spaces w/hookups, spt fac at CGTRACE, all activities on the beach.

SPACE-A: McGuire AFB, Bldg 1706, 24 hrs dly, Comm: 609-724-2749/3671, ATVN: 440-2749/3671, flts to CONUS and overseas.

ATTRACTIONS: Beaches, Atlantic City Resort, 45 mi N. Auto/Passenger ferry from Cape May to Lewes, DE, 1 hr, (20 min trip), operates year round.

EARLE NAVAL WEAPONS STATION (NJ11R1)

Colts Neck, NJ 07722-5000

Comm: 201-577-2000
ATVN: 449-2000

LOCATION: From New Jersey Turnpike, exit 8 to NJ-33 E through Freehold to NJ-537 E to NJ-34 S to main gate. Or exit Garden State Parkway to NJ-33 W to NJ-34 N to Station. RM: p-63, L/13. NMC: Newark, 50 mi N.

GENERAL INFORMATION: Naval Ammunition, Storage and Transshipment site with loading facilities near Sandy Hook.

TEMPORARY MILITARY LODGING: Very limited in bachelor type quarters. Off, EX-305, Enl, EX-434.

LOGISTICAL SUPPORT: Limited spt fac avail. Key phone numbers follow.

NX-349	**Caf-2479**	**Chap-405**
CPO C-306	**EM C-438**	**Gym-2119**
Legal-117	**Med-EX-300**	**Off C-306**
Pkg S-417	**Pol-EX-291**	**Spc Svcs-2167**
Theat-395 (wkend only)	**Waterfront C-EX-7100**	

HEALTH & WELFARE: Dispensary, Emerg, EX-300, Appoint, EX-300. Inpatient, Patterson Army Comm Hosp, Fort Monmouth, 10 mi N, Comm: 201-532-3205, ATVN: 992-2952, Hosp Duty Off, EX-2789. Chaplain serving all faiths, EX-405.

RECREATION: Bowling, EX-394, Gym, EX-351, Library, EX-103, Hobby Shops, EX-469, Youth Act, EX-351, Auto Hobby, EX-105, Spc Svcs, EX-351.

SPACE-A: McGuire AFB, Bldg 1706, 24 hrs dly, Comm: 609-724-2749/3671, ATVN: 440-2749/3671, flts to CONUS and overseas. Also, try Philadelphia IAP, Comm: 215-897-5630/5641/5642/5644.

ATTRACTIONS: Jersey Seashore, short drive. New York City, 1.5 hr NE.

FORT DIX ARMY TRAINING CENTER (NJ03R1)

Fort Dix, NJ 08640-5000

Comm: 609-562-1011
ATVN: 944-1110
FTS: 484-1011

LOCATION: From NJ Turnpike (I-95), exit 7, right onto NJ-206, short distance left on NJ-68, continue to General Cir & main gate. RM: p-63, 0/10. NMC: Trenton, 17 mi NW.

GENERAL INFORMATION: Army Training Center and Army training brigades.

TEMPORARY MILITARY LODGING: Biltg Ofc, ATTN: ATZDEH-H, Bldg 5255, Maryland Ave & First St, 0745-1630, other hrs desk of fac, Comm: 609-562-4849, ATVN: 944-4849. All ranks. DV/VIP, 5059/6293.

LOGISTICAL SUPPORT: Complete spt fac avail. Key phone numbers follow.

PX-723-6100	Caf-723-2671	Chap-562-4757
Chld C-723-1009	Commiss-724-2301	Conv S-723-6100
EM C-562-4896	Fam Svc-562-2767	Gas S-723-5744
Legal-562-3290	Loc-562-1011	Med-562-2695
NCO C-562-5984	Off C-723-7700	Pkg S-723-5488
Pol-562-6001	Snack B-723-6100	Theat-562-3986

HEALTH & WELFARE: Walson Army Community Hospital, clinics and depts, Emerg, EX-2695, 163 bed. Chapels serving all faiths, EX-4282/5146.

RECREATION: Arts/Crafts, EX-5691, Auto Hobby, EX-5746, Bowling, EX-6895, Golf, EX-5443, Library, EX-3587, Rod & Gun Club, EX-4290, Swimming, EX-2863, Teen Club, EX-4481, Youth Act, EX-4566. Brindle Lake Travel Camp, on Post, May-Sept, Comm: 609-562-2830, ATVN: 944-2830, FTS: 484-2830, 10 spaces w/o hookups, full fac and all activities. Also, Barnegat Rec Cottage, Long Beach, NJ, Cmdr 3rd CG Dist, Gov Island, NY, NY 10004, Comm: 212-668-7032, FTS: 664-7032, 5 unit beach house, modified CG Station, fully furn, beach and water act.

SPACE-A: McGuire AFB, Pax Term and Lounge, Bldg 1706, 24 hrs dly, Comm: 609-724-2749, ATVN: 440-2749, flts to CONUS and overseas.

ATTRACTIONS: Atlantic City and Philadelphia, short drives.

FORT MONMOUTH (NJ05R1)

Fort Monmouth, NJ 07703-5000

Comm: 201-532-1800
ATVN: 992-1800

LOCATION: Take New Jersey Turnpike to I-95; E to Garden State Pkway; N to exit 105 for Eatontown and Ft Monmouth. RM: p-63, L/14. NMC: New Brunswick, 23 mi NW.

GENERAL INFORMATION: Army Electronics Command, Army Electronics R&D Command, other supporting electronics org, US Military Academy Preparatory School, Chaplain Board & School.

TEMPORARY MILITARY LODGING: Biltg Ofc, ATTN: SELHI-EH-BH, Bldg 360, Allen & Barton Ave, 0745-2400 hrs dly, Comm: 201-532-1092/1611, ATVN: 992-1092/1611. All ranks. DV/VIP, EX-4015.

LOGISTICAL SUPPORT: Complete spt act avail. Key phone numbers follow.

PX-542-7235	Caf-542-7399	Chap-532-1366/4587
Chld C-532-2952	Commiss-532-3323	Conv S-542-5353
EM C-542-7267	Fam Svc-532-2076	Gas S-542-7417
Legal-532-4442	Loc-532-1492	Med-532-2952
NCO C-532-4520	Off C-532-4561	Pol-532-1112

Fort Monmouth, Cont'd

Snack B-532-3805 Theat-532-2737/2774

HEALTH & WELFARE: Patterson Army Community Hospital, clinics and depts, Emerg, EX-1818/2789, Appoint, EX-1151, 120 bed. Chapels serving all faiths, EX-1366/4587.

RECREATION: Bowling, EX-3805, Field House, EX-3275, Golf, EX-4307, Library, EX-3172, ITT, EX-3063, Morale Spt Ofc, EX-3281, Swimming, EX-3275, Outdoor Rec, EX-2202, Auto Hobby, EX-3301, Wood, EX-4150, Multi-Crafts, EX-2449, Photo, EX-3159, Community Ctr, EX-2695.

SPACE-A: Try ERADCOM Army Aviation, Lakehurst NAEC NJ, Comm: 201-323-2112, ATVN: 624-2112, flts via Army exec aircrft to CONUS locations. Also, Lakehurst NAEC, Base Ops, Comm: 201-323-2316, ATVN: 624-2316 for flts of opportunity.

ATTRACTIONS: New Jersey beaches nearby, Monmouth Park, New York City 1.5 hrs drive N.

LAKEHURST NAVAL AIR ENGINEERING CENTER (NJ08R1)

Comm: 201-323-2011
ATVN: 624-1110

Lakehurst, NJ 08733-5085

LOCATION: Take Garden State Parkway S to NJ-70 W to junction of NJ-547 right and proceed 1 mi to Base. RM: p-63, 0/11. NMC: Trenton, 30 mi NW.

GENERAL INFORMATION: Naval Air Engineering and Technical Air Training Centers, Army Aviation Flight Test Activity.

TEMPORARY MILITARY LODGING: Biltg Ofc, Navy Lodge, 0800-1630 hrs dly, Comm: 201-323-7442, ATVN: 624-7442. All ranks.

LOGISTICAL SUPPORT: Complete spt act avail. Key phone numbers follow.

NX-323-7684	**Caf-323-2459/7402**	**Chap-323-2272**
Chld C-323-2406	**Commiss-323-2516**	**EM C-323-2475**
Fam Svc-323-2680	**Gas-S-323-2909**	**Legal-323-2571**
Loc-323-2587/2308	**Med-323-2561**	**Off C-2340**
Pkg S-657-5100	**Pol-323-2457/2332**	**Snack B-323-2717**

HEALTH & WELFARE: Dispensary/Emerg, EX-2231. Inpatient, Walson Army Comm Hosp, Fort Dix, Comm: 609-562-2255/3142, ATVN: 944-2255/3142. Chaplain serving all faiths, EX-2272.

RECREATION: Golf, EX-7483, Bowling, EX-2027, Gym & Issue Rm, Library, Picnic Area, Swimming, Theater, Auto, for all of the above call EX-2468.

SPACE-A: No space avail at LNAEC, see McGuire AFB listing. PAX Term, Bldg 307, Comm: 201-323-2316, ATVN: 624-2316.

ATTRACTIONS: Atlantic City, resort, 60 mi S, New Jersey beaches, 10 mi E, and Great Adventure Amusement Park, 10 mi NE.

McGUIRE AIR FORCE BASE (NJ09R1)

Comm: 609-724-1110
ATVN: 440-0111

McGuire Air Force Base, NJ 08641-5000

LOCATION: From New Jersey Turnpike, exit 7 to NJ-68 SE to AFB. Adjacent to Fort Dix. Clearly marked. RM: p-63, 0/9. NMC: Trenton, 18 mi NW.

McGuire AFB, Cont'd

GENERAL INFORMATION: Military Airlift Command base. Airlift Wing, Air Refueling Group, ANG and AFRES units.

TEMPORARY MILITARY LODGING: Biltg Ofc, ATTN: SVH, Bldg 19-02, 24 hrs dly, Comm: 609-724-2954/3974/3167, ATVN: 440-2954/3974/3167. All ranks. DV/VIP, Reservations, 3336-3337.

LOGISTICAL SUPPORT: Complete spt fac avail. Key phone numbers follow.

BX-723-6100	Caf-723-5056	Chap-724-3811
Chld C-724-2966	Commiss-724-4155	Conv S-723-4466
Fam Svc-724-3145	Gas S-723-4704	Legal-724-3489
Loc-724-2345	NCO C-724-3188	Off C-724-3296
Pkg S-724-3888	Pol-724-2000	Theat-724-3240

HEALTH & WELFARE: Walson Army Community Hospital, Fort Dix, adjacent to AFB. Clinics and depts, Emerg and Appoint, 609-562-2695, 163 bed. Chapels serving all faiths, 724-3811.

RECREATION: Bowling, EX-3345, Aero Club, 723-4900, Library, EX-4319, Golf, EX-3330, Gym, EX-4645 Rec Ctr, EX-3736, Spc Svc Ofc, EX-3208, Hobby and Arts & Crafts, EX-3984, Youth Ctr, EX-3940. Fam-Camp, on Base, year round, Comm: 609-724-2176, ATVN: 440-2176, 6 camper spaces w/hookups, fac and act.

SPACE-A: Pax Term and Lounge, Bldg 17-06, Comm: 609-724-2749/3135, ATVN: 440-2749, flts to CONUS and overseas. Also from Philadelphia IAP MAC Terminal, Comm: 215-897-5642, ATVN: 443-5642.

ATTRACTIONS: Jersey seashore, Atlantic City, and Philadelphia, nearby.

NEW MEXICO

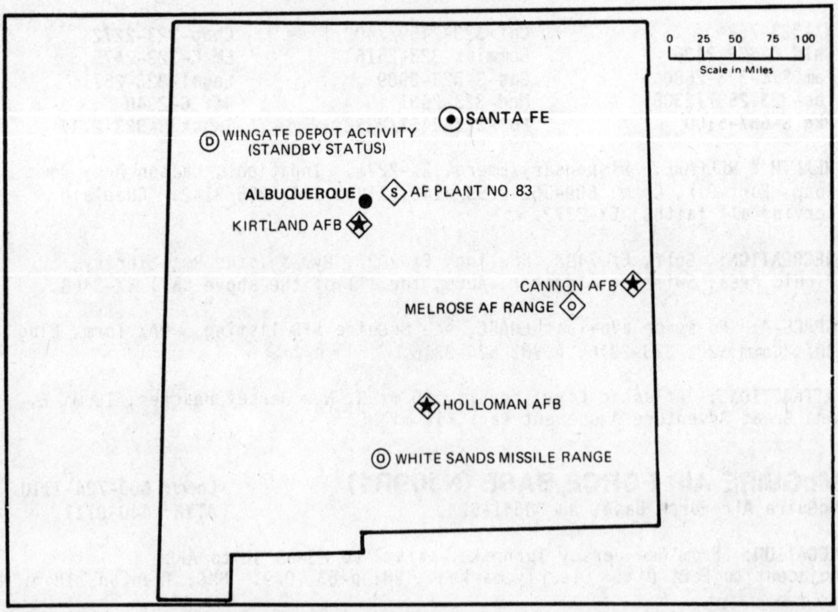

CANNON AIR FORCE BASE (NM02R3)

Cannon Air Force Base, NM 88103-5000

Comm: 505-784-3311
ATVN: 681-1110

LOCATION: From Clovis, W on US-60/84 to AFB. From NM-467 enter the Portales gate. RM: p-70, H/11. NMC: Clovis, 7 mi E.

GENERAL INFORMATION: Tactical Air Command Base. Tactical Fighter Wing, F111D Fighter Operations.

TEMPORARY MILITARY LODGING: Biltg Ofc, ATTN: 27/SVS/SVH, Bldg 1801B, Olympic St, 24 hrs dly, Comm: 505-784-2918/2935, ATVN: 681-2918/2935. All ranks. DV/VIP, EX-2769.

LOGISTICAL SUPPORT: Complete spt fac avail. Key phone numbers follow.

BX-784-4611	**Caf-784-2701**	**Chap-784-2507**
Chld C-784-3222	**Commiss-784-2160**	**Conv S-784-2035**
Fam Svc-784-2452	**Gas S-784-5177**	**Legal-784-2211**
Loc-784-2424	**Med-784-4038**	**Off C-784-2477**
Pkg S-784-2993	**Pol-784-4111**	**Snack B-784-4621**
Theat-784-2582		

HEALTH & WELFARE: USAF Hospital, clinics and depts, Emerg, EX-4033, Appoint, EX-4033, 35 bed. Chapels for all faiths, EX-2507.

RECREATION: Bowling, EX-2280, Golf, EX-2800, Gym, EX-2466, Hobby Arts & Crafts, EX-2563, Auto, EX-2170, Library, EX-2786, MWR Recording, EX-2010, Youth Ctr, EX-2747. Lake Conchas Rec Area, 119 mi NW of AFB at Lake Conchas, Comm: 505-784-2795, ATVN: 681-2795, 3 camper spaces w/water & elec, 15 camper spaces w/water only, 13 mobile homes, comp fac and all lake activities.

SPACE-A: Extremely ltd Space-A, Pax Term and Lounge, Bldg 135, 0700-2300 M-F, 0700-2100 other times, Comm: 505-784-2935, ATVN: 681-2935, flts to CONUS.

ATTRACTIONS: Carlsbad Caverns, Sierra Blanca ski resort, Clovis.

HOLLOMAN AIR FORCE BASE (NM05R3)

Holloman Air Force Base, NM 88330-5000

Comm: 505-479-6511
ATVN: 867-1110

LOCATION: Exit US-70/82, 8 mi SW of Alamogordo. Route to AFB clearly marked. RM: p-70, K/6. NMC: Las Cruces, 50 mi SW.

GENERAL INFORMATION: Tactical Air Command Base. Hq Air Div, tactical fighter wings and many support activities.

TEMPORARY MILITARY LODGING: Biltg Ofc, Bldg 342, 5th St, 24 hrs dly, Comm: 505-479-3311, ATVN: 867-3311. All ranks. DV/VIP, EX-4541.

LOGISTICAL SUPPORT: Complete spt fac avail. Key phone numbers follow.

BX-479-6146	**Caf-479-2698**	**Chap-479-7211/7024**
Chld C-479-7505	**Commiss-479-5127**	**Conv S-479-2381**
Fam Svc-479-3944	**Gas S-479-6961**	**Legal-479-7216**
Loc-479-7510/3498	**Med-479-3268/5477**	**NCO C-479-3226**
Off C-479-3611	**Pkg S-479-3226**	**Pol-479-7171**
Snack B-479-2779	**Theat-479-3286**	

HEALTH & WELFARE: USAF Hospital, clinics and depts, Emerg, EX-3268, Appoint, EX-5477, 35 bed. Chapels serving all faiths, EX-7211/7024.

RECREATION: Bowling, EX-3596, Golf, EX-3574, Gym, EX-3729, Auto Hobby, EX-

Holloman AFB, Cont'd

7438, Cons Hobby, EX-3760, Library, EX-3939, Rec Supply, EX-3728, Youth Ctr, EX-3753, Aero Club, EX-3752, Swimming, EX-3639. Fam-Camp, on Base, year round, Comm: 505-479-7476, ATVN: 867-7476, 12 camper spaces w/hookups, on Base fac.

SPACE-A: Pax Term and Lounge, Bldg 571, 24 hrs dly, Comm: 505-479-5411, ATVN: 867-5411, limited flts to CONUS locations.

ATTRACTIONS: Sierra Blanca ski area, 50 mi N. El Paso TX and Ciudad Juarez, Mexico easy drives S. International Space Hall of Fame in Alamogordo; White Sands National Monument, Alamogordo; Horse Racing, Ruidoso; Carlsbad Caverns, Whites City.

KIRTLAND AIR FORCE BASE (NM03R3)

Kirtland Air Force Base, NM 87117-5000

Comm: 505-844-0011
ATVN: 244-0011

LOCATION: From I-40 E, exit on Wyoming Blvd, S for 2 mi to Wyoming gate to AFB. RM: p-70, P/10. NMC: Albuquerque, 1 mi NE.

GENERAL INFORMATION: Military Airlift Command Base. Air Force Operational Test and Evaluation, Weapons Lab, Defense Nuclear Agency Field Command, ANG and AFRES units and other support units.

TEMPORARY MILITARY LODGING: Biltg Ofc, ATTN: 1606 ABW/SVH, Bldg 22010, 18th St, 24 hrs dly, Comm: 505-844-2936/0561, ATVN: 244-2936/0561. All ranks. DV/VIP, EX-1606.

LOGISTICAL SUPPORT: Complete spt fac avail. Key phone numbers follow.

BX-265-3679	**Caf-265-4304**	**Chap-844-2011**
Chld C-844-2314	**Commiss-844-1410**	**Conv S-265-7301**
EM C-844-5611	**Fam Svc-844-2591**	**Gas S-265-9093**
Legal-844-2943	**Loc-844-6688**	**Med-844-4611**
Off C-265-7512	**Pkg S-268-0639**	**Pol-844-4618**
Snack B-265-7301	**Theat-265-8374**	**NCO-844-7512**

HEALTH & WELFARE: USAF Hospital, clinics and depts, Emer, EX-4611, Appoint, EX-4507, 40 bed. Chapels serving all faiths, EX-2011.

RECREATION: Bowling, EX-4327, Library, EX-7320, Golf, EX-4427, Gym, EX-1818/9698, Auto Hobby, EX-2410, Arts & Craft, EX-0222, Sandia Skeet Club, 268-9816, Youth Ctr, EX-9975, Wood Hobby, EX-9127. Fam-Camp, on Base, year round, 20 camper spaces w/hookups, 20 camper spaces w/o hookups, Comm: 505-844-5420, ATVN: 244-5420.

SPACE-A: Pax Term and Lounge, Bldg 333, 0600-2200 hrs dly, Comm: 505-844-0889, ATVN: 244-0889, flts to CONUS locations.

ATTRACTIONS: Natl Atomic Museum. Albuquerque. World's longest freespan tram, Old Town founded in 1706, variety of Indian Pueblos, Albuquerque Intl Hot Air Balloon Fiesta ea October, State fair ea Sep - 4th largest in US, ski resorts, Santa Fe-60 mi N; Indian cliff dwellings, Santa Fe Opera.

WHITE SANDS MISSILE RANGE (NM04R3)

White Sands Missile Range, NM 88002-5047

Comm: 505-678-2121
ATVN: 258-2121

LOCATION: From Las Cruces, E on US-70, 30 miles to WSMR. From Alamogordo W on US-70, 45 miles to WSMR. RM: p-70, L/6. NMC: El Paso, TX, 45 mi S.

White Sands MR, Cont'd

GENERAL INFORMATION: Army Test and Evaluation Command, Naval Ordnance Missile Test Station, NASA, and other tenants and support activities.

TEMPORARY MILITARY LODGING: Biltg Ofc, 0745-1530 hrs dly, Comm: 505-678-4559/1838, ATVN: 258-4559/1838. All ranks. DVQ/VOQ/VEQ.

LOGISTICAL SUPPORT: Complete spt fac avail. Key phone numbers follow.

PX-678-3270	Caf-678-2081	Chap-678-2615
Chld C-678-2059	Commiss-678-3606	Conv S-678-4877
EM C-678-2061	Fam Svc-678-6789	Gas S-678-3400
Legal-678-1263	Loc-678-1630	Med-678-1231
NCO C-678-2061	Off C-678-2057	Pkg S-678-4022
Pol-678-1234	Snack B-678-2081	Theat-678-2483

HEALTH & WELFARE: Dispensary/Emerg, EX-2882. Inpatient, William Beaumont Army Med Ctr, Fort Bliss, TX, Comm: 915-568-2159, ATVN: 978-2159. Chapels serving all faiths, EX-2615.

RECREATION: Arts & Crafts, EX-5321, Auto Hobby, EX-5880, Bowling, EX-1394, Golf, EX-1759, Gym, EX-3374, Library, EX-5820, Swimming, EX-2860, Rec Equip, EX-1713, Rec Ctr, EX-4134.

SPACE-A: None.

ATTRACTIONS: White Sands Natl Monument and Space Center in Alamogordo.

NEW YORK

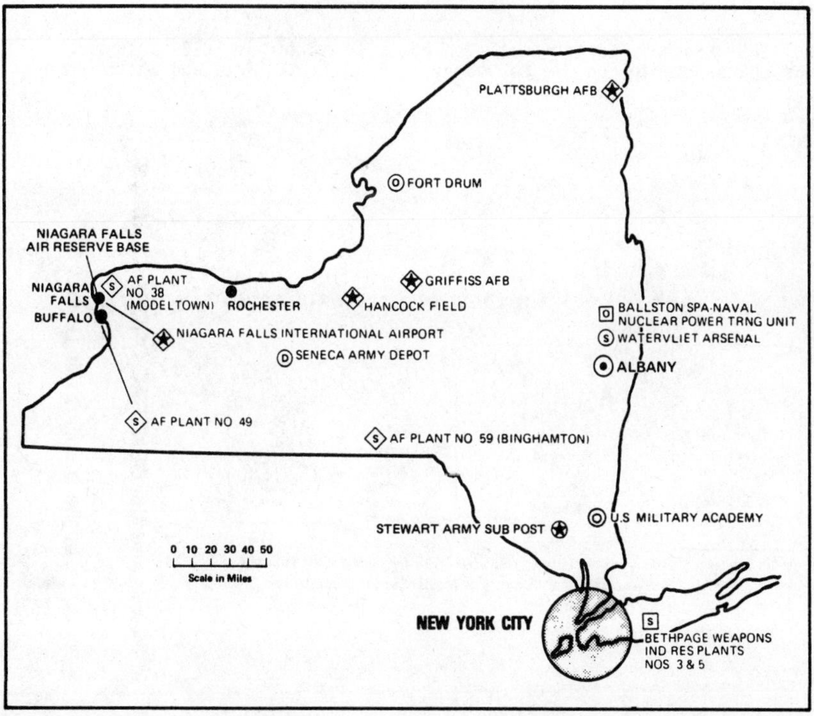

FORT DRUM (NY06R1)

Fort Drum, NY 13602-5000

Comm: 315-772-5647
ATVN: 341-5647

LOCATION: From Syracuse, take I-81 N to exit 48, past Watertown, and follow signs to Fort Drum. RM: p-67, EG/6. NMC: Watertown, 8 mi SW.

GENERAL INFORMATION: Hq 10th Mountain Division Lt Inf and supporting units.

TEMPORARY MILITARY LODGING: Biltg Ofc, ATTN: AFZS-EH-H, Bldg T-2227, Officers Loop, 0730-2300 hrs dly, Comm: 315-772-5437, ATVN: 341-5437. All ranks.

LOGISTICAL SUPPORT: Complete spt fac avail. Key phone numbers follow.

PX-772-6103	Caf-772-5960	Chap-772-5591
Commiss-772-5294	Conv S-773-5527	EM C-772-6201
Fam Svc-772-5475	Gas S-773-2343	Legal-772-5261
Loc-772-5869	Med-772-5236	NCO C-772-6201
Off C-772-6222	Pkg S-773-5885	Pol-772-5157
Theat-772-5571		

HEALTH & WELFARE: USA Hospital, clinics and depts, Emerg, EX-5333, Appoint, EX-5236, 25 bed. Chapels serving all faiths, EX-5591.

RECREATION: Rec Ctr, EX-5692, Crft Shop, EX-5606, Auto Hobby, EX-5785, Ice Skating, EX-5769, Bowl, EX-5086, Gym, EX-6163. Remington Pond Rec Area, on Post, Apr-Nov, wilderness area, Comm: 315-772-5169, ATVN: 341-5169, 6 camper spaces w/o hookups, other fac & activities incl sm ski slope on Post.

SPACE-A: Wheeler-Sack Army Airfield, flt via exec aircraft to CONUS locations, EX-5681.

ATTRACTIONS: Sackets Harbor Battleground, site of War of 1812, Thousand Islands, Lake Ontario and Canada nearby. Dry Hill ski area and white water rafting, Watertown.

NEW YORK CITY

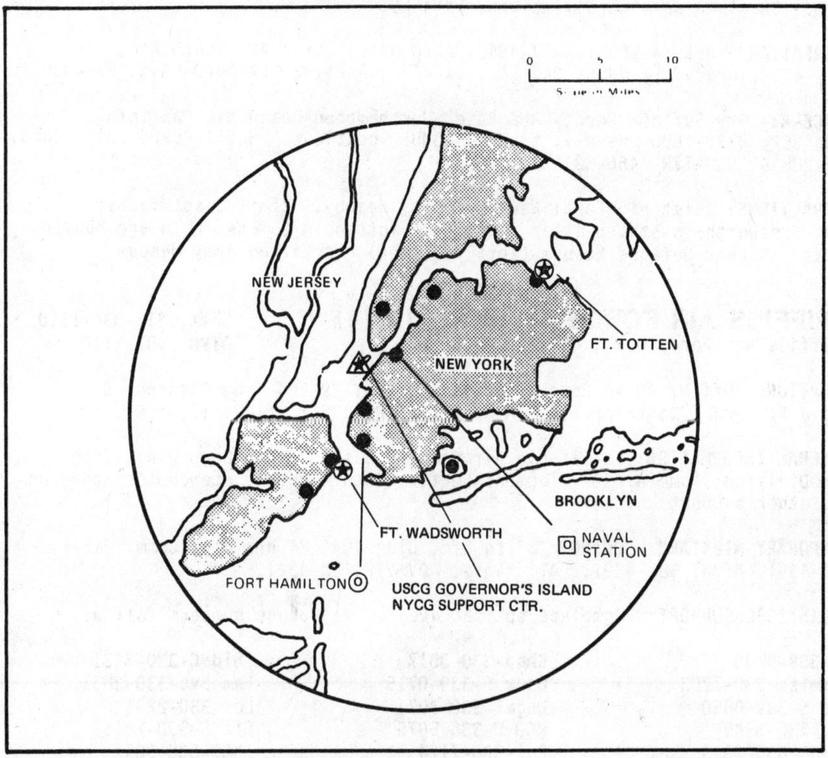

FORT HAMILTON (NY02R1)

Brooklyn, NY 11252-5330

Comm: 718-630-4101
ATVN: 232-1110

LOCATION: From Belt Parkway, exit 2 (Ft Hamilton Parkway) to 100th St, right to Ft Hamilton Parkway, right to main gate. RM: p-65, S/8. NMC: New York, in the city.

GENERAL INFORMATION: New York Area Command providing facilities and support for Army and other DOD activities in the New York City metro area.

TEMPORARY MILITARY LODGING: Biltg Ofc, Bldg 109, Schum Ave, 0800-1630 hrs dly, Comm: 718-630-4892/4937, ATVN: 232-4892/4564. All ranks. DV/VIP, EX-4324. Soldiers, Sailors and Airmen's Club, 283 Lexington Ave, New York City, club-type facilities, rooms, Comm: 212-683-4353.

LOGISTICAL SUPPORT: Complete spt fac avail. Key phone numbers follow.

PX-748-3440	Snack B-748-3440	Chap-630-4969
Commiss-630-4568	Theat-630-4840	EM C-630-4556
Fam Svc-630-4332	Gas S-748-3440	Legal-630-4562
Loc-630-4958	Med-630-4615	CDC-630-4079
NCO C-630-4556	Off C-630-4361	Pkg S-630-4049

HEALTH & WELFARE: Clinic/Emer, EX-4555, Appoint, EX-4268. Inpatient, Uni-

Fort Hamilton, Cont'd

formed Services Med Treatment Fac, Bayley-Seton Hosp, Staten Island, 212-447-3010. Chaplain serving all faiths, EX-4969.

RECREATION: Arts & Crafts, EX-4942, Auto Hobby, EX-4149, Youth Act, EX-4442, Library, EX-4875, Rec Ctr, Ex-4476, Bowling, EX-4440, Gym, EX-4830.

SPACE-A: Try Suffolk County ANG Base, Westhampton Beach NY, Pax Info, Bldg 362, 0730-1600 hrs dly, flts to CONUS locations. Nat'l Guard Unit, Comm: 516-288-4200, ATVN: 456-7216.

ATTRACTIONS: Historic Fort, New York City nearby. Limited spt fac at Fort Wadsworth in Staten Island, and Fort Totten in Queens, both are housing areas. Harbor Defense Museum (Fort Hamilton), certified Army Museum.

GRIFFISS AIR FORCE BASE (NY11R1)

Griffiss Air Force Base, NY 13441-5000

Comm: 315-330-1110
ATVN: 587-1110

LOCATION: Off NY-49 in Rome. Entrance off NY-49 and from Chestnut St, Floyd St, and E Dominick St. RM: p-67, EK/8. NMC: Utica, 17 mi SE.

GENERAL INFORMATION: Strategic Air Command Base. 416th Bomb Wing, 24th Air Division, Rome Air Development Center, 49th Fighter Interceptor Squadron and other support units.

TEMPORARY MILITARY LODGING: Biltg Ofc, Bldg 704, 24 hrs dly, Comm: 315-330-4391, ATVN: 587-4391. All ranks. DV/VIP, EX-4391.

LOGISTICAL SUPPORT: Complete spt fac avail. Key phone numbers follow.

BX-339-0700	**Chap**-330-3017	**Chld C**-330-7931
Commiss-330-7203	**Conv** S-339-0715	**Fam Svc**-330-2031
Gas S-339-0850	**Legal**-330-3071	**Loc**-330-2231
Med-330-3545	**NCO** C-330-5075	**Off** C-330-7915
Pkg S-330-7913	**Pol**-330-7174	**SNCO**-330-3057
Theat-330-2993		

HEALTH & WELFARE: USAF Hospital, clinics and depts, Emer, EX-4108, Appoint, EX-3671, 70 bed. Chapels serving all faiths, EX-3017.

RECREATION: Bowling, EX-7097, Golf, EX-2647, Arts & Crafts, Auto Hobby, Swimming, Rec Ctr, EX-3163, Rec Supply, EX-7020, Gym, Library, Youth Act, Aero Club, EX-1110. Fam-Camp, on Base, May-Oct, Comm: 315-330-4391, ATVN: 587-4391, 10 camper spaces w/elec only, other fac and activities.

SPACE-A: Pax Term and Lounge, Base Ops, 24 hrs dly, Comm: 315-330-7400, ATVN: 587-7400, flts to CONUS and overseas via KC-135 aircraft.

ATTRACTIONS: Erie Canal Village, Fort Stanwix, Fort Rickey Game Farm, Battle of Oriskany Battlefield, Sports Hall of Fame.

HANCOCK FIELD (NY10R1)

Syracuse, NY 13211-7099

Comm: 315-470-6100
ATVN: 587-9100

LOCATION: From Syracuse, take I-81 N to exit 28, Taft Road E, 1 mi to north side entrance. From I-90 (New York Throughway), exit 35 to Thompson Road, 1 mi to S entrance. RM: p-69, WJ/6. NMC: Syracuse, 4 mi SW.

GENERAL INFORMATION: 174 Tactical Fighter Wing, New York Air National Guard (south side); 152 Tactical Control Group, New York Air National Guard

Hancock Field, Cont'd

(north side). Rest of USAF Base closed 30 June, 1984.

TEMPORARY MILITARY LODGING: None.

LOGISTICAL SUPPORT: Little spt fac avail. Key phone number-**470-6440.**

HEALTH & WELFARE: Inpatient, USAF Hosp, Griffiss AFB, Comm: 315-330-4108, ATVN: 587-4108.

RECREATION: Griffiss AFB is the nearest USAF Base.

SPACE-A: None.

ATTRACTIONS: New York Fairgrounds, Carrier Dome, Syracuse U.

NEW YORK COAST GUARD SUPPORT CENTER (NYO1R1)

Governors Island, NY, NY 10004-5000

Comm: 212-668-7000
FTS: 664-7000

LOCATION: Take free Governors Island Ferry from Battery Park area of Manhattan. RM: p-65, P/8. NMC: New York, 1 mi NW.

GENERAL INFORMATION: Third CG District, CG Training Ctr, CG Support Ctr.

TEMPORARY MILITARY LODGING: Biltg Ofc, Bldg 702, 0800-1600 hrs dly, Comm: 212-668-3452, FTS: 664-3452. All ranks. DV/VIP, EX-7251.

LOGISTICAL SUPPORT: Complete spt fac avail. Key phone numbers follow.

BX-668-7311	Burg K-825-0465	Chap-668-7313
Chld C-344-1942	Commiss-668-7338	Conv S-952-1347
CPO C-668-7266	EM C-668-7270	Fam Svc-668-6444
Gas S-425-2970	Legal-668-7159	Loc-668-7000
Med-668-7167	Off C-668-7400	Pkg S-668-7394
Pol-668-7474	Theat-668-7401	

HEALTH & WELFARE: Clinic/Emerg, EX-7123. Inpatient, Uniformed Services Med Treat Fac, Bayley-Seton Hosp, Staten Island, 212-447-3010. Chapels, serving all faiths, EX-7313.

RECREATION: Auto Crafts, EX-7231, Bowling, EX-7315, Crafts Shop, EX-7248, Golf, EX-7319, Gym, EX-7255, Library, EX-7394, Picnic Grounds, EX-7255, Swimming, EX-7255, Tennis, EX-3401, Tickets, EX-7255.

SPACE-A: McGuire AFB, Bldg 1706, 24 hrs dly, Comm: 609-724-2749/3671, ATVN: 440-2749/3671, flts CONUS and overseas.

ATTRACTIONS: New York City, Action Park.

NEW YORK NAVAL STATION (NY07R1)

Brooklyn, NY 11251-5000

Comm: 718-834-2000
ATVN: 456-2000

LOCATION: In Brooklyn, .25 mi W of I-278, at Flushing & Washington Avenues (207 Flushing Avenue). RM: p-65, T/12. NMC: New York, in the city.

GENERAL INFORMATION: Naval Station New York, support activities for Navy in New York area.

New York NS, Cont'd

TEMPORARY MILITARY LODGING: Extremely limited. Officers & chiefs, Comm: 718-834-2298, ATVN: 456-2298; EM, Comm: 718-834-2410, ATVN: 456-2410.

LOGISTICAL SUPPORT: Most spt fac avail. Key phone numbers follow.

NX-834-2432	Caf-834-2432	Chap-834-2791
EM C-834-2430	Snack B-834-2430	Legal-834-2316
Loc-834-2609	Med-834-2222	Off C-834-2215
Pol-834-2323		

HEALTH & WELFARE: Dispensary/Emerg, EX-2222. Inpatient, Uniformed Services Med Treat Fac, Bayley-Seton Hosp, Staten Island, 212-447-3110. Chapel serving all faiths, EX-2791.

RECREATION: Limited.

SPACE-A: See McGuire AFB, NJ, listing.

ATTRACTIONS: New York City.

NIAGARA FALLS AIR RESERVE BASE (NY12R1)

Comm: 716-236-2000
ATVN: 489-3011

Niagara Falls International Airport, NY 14304-5000

LOCATION: Take I-190 to Niagara Falls, exit Lockport Rd to Base. Take US-62 to Walmore Rd N to Base. RM: p-69, WD/12. NMC: Niagara Falls, 6 mi W.

GENERAL INFORMATION: Tactical Airlift Group, Fighter Interceptor Group.

TEMPORARY MILITARY LODGING: Biltg Ofc, Bldg 312, Flint Ave, 0730-1600 hrs dly, Comm: 716-236-2014, ATVN: 489-2014. All ranks. DV/VIP, EX-2014/2000.

LOGISTICAL SUPPORT: Most spt fac avail. Key phone numbers follow.

BX-236-2100	Chap-236-2096/7	Cons C-236-2027
Gas S-236-2333	Legal-236-2131	Loc-236-2002
Med-236-2086/7	Pkg S-236-2009	Pol-236-2279
Snack B-236-2328	Theat-236-2311	

HEALTH & WELFARE: Dispensary/Emerg, EX-2086/7. Inpatient, USAF Hosp, Griffiss AFB, Rome, 200 mi E, 315-330-4108. Chapels serving all faiths, EX-2096/7.

RECREATION: Limited on Base.

SPACE-A: Pax Term and Lounge, Bldg 800, 0730-1600 M-F, Comm: 716-236-2150, ATVN: 489-2150, flts to CONUS via AFRES aircraft.

ATTRACTIONS: Beautiful Niagara Falls.

PLATTSBURGH AIR FORCE BASE (NY08R1)

Comm: 518-565-5000
ATVN: 689-5000

Plattsburgh Air Force Base, NY 12903-5000

LOCATION: From I-87 take exit 36. Directions to AFB are clearly marked. RM: p-67, EC/15. NMC: Burlington, 20 mi SE.

GENERAL INFORMATION: Strategic Air Command Base. Bomb Wing, Combat Crew Training, Aerospace Rescue and Recovery Squadron.

Plattsburgh AFB, Cont'd

TEMPORARY MILITARY LODGING: Biltg Ofc, Bldg 381, Club Rd, 24 hrs dly, Comm: 518-565-7614, ATVN: 689-7614. All ranks. DV/VIP, EX-5171.

LOGISTICAL SUPPORT: Complete spt fac avail. Key phone numbers follow.

BX-561-6622	**Chap-565-5211**	**Chld C-565-5570**
Commiss-565-5553	**Conv S-561-2360**	**Fam Svc-565-5732**
Gas S-561-5010	**Legal-565-5476**	**Loc-565-5579**
Med-565-7223	**NCO C-563-3956**	**Off C-563-4780**
Pkg S-565-7657	**Pol-565-7191**	**Theat-565-5519**

HEALTH & WELFARE: USAF Hospital, clinics and depts, Emerg, EX-7223, Appoint, EX-7524. Chapels serving all faiths, EX-5211.

RECREATION: Arts & Crafts, EX-7360, Gym, EX-7028, Auto Hobby, EX-5269, Rec Ctr, EX-7036, Library, EX-7046, Boating/Swimming, EX-5244, Youth Ctr, EX-5412, Rec Supply, EX-5287, Wood Hobby, EX-5155, Bowling, EX-5558, Golf, EX-7098.

SPACE-A: Base Ops, Comm: 518-565-5442, ATVN: 689-5442, flts to CONUS and overseas.

ATTRACTIONS: Lake Champlain, ski areas, Canada short drive. Lake Placid, home of 1932 and 1980 Winter Olympics. Montreal, Quebec, second-largest French speaking city in the world, 60 mi north, Old World charm close to home.

SENECA ARMY DEPOT (NY03R1)

Romulus, NY 14541-5000

Comm: 607-869-1110
ATVN: 489-5110

LOCATION: Fifty-five mi SE of Rochester, 12 mi S of Geneva, on NY-96A. RM: p-69, WF/21. NMC: Rochester, 55 mi NW.

GENERAL INFORMATION: Army Readiness Group, Coast Guard Loran Station.

TEMPORARY MILITARY LODGING: Biltg Ofc, ATTN: SDSSE-AH, Bldg B-116, 0830-1630 hrs dly, Comm: 607-869-1314, ATVN: 489-5314. All ranks.

LOGISTICAL SUPPORT: Complete spt fac avail. Key phone numbers follow.

PX-869-0294	**Chap-869-0340**	**Chld C-868-0401**
Commiss-869-0256	**EM C-869-0431**	**Fam Svc-869-0388**
Gas S-869-0284	**Legal-869-0230**	**Loc-869-1398**
Med-869-1243	**NCO C-869-1438**	**Off C-869-1666**
Pkg S-869-0296	**Pol-869-0448**	**Theat-869-0425**

HEALTH & WELFARE: Dispensary/Emerg, EX-1117. Inpatient, USAF Hosp, Griffiss AFB, Rome, 90 mi E, 315-330-4108. Chapels serving all faiths, EX-0340.

RECREATION: Rec Ctr, Rec Supply, Bowling, Auto Hobby, Multi-Craft Hobby, EX-18-512. Army Travel Camp, on Post, year round, Comm: 607-869-1221, ATVN: 489-5211, 6 camper spaces w/hookups, 2 camper spaces w/o hookups, 13 mobile homes, all fac and act, on E shore of Seneca Lake, largest of the Finger Lakes.

SPACE-A: See Hancock Field NY listing.

ATTRACTIONS: Finger Lakes region of NY.

STEWART ARMY SUB-POST (NY09R1)

Newburgh, NY 12550-9999

Comm: 914-564-7000
ATVN: 247-1110

Stewart Army Sub-Post, Cont'd

LOCATION: From I-87 take Newburgh exit to Union Ave, S to NY-207. Follow signs to Stewart Airport and Sub-Post. RM: p-67, EU/13. NMC: New York City, 60 mi S.

GENERAL INFORMATION: Support activity for US Military Academy, West Point.

TEMPORARY MILITARY LODGING: Biltg Ofc, Bldg 2605, Five Star Inn, 6th and D Sts, 24 hrs dly, Comm: 914-564-6309, ATVN: 247-3311/3524. All ranks. DV/VIP, EX-6309.

LOGISTICAL SUPPORT: Complete spt fac avail. Key phone numbers follow.

PX-564-7601	**Chap-564-5291**	**Chld C-563-3522**
Commiss-563-3421	**Conv S-564-7601**	**EM C-564-7590**
Fam Svc-563-3257	**Legal (West Pt)-938-4570**	**Loc-564-7000**
Med-564-7788	**NCO C-564-7590**	**Off C-564-6661**
Pkg S-564-7730	**Pol-564-3602**	

HEALTH & WELFARE: Clinic/Emerg, 563-3440/3430, Appoint, 563-3440. Inpatient, William L. Keller Army Comm Hosp, USMA, West Point, Comm: 914-938-4004, ATVN: 688-4004. Chapels serving all faiths, 563-3310/3429.

RECREATION: Library, 563-3501, Youth Act Ctr, 563-3544, Swimming, 563-3485, Bowling, 563-3447, Gym, 563-3565, Picnic & Archery, 563-3485, Multi-Crafts, 563-3584, Auto Hobby, 563-3420, Morale Spt Ofc, 563-3485.

SPACE-A: Stewart IAP, ANG Ops, Comm: 914-564-3376, ATVN: 247-3376, flts to CONUS and overseas.

ATTRACTIONS: Historic USMA, West Point, and Hudson River Valley.

UNITED STATES MILITARY ACADEMY, WEST POINT (NY16R1)

Comm: 914-938-4011
ATVN: 688-1110

West Point, NY 10996-5000

LOCATION: Off I-87 or US-9W. Clearly marked. RM: p-67, EU/14. NMC: New York City, 50 mi S.

GENERAL INFORMATION: Military Educational Institution, Corps of Cadets and support units.

TEMPORARY MILITARY LODGING: Bldg 674, Hotel Thayer, 24 hrs, Comm: 800-247-5047, 914-446-4731, ATVN: 688-2632. All ranks. DV/VIP, 4021.

LOGISTICAL SUPPORT: Complete spt fac avail. Key phone numbers follow.

PX-446-5406	**Chld C-938-4523**	**Commiss-938-3663**
Fam Svc-938-4506	**Gas S-446-5556**	**Legal-938-4541**
Med-938-4004	**NCO C-446-5507**	**Off C-446-5506**
Pol-938-3333		

HEALTH & WELFARE: William L. Keller Army Community Hospital, clinics and depts, Emerg, EX-4004, Appoint, EX-4091, 65 bed. Chapels serving all faiths, EX-3412.

RECREATION: Golf, EX-2435, Ski Slope, EX-3726, Youth Ctr, EX-3208, Swim, EX-2946/5158, Lodge, EX-4335, Gym, EX-2338, Smith Rink, EX-4236, Equip Rental, EX-3481/2428, Bowling, EX-2140, Library, EX-2974, Rec Adm, EX-3809/4455, Rec Ctr, EX-3601/2070, Craft Shop, EX-4812/2074. Operates Round Pond Rec Area,

US Mil Academy WP, Cont'd

5 mi W, Apr-Nov, Comm: 914-938-3809, ATVN: 688-3809, 14 camper spaces w/elec, 20 tent spaces, 6 mobile homes, 3 travel trailers, all fac/full activities.

SPACE-A: See Stewart Army Sub-Post listing.

ATTRACTIONS: Walking/bus tours historic USMA, West Point, Hudson River Valley.

NORTH CAROLINA

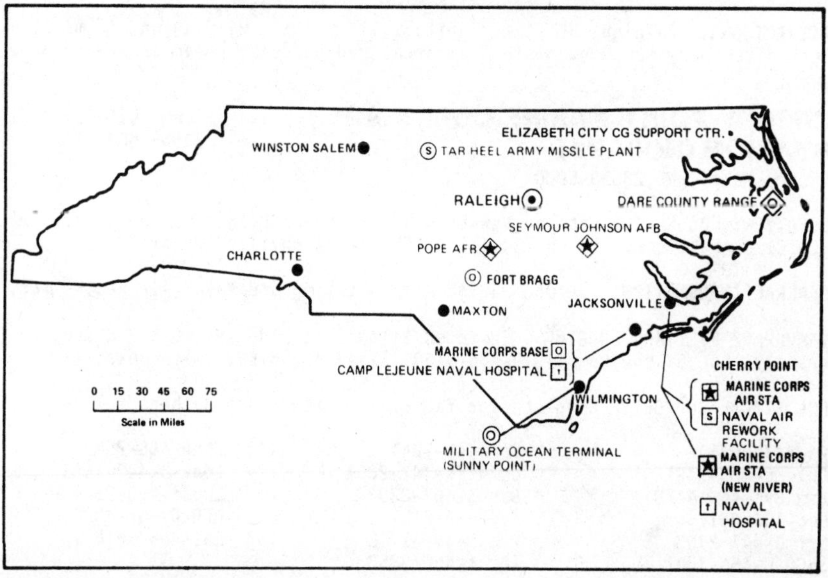

CAMP LEJEUNE MARINE CORPS BASE (NC10R1)

Camp Lejeune Marine Corps Base, NC 28542-5000

Comm: 919-451-1113
ATVN: 484-1113
FTS: 676-1113

LOCATION: Main gate is 6 mi E of junction of US-17 and NC-24, off NC-24. RM: p-73, J/19. NMC: Jacksonville, 3 mi NW.

GENERAL INFORMATION: Marine Division, Support Group, Marine Amphibious Brigade, Naval Hospital.

TEMPORARY MILITARY LODGING: Biltg Ofc, Bldg 2617, Paradise Point, 24 hrs dly, Comm: 919-451-1856/1385, ATVN: 484-1856/1385. All ranks. DV/VIP, EX-2526.

BX-451-2481	**Caf-451-2591**	**Chap-451-0794/5646**
Commiss-451-2172	**Conv S-451-5491**	**EM C-451-2872**
Fam Svc-451-1362	**Gas S-451-2443**	**Legal-451-1903**
Loc-451-3074	**Med-451-4551/4300**	**NCO C-451-2839**
Off C-451-2777	**Pol-451-2555**	**Theat-451-1759**

HEALTH & WELFARE: Naval Hospital, clinics and depts, Emerg, EX-4551/4300, Appoint, EX-4630/4644, 236 bed. Chapels serving all faiths, EX-0794/5775.

RECREATION: Arts/Crafts, EX-2077/6711/5191/6690, Auto Hobby, EX-7368/1550/

Camp Lejeune MC Base, Cont'd

6709, Intramural Sports, EX-3125/6714, Golf, EX-5445/1668, Gym, EX-3516/1612/3768/5288/1879, Boating, EX-1956/7386/6578, Riding Stables, EX-2238, Library, EX-5724/0844/6715, Swimming, EX-2024/6281/1441 and EX-4149/2465/6424, Tickets, EX-5694/6704, Trap/Skeet Range, EX-3889, Beach, EX-5694, Fishing/Hunting, EX-5226. Onslow Beach Camp Sites & Rec Area, on Base, year round, Comm: 919-451-5694, ATVN: 484-5694, 59 camper spaces w/hookups, 15 Burro Campers w/hookups, comp fac at beach and all activities.

SPACE-A: Cherry Point MCAS, Pax Term, ATC tower, 0700-2300 hrs dly, Comm: 919-466-2379, ATVN: 582-2379, flts to CONUS locations.

ATTRACTIONS: Raleigh, NC state capital, 116 mi NE. Wilmington, 45 mi S. Both offer many museums, historical areas and gardens. Beaches are on Base.

CHERRY POINT MARINE CORPS AIR STATION (NC02R1)

Comm: 919-466-4232
ATVN: 582-4232

Cherry Point, NC 28533-5000

LOCATION: On NC-101 between New Bern and Morehead City. US-70 connects with NC-101 at Havelock. RM: p-73, I/21. NMC: Jacksonville, 30 mi SW.

GENERAL INFORMATION: Second Marine Aircraft Wing and Naval Air Rework Fac.

TEMPORARY MILITARY LODGING: Enlisted Biltg Ofc, Bldg 294, C & 3rd Ave, 24 hrs dly, BOQ Biltg, Comm: 919-466/5169, ATVN: 582-5169. All ranks.

LOGISTICAL SUPPORT: Complete spt fac avail. Key phone numbers follow.

PX-447-7041	**Caf-447-7041**	**Chap-466-2585**
Chld C-466-3491	**Commiss-447-2061**	**Conv S-447-7041**
Fam Svc-466-4470	**Gas S-447-2392**	**Legal-466-2361**
Loc-466-2917	**Med-466-5751**	**NCO C-466-3020**
Off C-447-2303	**Pkg S-447-1718**	**Pol-466-3615**
SNCO C-466-3087	**Theat-466-3850**	

HEALTH & WELFARE: Naval Hospital, clinics & depts, Emer EX-4410, Appoint, 466-4216, 52 bed. Chapels serving all faiths, EX-2585.

RECREATION: Arts/Crafts, EX-3965, Bowling, EX-3910, Stables, EX-2753, Boat Docks, EX-5812, Library, EX-3532, Swimming, EX-2302, Rec Ctr, EX-5184, Picnic Area, EX-2197, Skeet Range, EX-3263. Travel Park, on Base, year round, Comm: 919-466-2197, ATVN: 582-2197, 50 camper spaces w/hookups, 5 mobile homes complete fac and all act on the Neuse Waterway of the Outer Banks area.

SPACE-A: Pax Term and Lounge, ATC tower, 0700-2300 hrs dly, Comm: 919-466-2379/3225, ATVN: 582-2379/3225, flts to CONUS locations.

ATTRACTIONS: Greater Outer Banks of NC, hunting, fishing, swimming, beaches and outdoor recreation of all types. New Bern, first colonial capital of NC, 17 mi N Morehead City, is a large commercial fishing port.

ELIZABETH CITY COAST GUARD SUPPORT CENTER (NC03R1)

Comm: 919-338-3941
FTS: 931-0001

Elizabeth City, NC 27909-5000

LOCATION: Take I-65 E to VA-104S to US-17 S to Elizabeth City, left on Halstead Blvd, 3 mi to main gate of Ctr. RM: p-73, C/23. NMC: Elizabeth City, in the city.

Elizabeth City CGSC, Cont'd

GENERAL INFORMATION: Air Station, Aviation Technical Training and Repair Center for the Coast Guard.

TEMPORARY MILITARY LODGING: Biltg Ofc, Bldg 5, 0800-1630 dly, Comm. 338-3941, ATVN: 335-6398.

LOGISTICAL SUPPORT: Most spt fac avail. Key phone numbers follow.

CGX/Pkg S-335-6269	Chap-335-6202	Conv S-335-6187
EM C-335-6388	Gas S-335-6187	Loc-335-6130
Med-335-6460	NCO C-335-6301	Off C-335-6226
Pol-335-6398	Beauty Shop-335-6382	Barber Shop-335-6284

HEALTH & WELFARE: Dispensary/Emerg, EX-6460, Appoint, EX-6460. Inpatient, Naval Hosp, Portsmouth, VA, 804-398-5111. Chapels serving all faiths, EX-6202.

RECREATION: Gym, Swimming, Auto Hobby, Boat Ramp, Ceramics Shop, Racquetball Courts, Tennis Courts, Beaches. Cape Hatteras CG Rec Area, Buxton, NC, year round, operated by 5th CG District, Comm: 804-398-6475, FTS: 827-9257, 2 suites and 6 single rms, common kitchen, most fac and all beach activities.

SPACE-A: Pax Term, Air Ops Ctr, 24 hrs dly, Comm: 919-338-3941, ATVN: 723-1540, FTS: 931-0333, flts to CONUS and overseas.

ATTRACTIONS: North Carolina Outer Banks, fresh seafood, beaches.

FORT BRAGG (NC05R1)

Fort Bragg, NC 28307-5000

Comm: 919-396-0011
ATVN: 236-0011

LOCATION: From I-95, exit to NC-24 W for 15 mi. NC-24 runs through Post as Bragg Blvd. From US-401 (Fayetteville Bypass) exit to All American Expressway, 5 mi to Fort. RM: p-73, H/14. NMC: Fayetteville, 10 mi SE.

GENERAL INFORMATION: Airborne Corps, Airborne Division, Special Operations Command, Field Artillery Command, Corps Support Command, other support units.

TEMPORARY MILITARY LODGING: Biltg Ofc, Bldg D-3601 (Moon Hall), Rm 101, off Bastogne Dr, 0730-1600 M-F, other hrs SDO, Comm: 919-396-4811/9574, ATVN: 236-4811/9574. All ranks. DV/VIP, EX-2409.

LOGISTICAL SUPPORT: Complete spt fac avail. Key phone numbers follow.

PX-436-4888	Caf-436-5202	Chap-396-1121/3836
Commiss-396-2428/8191	Conv S-436-3950/0800	EM C-436-3200
Fam Svc-396-5601	Gas S-436-3402/497-3985	Legal-396-6113
Loc-396-3794/6972	Off C-436-1700/1945	Pkg S-436-4646/0404
Pol-396-0391	Snack B-436-0217	Theat-396-4476

HEALTH & WELFARE: Womack Army Community Hospital, clinics and depts, Emerg, EX-0301, Appoint, EX-0341, 325 bed. Chapels serving all faiths, EX-3836/1121.

RECREATION: Arts/Crafts, EX-7894/1093/9393/7919/2931, Auto Hobby, EX-4397/4132/5696, Bowling, EX-4128, ITT, EX-8670/3625, Golf, EX-4325 or 436-3390, Field Hs, EX-3037/4468/3573/6489/2231/6493, Sports Rntal, EX-7520, Riding Stables, EX-4510, Ice Rink, EX-5127, Library, EX-3526/3923/6111/2715, Swimming, EX-6152/9941/6762 & EX-0883/7266/2944, Playhouse, EX-7555, Music Ctr, EX-4373, Fishing Club, EX-6717, Sport Parachute Club, EX-1010/2919/4056. Operates Rogers Bay Campway, 6 mi N of Surf City, year round, Comm: 919-396-7520, ATVN: 236-7520, 8 campers (self-contained), 7 pop-ups w/o hookups, full fac/water sport act. Also operates Smith Lake Army Travel Camp, on Post, Apr-Nov, Comm: 913-369-7520, 24 camper spaces w/hookups, all fac and act.

Fort Bragg, Cont'd

SPACE-A: Simmons Army Airfield, flts via exec aircraft to CONUS locations, EX-0011. Pope AFB, Pax Term, Bldg 708, 24 hrs dly, Comm: 919-394-4429, ATVN: 486-4429, flts to CONUS locations.

ATTRACTIONS: Pinehurst Resort is nearby. JFK Special Warfare Museum and 82nd Airborne Division Museum on Post.

NEW RIVER MARINE CORPS AIR STATION (NC06R1)

Comm: 919-451-6197
ATVN: 484-6197

Jacksonville, NC 28545-5000

LOCATION: Off US-17, 2 mi S of Jacksonville. Clearly marked. RM: p-73, J/19. NMC: Jacksonville, 2 mi NE.

GENERAL INFORMATION: Marine Aircraft Groups, Marine Air Traffic Control Squadron, and Marine Wing Support Group Detachment.

TEMPORARY MILITARY LODGING: Biltg Ofc, Bldg 122, dty hrs, Comm: 919-451-6111/2/3, ATVN: 484-6111/2/3. All ranks. DV/VIP, EX-6197.

LOGISTICAL SUPPORT: Most spt fac avail. Key phone numbers follow.

MX-347-2168	**Chld C-451-6712**	**Commiss-347-6395**
EM C-451-0589	**Gas S-347-6092**	**Legal-451-6169**
Loc-451-6169	**Med-451-6001**	**NCO C-451-6707**
Off C-347-4149	**Pol-451-6197**	**Theat-347-6292**

HEALTH & WELFARE: Dispensary/Emerg, EX-6512. Inpatient, Camp Lejeune MCB, Comm: 919-451-4300, ATVN: 484-4300. Chapels serving all faiths, Camp Lejeune MCB, EX-2113.

RECREATION: Arts & Crafts, EX-6711, Auto Hobby, EX-6709, Gym, EX-6436, Library, EX-6942, Bowling, EX-6582, Gear Issue, EX-67387, Ceramics, EX-6711, Wood, EX-6690, Marina, EX-6578, Swimming, EX-6346. Rec Beach Area, on Base, year round, Comm: 919-451-6704, ATVN: 484-6704, 12 camper spaces w/o hookups, all fac and water sport activities.

SPACE-A: Helicopter aircraft here. Try Cherry Pt MCAS, see that listing.

ATTRACTIONS: Great NC beaches. Cape Lookout Natl Seashore easy drive NE.

POPE AIR FORCE BASE (NC01R1)

Comm: 919-394-0111
ATVN: 486-0111

Pope Air Force Base, NC 28308-5000

LOCATION: Take I-95, exit to NC-87/24 W. Signs point the direction to AFB and Fort Bragg. RM: p-73, H/14. NMC: Fayetteville, 12 mi SE.

GENERAL INFORMATION: Military Airlift Command Base, Airlift Center, Tactical Airlift Wing, Aeromedical Evacuation Squadron, and AFRES units.

TEMPORARY MILITARY LODGING: Biltg Ofc, Bldg 235, Ethridge St, 24 hrs dly, Comm: 919-394-4404, ATVN: 486-4404. All ranks. DV/VIP, EX-4739.

LOGISTICAL SUPPORT: Complete spt fac avail. Key phone numbers follow.

BX-497-2111	**Caf-497-6670**	**Chap-394-2676**
Chld C-394-4323	**Commiss-396-2428**	**Conv S-497-2112**
EM C-394-2779	**Fam Svc-394-2119**	**Gas S-497-6615**

Pope AFB, Cont'd

Legal-394-2341	**Loc-394-1110**	**Med-394-2241**
NCO C-497-4051	**Off C-497-4031**	**Pkg S-497-4031**
Pol-394-2801	**Snack B-497-6670**	**Theat-694-2679**

HEALTH & WELFARE: Clinic/Emerg, EX-2327. Inpatient, Fort Bragg, Comm: 919-396-0301/1123, ATVN: 236-0301/1123. Chapels serving all faiths, EX-2111.

RECREATION: Auto Craft, EX-4234, Bowling, EX-2891, Golf, EX-2325, Sport Rental, EX-2135, Hobby Shop, EX-4702, Library, EX-4349, Rec Ctr, EX-2779, Playhs, EX-2679. Fam-Camp, operated by Fort Fisher AFS, on NC-421 at Kure Beach, year round, Comm: 919-458-8251, ATVN: 656-2212, 9 camper spaces w/hookups, dump station, other fac, and beach activities.

SPACE-A: Pax Term and Lounge, Bldg 708, 24 hrs dly, Comm: 919-394-4429, ATVN: 486-4429, flts to CONUS locations. Also, try Charlotte/Douglas Municipal Airport, Comm: 704-399-6363, ATVN: 583-9210, Tactical Airlift Group, ANG, training flts to CONUS locations.

ATTRACTIONS: Raleigh, Durham, Chapel Hill, easy drives N. Myrtle Beach, SC, is 2 1/2 hrs S.

SEYMOUR JOHNSON AIR FORCE BASE (NC11R1)

Comm: 919-736-0000
ATVN: 488-1110

Seymour Johnson Air Force Base, NC 27531-5000

LOCATION: From US-70 take Seymour Johnson AFB exit onto Berkeley Blvd which leads to main gate. Clearly marked. RM: p-73, G/17. NMC: Raleigh, 50 mi W.

GENERAL INFORMATION: Tactical Air Command Base. Tactical Fighter Wing, Air Refueling Group, and other support units.

TEMPORTARY MILITARY LODGING: Biltg Ofc, Bldg 3804, 24 hrs dly, Comm: 736-6705/6168, ATVN: 488-6705/6168.

LOGISTICAL SUPPORT: Complete spt fac avail. Key phone numbers follow.

BX-735-8511	**Chap-736-5211**	**Commiss-736-6198**
Fam Svc-736-5404	**Gas S-734-7358**	**Legal-736-6256**
Loc-736-5584	**Med-736-5577**	**Pol-736-5522**
NCO C-734-2753	**Off C-735-8546**	**Pkg S-734-2948**
Shoppette-735-9364	**Theat-736-6164**	

HEALTH & WELFARE: USAF Hospital, clinics and depts, Emerg, EX-5577, Appoint, EX-5323, 35 bed. Chapels serving all faiths, EX-5211.

RECREATION: Arts and Crafts Hobby Shop, EX-5423, Auto Hobby, EX-5284, Bowling, EX-6669, Golf, EX-6249, Gym, EX-6643, Rod & Gun Club, 736-0786, Libraries, EX-5707, Rec Ctr, EX-6745. Fam-Camp, on Base, year round, Comm: 919-736-6705, ATVN: 488-6705, 8 camper spaces w/hookups, all spt fac and act. Also operates, Rogers Bay AF Family Campgrounds, off Base, 6.5 mi N of Surf City, Mar-Nov, Comm: 919-736-5263, ATVN: 488-5263, 5 mobile homes.

SPACE-A: Bldg 4507, Comm: 919-736-6161, ATVN: 488-6161, flts to CONUS and overseas.

ATTRACTIONS: NC research triangle of Raleigh, Durham and Chapel Hill.

NORTH DAKOTA

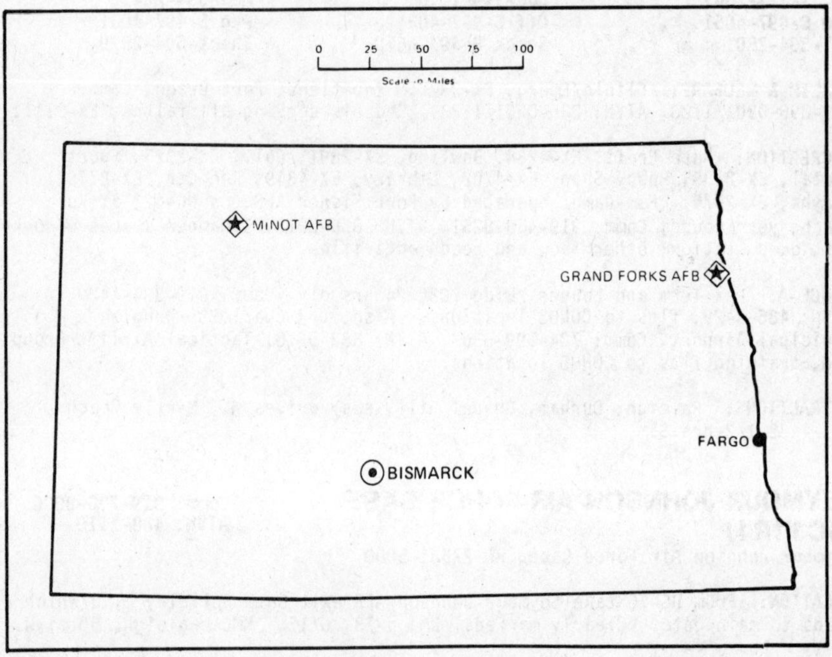

GRAND FORKS AIR FORCE BASE (ND04R3)

Grand Forks Air Force Base, ND 58205-5000
Comm: 701-594-6113
ATVN: 362-1110

LOCATION: From I-29 take US-2 W exit for 14 mi to Grand Forks, County Rd B-3 (Emerado/Air Base) 1 mi to AFB. RM: p-71, D/15. NMC: Grand Forks, 15 mi E.

GENERAL INFORMATION: Strategic Air Command Base. Bomb Wing and Strategic Missile Wing.

TEMPORARY MILITARY LODGING: Biltg Ofc, ATTN: CSG/SVH, Bldg 117, I St & 6th Av, 24 hrs dly, Comm: 701-594-6541, ATVN: 362-6541. All ranks. DV/VIP, EX-6310

LOGISTICAL SUPPORT: Complete spt fac avail. Key phone numbers follow.

BX-594-5542	Caf-594-6729	Chap-594-6235
Chld C-594-3831	Commiss-594-6481	Fam Svc-594-6221
Gas S-594-5951	Legal-594-6301	Loc-594-6761
Med-594-6771	NCO C-594-3897	Off C-594-3896
Pkg S-594-5206	Pol-594-3777	Rec Ctr-594-6527
Snack B-594-5941	Theat-594-5587	

HEALTH & WELFARE: USAF Hospital, clinics and depts, Emer, EX-6771, Appoint, EX-3716, 35 bed. Chapels serving all faiths, EX-6235.

RECREATION: Arts/Crafts, EX-3851, Auto Hobby, EX-3319, Bowling, EX-6655, Gym, EX-6455, Library, EX-6725, Riding Club, EX-4041. Fam-Camp, on Base, May-Sept, 10 camper spaces w/elec & sewer, Comm: 701-594-3703, other fac and act.

SPACE-A: Pax Term and Lounge, Bldg 528, 24 hrs dly, Comm: 701-594-6778,

Grand Forks AFB, Cont'd

ATVN: 362-6778, flts to CONUS and overseas.

ATTRACTIONS: Outdoor sports and recreation. Canada easy drive N via I-29.

MINOT AIR FORCE BASE (ND02R3)

Minot Air Force Base, ND 58705-5000

Comm: 701-727-4761
ATVN: 344-1110

LOCATION: On US-83, N of Minot. RM: p-71, B/7. NMC: Minot, 15 mi S.

GENERAL INFORMATION: Strategic Air Command Base, Air Division, Strategic Missile Wing, Bomb Wing, Fighter Interceptor Squadron.

TEMPORARY MILITARY LODGING: Biltg Ofc, Bldg 173, Summit Dr, 24 hrs dly, Comm: 701-727-2184/6161, ATVN: 344-2184/6161. All ranks. DV/VIP, EX-3818.

LOGISTICAL SUPPORT: Complete spt fac avail. Key phone numbers follow.

BX-727-4717	Caf-727-4625	Chap-727-2111/2456
Chld C-727-3750	Commiss-727-2464	Conv S-727-4973
EM C-727-3806	Fam Svc-727-2471	Gas S-727-4876
Legal-727-3026	Loc-727-2481	Med-727-3999
NCO C-727-6156	Off C-727-3731	Pol-727-3096
Snack B-727-4625	Theat-727-3802	

HEALTH & WELFARE: USAF Hospital, located off Base in Minot, clinics and depts, Emer, EX-5304, Appoints, 727-5307. Chapels serving all faiths, EX-2111.

RECREATION: Arts/Crafts, EX-3640, Auto Hobby, EX-2127, Bowling, EX-2610, Golf, EX-3164, Gym, EX-2174/2124, Sports Rental, EX-2175, Handball, EX-2174, Picnic Area, EX-3164, Library, EX-3344, Rec Ctr, EX-3352, Tennis, EX-2174, Tickets, EX-3352, Outdoor Rec, EX-3840, Swimming, EX-3388 (Summer only). Riverdale Rec Area has been sold.

SPACE-A: Pax Term and Lounge, Bldg 746, 24 hrs dly, Comm: 701-727-3226, ATVN: 344-3226, flts to CONUS and overseas.

ATTRACTIONS: Trestle Valley Ski Area, outdoor recreation. Canada easy drive N via US-83.

- NOTES -

OHIO

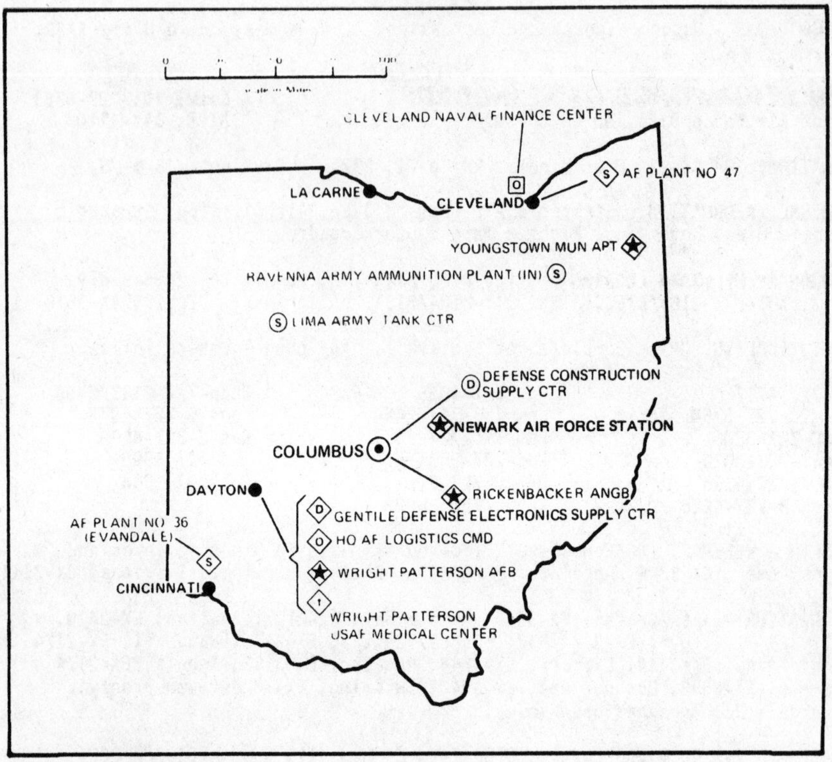

DEFENSE CONSTRUCTION SUPPLY CENTER (OHO5R2)

Columbus, OH 43216-5000

Comm: 614-238-2121
ATVN: 850-3131

LOCATION: From I-270 (Beltway) take exit 39 to Broad Street W, main gate at 3990 Broad Street. Mailing address: PO Box 3990. RM: p-77, SJ/25. NMC: Columbus, in the city.

GENERAL INFORMATION: Logistical Support for construction supplies and equipment for DOD.

TEMPORARY MILITARY LODGING: Ltd facility near Officers' Club, EX-3131.

LOGISTICAL SUPPORT: Very limited. Key phone numbers follow.

PX-231-0976 Caf-238-2121/3131 Loc-238-3211/3131
Off C-239-0482

HEALTH & WELFARE: Rickenbacker ANGB, Medical, Emerg, 492-4542. Chapels serving all faiths, 492-4404.

RECREATION/SPACE-A: See Rickenbacker ANGB listing.

ATTRACTIONS: State Fairgrounds and Ohio Historical Center.

NEWARK AIR FORCE STATION (OHO4R2)

Newark, OH 43057-5000

Comm: 614-522-2171
ATVN: 580-1110

LOCATION: From I-70 exit to OH-79 S to Heath, main entrance from Irving-Wick Drive. RM: p-77, SA/15. NMC: Newark, 3 mi N.

GENERAL INFORMATION: Air Force Logistics Command Base. The Air Force Aerospace Guidance and Metrology Center.

TEMPORARY MILITARY LODGING: None.

LOGISTICAL SUPPORT: Limited spt fac avail. Key phone numbers follow.

BX-522-7213	Caf-522-7666	Cons C-522-7239
Loc-522-7220/7281	Med-522-7333	Pol-522-7200

HEALTH & WELFARE: Dispensary/Emerg, EX-7333/117. Inpatient, USAF Hosp, Wright-Patterson AFB, Dayton, Comm: 513-257-2968, ATVN: 787-2968. No chapel.

RECREATION: Jogging, Track, Handball/Racquetball, EX-7220.

SPACE-A: None. Try Mansfield Lahm Airport (ANG Base). Pax Term, Bldg 101, 0745-1630 M-F, Comm: 419-522-9355, ATVN: 696-6124, flts via C-130 aircraft to CONUS locations.

ATTRACTIONS: Heisey Museum, Moundbuilders and Flint Ridge Parks, Buckeye Lake, Denison and Newark Campus of Ohio St Univ. 1/2 hr drive to Columbus.

RICKENBACKER AIR NATIONAL GUARD BASE (OHO2R2)

Comm: 614-492-8211
ATVN: 950-1110

Rickenbacker Air National Guard Base, OH 43217-5000

LOCATION: From I-270 take Alum Creek exit S to ANGB. Also, accessible off US-23 onto OH-317. Clearly marked. RM: p-77, SC/12. NMC: Columbus, 13 mi NW.

GENERAL INFORMATION: Air National Guard Base. Tactical Fighter Wing, Tactical Airlift Group, Air Refueling Group, Naval and AFRES units.

TEMPORARY MILITARY LODGING: Biltg Ofc, Bldg 92, 24 hrs dly, Comm: 614-492-4545/6, ATVN: 950-4545/6. All ranks. DV/VIP, EX-8211/4545.

LOGISTICAL SUPPORT: Complete spt fac avail. Key phone numbers follow.

BX-492-4603	Loc-492-8211/1110	Pkg S-492-4473
Pol-492-4727		

HEALTH & WELFARE: Dispensary/Emerg, EX-4542. Inpatient, USAF Hosp, Wright-Patterson AFB, Comm: 513-257-2968, ATVN: 787-2968. Chapels serving all faiths, EX-4404, Unit Training Assembly, weekends only.

RECREATION: Gym, EX-4158, racquetball, sauna, track, weights.

SPACE-A: Pax Term and Lounge, Bldg 500, 1st floor, 0730-2330 Tues-Fri, 0730-1930 Sat, 0800-1600 Sun-Mon, Comm: 614-492-4595, ATVN: 950-4595, flts to CONUS and overseas.

ATTRACTIONS: Columbus, OH capital, 13 mi NW. Columbus Clippers (New York Yankee Farm Club, baseball). Ohio State University, state capital, DOD Aerial Spray Branch, C-130A aircraft at Rickenbacker.

WRIGHT-PATTERSON AIR FORCE BASE (OH01R2)

Comm: 513-257-1110
ATVN: 787-1110(A & C)
ATVN: 785-1110(B)

Wright-Patterson Air Force Base, OH 45433-5000

LOCATION: South of I-70, off I-675 at Fairborn. Also, access from OH-4. AFB clearly marked. RM: p-77, SN/12. NMC: Dayton, 10 mi SW.

GENERAL INFORMATION: Air Force Logistics Command Base. Test Wing, Foreign Technology Div, Air Force Institute of Technology, USAF Medical Center, and Aeronautical Systems Division.

TEMPORARY MILITARY LODGING: Biltg Ofc, ATTN: 2750 ABW/SVH, Bldg 825, 24 hrs dly, Comm: 513-257-3810, ATVN: 787-3451. All ranks. DV/VIP, EX-3810.

LOGISTICAL SUPPORT: Complete spt fac avail. Key phone numbers follow.

BX-879-5730	Chap-257-2909	Chld C-257-2173
Commiss-257-4601	Conv S-878-2103	Day Care Ctr-257-3107
Fam Svc-257-6934	Gas S-257-3760	Legal-257-6142
Loc-257-3231	Med-257-2968	NCO C-257-3767
Off C-257-2216	Pol-257-6516	Theat-257-4697 after 7 pm only

HEALTH & WELFARE: USAF Medical Center, Emerg 257-2968, Appoint, call Med Ctr Info, 257-3541, 285 beds. Chapels serving all faiths, EX-2909.

RECREATION: Arts & Crafts Store, EX-77025, Auto Hobby, EX-74937, Bowling, EX-77796, Gym, EX-74225, 55175, 73607, Sports Rental, EX-74222, Aero Club, EX-77950, Fishing/Hunting, EX-72001, Library, EX-74815, Swimming, EX-74225/72977/74254, Rec Ctr, EX-72001, Tennis, EX-77248, Youth Ctr, EX-55053, Picnic Area, EX-72001, Riding Stables, 253-3810. Fam-Camp, on Base, year round, Comm: 513-257-2001, ATVN: 787-2001, 16 camper spaces w/hookups (no water in winter), other fac and act.

SPACE-A: Pax Term and Lounge, Bldg 142, Area C, Skeel Ave, 0530-2000 M-F, other dys, 0800-1600, Comm: 513-257-6235/7741, ATVN: 787-6235/7741, flts to CONUS locations. Also, flts to CONUS locations from Youngstown Municipal Airport, AFRES units, Comm: 216-856-1645, ATVN: 346-9211.

ATTRACTIONS: Air Force Museum on Base. Wright Brothers Mem'l, Dayton. Other aviation attractions in Dayton area, contact Dayton Ch of Commerce, 226-1444, or Dayton Convention Bureau.

- NOTES -

OKLAHOMA

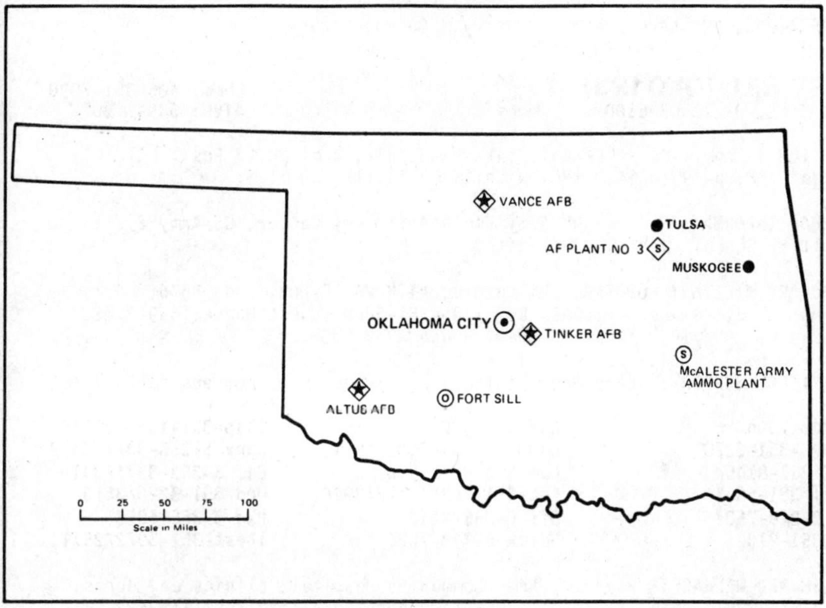

ALTUS AIR FORCE BASE (OK02R3)

Altus Air Force Base, OK 73523-5000

Comm: 405-481-8100
ATVN: 866-1110

LOCATION: Off US-62 S of I-40 and W of I-44. From US-62 traveling W from Lawton, turn R at 1st traffic light in Altus and follow road to main gate NE of Falcon Rd. RM: p-79, J/12. NMC: Lawton, 56 mi E.

GENERAL INFORMATION: Military Airlift Command Base. 443 MAW training units and support activities. 340 Air Refueling Wing/SAC tenant unit.

TEMPORARY MILITARY LODGING: Biltg Ofc, Bldg 82, 24 hrs dly, Comm: 405-481-7356/7556, ATVN: 866-7356. All ranks. DV/VIP, EX-7554.

LOGISTICAL SUPPORT: Complete spt act avail. Key phone numbers follow.

BX-481-8733	Chap-481-7485	Commiss-481-7533
Conv S-481-7095	Fam Svc-481-7460	Gas S-481-7002
Legal-481-7294	Loc-481-6402	Med-481-5222/6224
Nurse-481-5293	NCO C-481-7034	Off C-481-6780/6224
Pkg S-481-6276	SP's-481-7444/7709	Theat-481-7341

HEALTH & WELFARE: USAF Hospital, clinics and depts, Emerg EX-5222/6666, Appoint EX-5970, 30 bed. Chapels serving all faiths, EX-7485.

RECREATION: Arts & Crafts, EX-7048, Auto Hobby, EX-6326, Bowling, EX-6420, Golf, EX-7207, Gym, EX-7440, Library, EX-6302, Base Pool, EX-6377, O Club Pool, EX-6281. Fam-Camp, on Base, year round, Comm: 405-482-6420, ATVN: 866-6420, 4 camper spaces w/hookups, 3 camper spaces w/o hookups, full fac and act.

SPACE-A: Pax Term and Lounges, Bldg 185, 0730-1630 M-F, Comm: 405-481-

Altus AFB, Cont'd

6350/6428, ATVN: 866-6350/6428, flts to CONUS and via KC-135 overseas.

ATTRACTIONS: Western prairie country. Outdoor sports.

FORT SILL (OK01R3)

Fort Sill, OK 73503-5100

Comm: 405-351-7090
ATVN: 639-7090

LOCATION: From I-44 at Lawton, take US-62/277, 4 mi NW to Post. Clearly marked. RM: p-79, J/14. NMC: Wichita Falls, TX, 50 mi S.

GENERAL INFORMATION: Hq, US Army Field Artillery Center, US Army Field Artillery School, and support units.

TEMPORARY MILITARY LODGING: Biltg Ofc, ATTN: ATZR-EHB, Bldg 5676, Fergusson Rd, 24 hrs dly, BAQ, BOQ: 405-351-5000, Guest House: 639-3762. All Ranks. DV/VIP, EX-5511. Guest House, 353-5007.

LOGISTICAL SUPPORT: Complete spt fac avail. Key phone numbers follow.

PX-248-7504	**Caf-353-6209**	**Chap-351-1393-2701**
Chld C-351-2320	**Commiss-351-3601**	**Conv S-353-3371-3419**
EM C-357-8145	**Fam Svc-351-5018**	**Gas S-353-3371/3419**
Legal-351-6961	**Loc-351-3639/6172/3924**	**Med-351-3226/3616**
NCO C-248-7604	**Off C-355-9112**	**Pkg S-355-4053**
Pol-351-2102	**Snack B-248-7025**	**Theat-351-3972/2571**

HEALTH AND WELFARE: Reynolds Army Community Hospital, clinics and depts, Emerg EX-3226/7/3616, 214 beds. Chaplain serving all faiths, EX-2423.

RECREATIONS: Arts/Crafts, EX-5687, Auto Hobby, EX-1347, Bowling, EX-2882, ITT, EX-6211, Golf, EX-2723, Gym, EX-3781/4670, Outdoor Equip Checkout Ctr, EX-3342, Library, EX-1340/4477/3021, Rec Ctr, EX-6211, Youth Ctr, Ex-3065. Lake Elmer Thomas Rec Area, on Post, year round, Comm: 405-351- 2025, ATVN: 639-2025, 14 camper spaces w/hookups, 20 camper spaces w/o hookups, complete fac and all act. Also operates Possum Kingdom Rec Area, 75 mi W of Fort Worth, Comm: 405-351-2025, ATVN: 639-2025, 4 camper spaces w/hookups, 20 family units, 16 bachelor units, all fac and act on reservoir.

SPACE-A: Henry Post Army Airfield, flts via exec aircraft to CONUS locations, EX-3508. Also, Will Rogers ANG Base, Oklahoma City, Bldg 1011, 0630-1700 M-Thurs, Comm: 405-686-5251, ATVN: 956-8251, flts to CONUS location.

ATTRACTIONS: Lawton, third largest city in OK.

McALESTER ARMY AMMUNITION PLANT (OK09R3)

McAlester, OK 74501-5000

Comm: 918-421-2539
ATVN: 956-6539

LOCATION: Off OK-31, W of US-69. This SE Oklahoma area is mostly rolling pasture with timber-covered hills and creek bottoms. RM: p-79, I/21. NMC: Tulsa, 90 mi N.

GENERAL INFORMATION: Receives, stores, renovates and issues ammunitions, explosives and ordnance items to the Dept of the Army.

TEMPORARY MILITARY LODGING: Limited.

LOGISTICAL SUPPORT: PX and Commiss-421-2388; limited support facilities, full range of support at Tinker AFB, 116 mi NW.

McAlester Army Ammunition Plant, Cont'd

HEALTH & WELFARE: Dispensary, 421-2496. Inpatient, USAF Hosp, Tinker AFB, Oklahoma City, 405-734-8249.

RECREATION: Arts & Crafts, Auto Shop, Bowling, Flying Club, Intermural Sports, Golf, Gymnasium, Sports Rental, Library, Pool, Rec Center, Theater, Youth Services, Tennis, Boat Rental. Murphy's Meadow Campgrounds, on Post, on 625 acre Brown Lake, Jan-Sept, 2 camper spaces w/hookups, 15 camper spaces w/o hookups, Comm: 918-421-2780, ATVN: 956-2780.

SPACE-A: Avail out of Tinker AFB, 116 mi NW. PAX Term, bldg 244, Comm: 405-739-4360, ATVN: 735-4360.

ATTRACTIONS: Lake Eufaula.

TINKER AIR FORCE BASE (OK04R3)

Tinker Air Force Base, OK 73145-5000

Comm: 405-732-7321
ATVN: 884-1110

LOCATION: SE Oklahoma City, off I-40. Use gate 1 off Air Depot Blvd. Clearly marked. RM: p-79, 0/6. NMC: Oklahoma City, 12 mi NW.

GENERAL INFORMATION: Air Force Logistics Command Base. Major Logistics Center for the Air Force. Also AFRES units here.

TEMPORARY MILITARY LODGING: Biltg Ofc, ATTN: 2854 ABG/SVH, Bldg 5604, 24 hrs dly, Comm: 405-734-5095, ATVN: 884-5059. Billeting in-dial: 737-1475. All ranks. DV/VIP, 734-5511.

LOGISTICAL SUPPORT: Complete spt fac avail. Key phone numbers follow.

BX-734-3035	**Caf-734-3161**	**Chap-734-2111**
Chld C-734-3116	**Commiss-734-5085**	**Conv S-733-3445**
Fam Svc-739-2505	**Gas S-734-3040**	**Legal-739-5811**
Loc-734-2456	**NCO C-734-3435**	**Off C-734-3418**
Pkg S-734-5617	**Pol-734-3737/2000**	**Rec Ctr-734-2077**
Snack B-734-3486	**Theat-734-3400**	

(ATVN equivalants: 734 is 884, 736 is 336, and 739 is 339)

HEALTH & WELFARE: USAF Hospital, clinics and depts, Emerg, 734-8249, Appoint, 734-8366, 30 bed. Chapels serving all faiths, 734-2111.

RECREATION: Hobby Shop, EX-47234, Auto Hobby, EX-45616, Bowling, EX-43484, Golf, EX-42909, Gym, EX-43651, Handball, EX-43651, Boat Rental, EX-42289, Fishing, EX-42077, Library, EX-43083, Rec Ctr, EX-42077, Swimming, EX-43651, Riding Club, 737-4775, Rod/Gun Club, EX-42761, Youth Ctr, EX 47866. Fam-Camp, on Base, year round, Comm: 405-734-2077, ATVN: 884-2077, 29 camper spaces w/hookups, other fac and activities. Also, Murphy's Meadow Campgrounds, McAlester Army Ammunition Plant, on Post, Jan-Sept, Comm: 918-421-2780/2371, ATVN: 956-2780/2371, 2 camper spaces w/hookups, 15 camper spaces w/o hookups.

SPACE-A: Pax Term and Lounge, Bldg 244, 0715-2400 M-F, other days, 0800-1600, Comm: 405-739-4360/4339, ATVN: 339-4360/4339, flts to CONUS locations.

ATTRACTIONS: Oklahoma City, state capital. Kirkpatrick Planetarium, Oklahoma Air Space Museum, Omniplex Science Museum, National Cowboy Hall of Fame and Western Heritage Center, Oklahoma City Zoo, arts and entertainment.

VANCE AIR FORCE BASE (OK05R3)

Vance Air Force Base, OK 73705-5000

Comm: 405-237-2121
ATVN: 962-7110

Vance AFB, Cont'd

LOCATION: Off US-81 S of Enid. Clearly marked. RM: p-79, D/15. NMC: Oklahoma City, 90 mi SE.

GENERAL INFORMATION: Air Training Command Base. Flying Training Wing, Undergraduate Pilot Training.

TEMPORARY MILITARY LODGING: Biltg Ofc, Bldg 705, Fisher Rd, 24 hrs dly, Comm: 405-249-7358/7724, ATVN: 962-7358/7724. All ranks. DV/VIP, 249-7102.

LOGISTICAL SUPPORT: Complete spt fac avail. Key phone numbers follow.

BX-237-6765	**Chap-249-7211**	**Chld C-249-7310**
Commiss-249-7428	**Conv S-237-6765**	**Fam Svc-249-7322**
Gas S-237-7445	**Legal-249-7404**	**Loc-249-7791/7358**
Med-249-7416	**NCO C-237-7311**	**Off C-237-2326**
Pkg S-237-1276	**Pol-249-7200**	**Theat-249-7738**

HEALTH & WELFARE: Dispensary/Clinic, Emerg, 249-7200, Appoint, 249-7416. Inpatient, USAF Hosp, Tinker AFB, Oklahoma City, 405-734-8249. Chapels serving all faiths, 249-7211.

RECREATION: Arts/Crafts, 249-7402, Auto Hobby, 249-7508, Bowling, 249-7331, Gym, 249-7830/7670, Rec Ctr, 249-7251, Lib, 249-7368, Yth Ctr, 249-7474.

SPACE-A: Limited Space-A, Base Ops, Comm: 405-249-7424, ATVN: 962-7424.

ATTRACTIONS: Oklahoma City, capital, 90 mi SE. National Cowboy Hall of Fame, Kirkpatrich Planetarium, National Softball Hall of Fame, OKC Zoo and Frontier City.

OREGON

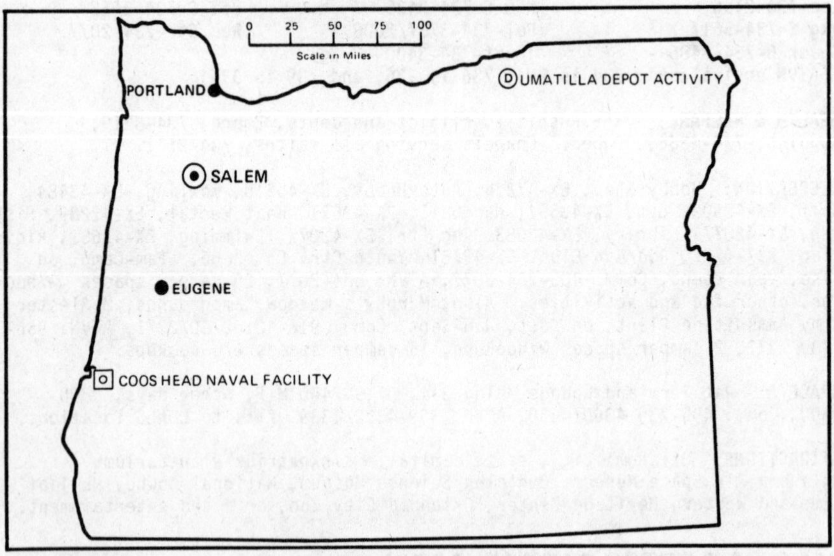

COOS HEAD NAVAL FACILITY (OR01R4)

Charleston, OR 97420-0630

Comm: 503-888-3221
ATVN: 355-2011

Coos Head NF, Cont'd

LOCATION: From US-101 N exit to Charleston and state parks, follow signs to Coos Bay, 8 mi E. RM: p-81, K/2. NMC: Eugene, 100 mi NE.

GENERAL INFORMATION: Oceanographic Research.

TEMPORARY MILITARY LODGING: Biltg Ofc, Bldg 1, 0715-1530 hrs dly, Comm: 503-888-3221. Officer billeting only (active and retired).

LOGISTICAL SUPPORT: Limited spt fac avail. Key phone numbers follow.

BX-882-3221-EX-224 **EM C-882-3221-EX-242** **Med-888-3221-EX-231**
Pol-888-3221

HEALTH & WELFARE: Dispensary/Emerg EX-231. Inpatient, Madigan Army Med Ctr, Tacoma, WA, 300 mi N, 206-967-6377. No chapel facilities.

RECREATION: Gym, two 16 ft Boston Whalers, 1 bass boat, 4 canoes, EX-246/235.

SPACE-A: None.

ATTRACTIONS: Wonderful Oregon coast and outdoor sports.

PENNSYLVANIA

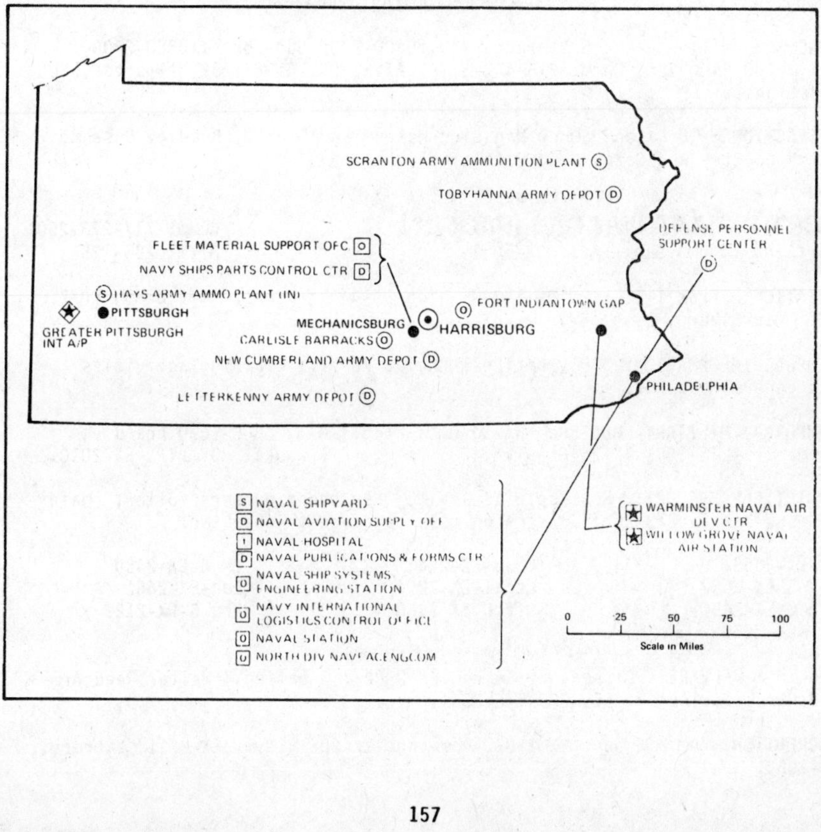

CARLISLE BARRACKS (PA08R1)

Carlisle Barracks, PA 17013-5002

Comm: 717-245-3131
ATVN: 242-4141

LOCATION: From I-81 exit 16 to US-11, 2 mi SW to Carlisle, signs clearly marked to Barracks and Army War College. RM: p-85, ES/3. NMC: Harrisburg, 18 mi NE.

GENERAL INFORMATION: Army War College and Army Military History Institute.

TEMPORARY MILITARY LODGING: Biltg Ofc, ATTN: ATZE-DIS, Bldg 7, Ashburn, 0700-1800 hrs wkdys, 0900-1700 other days, Comm: 717-245-3817/4245, ATVN: 242-3817/4245. All ranks. DV/VIP, EX-4818.

LOGISTICAL SUPPORT: Complete spt fac avail. Key phone numbers follow.

PX-243-2463	Chap-245-3318	Chld C-245-3701
Commiss-245-3409	EM C-245-4961	Fam Svc-245-4357
Gas S-243-2666	Legal-245-4940	Loc-245-3131
Med-245-3915/3400	NCO C-245-4961	Off C-245-3215
Pkg S-245-4325	Pol-245-4315	Theat-245-4108

HEALTH & WELFARE: Army Health Clinic, Emerg EX-3400. Inpatient, Walter Reed Army Medical Center, Washington, DC, Comm: 202-576-1199, ATVN: 291-1199. Chapels serving all faiths, EX-3318.

RECREATION: Art Center, 245-3319, Morale Spt Ofc, 245-4343, Golf, 243-3262, Bowling, 245-4375, New Gym, 245-3418, Old Gym, 245-3418, Woodworking Shop, 245-3466, Library, 245-3718, Swimming, 245-3560.

SPACE-A: Willow Grove NAS, Hangar 80, 0800-2300 Sun-Thurs, 0800-2400 Fri, 0700-2400 Sat, Comm: 215-443-6196, ATVN: 991-6196/1252, flts to CONUS locations.

ATTRACTIONS: Hessian Powder Magazine Museum and Omar N. Bradley Museum; Amish Country and Gettysburg Battlefield nearby.

FORT INDIANTOWN GAP (PA04R1)

Fort Indiantown Gap, PA 17003-5011

Comm: 717-273-2601
ATVN: 235-2252

LOCATION: From I-81 take exit 29W, N on PA-443 to Post. RM: p-85, EQ/7. NMC: Harrisburg, 20 mi SW.

GENERAL INFORMATION: Army National Guard, Reserve training activities and general support.

TEMPORARY MILITARY LODGING: Biltg Ofc, Bldg T-0-1, 0800-1630 hrs dly, Comm: 717-273-2601 EX-2540, ATVN: 235-2540. All ranks. DV/VIP, EX-2010.

LOGISTICAL SUPPORT: Most spt fac avail. Key phone numbers follow. Main Post number is 273-2601, extension must be given to operator.

PX-EX-2058	Chap-EX-2241	EM C-EX-2450
Gas S-EX-2032	Legal-EX-2294	Loc-EX-2601
NCO C-EX-2450	Off C-EX-2460	Pkg S-EX-2129
Pol-EX-2160		

HEALTH & WELFARE: Dispensary/Emerg, EX-2888. Inpatient, Walter Reed Army Med Ctr, Washington, DC, 202-576-3501.

RECREATION: Auto Hobby, EX-2702, Bowling, EX-2061, Gym, EX-2711, Library, EX-2063.

Fort Indiantown Gap, Cont'd

SPACE-A: Muir Army Airfield, occasional transient aircraft to CONUS locations, EX-2151/2813.

ATTRACTIONS: Hershey Amusement Park, 15 mi SW, Lebanon and origin of Lebanon Bologna, 14 mi S.

LETTERKENNY ARMY DEPOT (PA03R1)

Chambersburg, PA 17201-4150

Comm: 717-267-9111
ATVN: 570-9111

LOCATION: From I-81 exit 8 W to PA-997 to PA-233 left and enter depot at gate 6. RM: p-83, WT/17. NMC: Harrisburg, 45 mi NE.

GENERAL INFORMATION: Army Depot that maintains Army materiel, ammunition and general supplies.

TEMPORARY MILITARY LODGING: Biltg Ofc, ATTN: SDSLE-S, Bldg 500, 0800-1600 hrs dly, Comm: 717-267-8890, ATVN: 570-8890. All ranks. DV/VIP, EX-8659.

LOGISTICAL SUPPORT: Most spt fac avail. Key phone numbers follow.

PX-264-1713	ACS-267-9051	Caf-264-1521
Chap-267-8628	Chld C-267-8846	Legal-267-9889
Loc-264-1413	Med-267-8600	NCO C-267-9238
Off C-264-1413	Pkg S-264-1113	Pol-267-8800
Snack B-264-1931	Theat-267-9096	

HEALTH & WELFARE: Dispensary/Emerg EX-17. Inpatient, Walter Reed Army Med Ctr, Washington, DC, 202-576-3501. Chapels serving all faiths, EX-8628.

RECREATION: Hobby Shop, EX-8675, Auto Hobby, EX-8649, Gym, EX-8666, Lib, EX-6875, Swimming, EX-8956. Army Travel Camp, on Post, Apr-Oct, Comm: 717-267-9600, ATVN: 570-9600, 6 camper spaces w/o hookups, other fac and activities.

SPACE-A: Warfield ANGB, Glenn L. Martin State Aprt, Baltimore, Bldg 3198, 0600-2400 hrs dly, Comm: 301-981-2740/2744, ATVN: 858-2740/2744, flts to CONUS locations. Also, try Greater Pittsburgh IAP, AFRES units, Comm: 412-269-8300, ATVN: 277-8300, ANG and Comm: 412-269-8000, ATVN: 277-8000, AFRES, training flts to CONUS and overseas.

ATTRACTIONS: Gettysburg Battlefield.

MECHANICSBURG NAVY SHIPS PARTS CONTROL CENTER (PA07R1)

Mechanicsburg, PA 17055-0788

Comm: 717-790-2900
ATVN: 430-2900

LOCATION: From I-83 exit 20 to US-11, 4 mi. Or from PA Turnpike (I-76), exit 16 (Carlisle), 8 mi to Ctr. RM: p-75, EY/1. NMC: Harrisburg, 10 mi NE.

GENERAL INFORMATION: Logistical Activity for Naval ships parts.

TEMPORARY MILITARY LODGING: Navy Lodge, 0800-1600 hrs dly, Comm: 717-790-2608, ATVN: 430-2608. All ranks.

LOGISTICAL SUPPORT: Most spt fac avail. Key phone numbers follow.

NX-790-2608	Caf-790-3537	Nur-790-5683
Conv S-790-2608	CPO C-790-3104	Loc-790-2000
Med-790-3461	Off C-790-3505	Pkg S-790-2055
Pol-790-3351		

Mechanicsburg NSPCC, Cont'd

HEALTH & WELFARE: Dispensary/Emerg, EX-3461. Inpatient, Naval Hosp, Bethesda, MD, 202-295-2206. No chapel facilities.

RECREATION: Limited Base facilities, see Carlisle Barracks. One 3 bdrm mobile home reserved at Solomons Navy Recreation Area, Comm: 717-790-2720, ATVN: 430-2720.

ATTRACTIONS: Gettysburg Battlefield.

NEW CUMBERLAND ARMY DEPOT (PA06R1)

New Cumberland, PA 17070-5001

Comm: 717-782-6011
ATVN: 977-6011
FTS: 589-6011

LOCATION: From I-83 take exit 18 to PA-114, E for 1 mi to Old York Road, left .75 mi to Ross Ave, right for 1 mi to main gate of Depot. RM: p-85, EZ/4. NMC: Harrisburg, 7 mi NE.

GENERAL INFORMATION: Army Depot Activity.

TEMPORARY MILITARY LODGING: Biltg Ofc, ATTN: SDSNC-AH, Bldg 268, 0700-1530 hrs dly, Comm: 717-782-7035, ATVN: 977-7035. All ranks. DV/VIP, EX-7192.

LOGISTICAL SUPPORT: Complete spt fac avail. Key phone numbers follow.

PX-774-4482	Caf-782-7165/6973	Chap-782-6111
Chld C-782-7360	Commiss-782-6540	Comm Svc-782-6203/7737
Conv S-774-4482/4522	Gas S-774-4066	Legal-782-7971/6310
Loc-782-6011	Med-782-7281	NCO C-774-4240
Off C-774-4894	Pkg S-774-3623	Pol-782-6222
Snack B-774-4680	Theat-782-6765	

HEALTH & WELFARE: Dispensary/Emerg EX-6333. Inpatient, Walter Reed Army Med Ctr, Washington, DC, 202-576-3501. Chapels serving all faiths, EX-6111/7176.

RECREATION: Auto Hobby, EX-6664, Golf, 774-4686, Gym, EX-6473, Sports Rental, EX-7625, Library, EX-7121, Swimming, EX-6476, Rec Ctr, EX-7625, Youth Ctr, EX-6768.

SPACE-A: Warfield ANGB, Glenn L. Martin State Arprt, Baltimore MD, Bldg 3198, 0600-2400 hrs dly, Comm: 301-981-2740/2744, ATVN: 858-2740/2744, flts to CONUS locations via ANG and AFRES units.

ATTRACTIONS: Hershey Park, Gettysburg Battlefield, Dutch Country.

PHILADELPHIA NAVAL BASE (PA09R1)

Philadelphia, PA 19112-5098

Comm: 215-465-1000
ATVN: 443-1000

LOCATION: From I-95 N exit at Broad Street, Base gate is 200 yds S of exit. RM: p-87, I/6. NMC: Philadelphia, within city limits.

GENERAL INFORMATION: Base includes Naval Shipyard, Naval Station, Naval Hospital, Naval training and reserve units.

TEMPORARY MILITARY LODGING: Biltg Ofc, Bldg 889, Patrol Rd & 2nd St E, 24 hrs dly, Comm: 215-897-5160/1/5107/3090, ATVN: 443-5160/1/5107/3090. All ranks. DV/VIP, EX-2176/7622.

LOGISTICAL SUPPORT: Complete spt fac avail. Key phone numbers follow.

Philadelphia NB, Cont'd

NX-897-5044	Caf-952-8340	Chap-897-5103
Chld C-897-5288	Commiss-897-6675	Conv S-952-8312
CPO C-897-5274	EM C-897-5286	Fam Svc-897-5130
Gas S-952-8361	Legal-897-6601	Loc-897-5550
Med-897-6836	Off C-467-1810	Pkg S-467-3042
Pol-897-3740	Snack B-952-6847	Theat-897-5270

HEALTH & WELFARE: Naval Hospital, clinics and depts, Emerg, EX-6836, Appoint, EX-8220, 123 bed. Chapels serving all faiths, EX-3054/2896.

RECREATION: Archery Range, Athletic Field, Picnic Area, Bowling, Golf Driving Range, Gym, Hobby Shop, Library, Pistol Range, Rec Area (Camp Spearhead), Skeet Range, Spc Svc, Swimming, EX-1000.

SPACE-A: Philadelphia IAP, Concourse D, International Terminal, 24 hrs dly, Comm: 215-897-5630/5641/2/4, Recording: 215-897-5645, flts via commercial type aircraft to CONUS and overseas.

ATTRACTIONS: Liberty Bell, many historic exhibits and museums.

TOBYHANNA ARMY DEPOT (PA05R1)

Tobyhanna, PA 18466-5000

Comm: 717-894-7000
ATVN: 795-7110
FTS: 590-7000

LOCATION: I-80 E or W to I-380 N, exit 7 to Depot. RM: p-85, EK/13. NMC: Scranton, 24 mi NW.

GENERAL INFORMATION: Army Communications-Electronics Depot.

TEMPORARY MILITARY LODGING: Biltg Ofc, ATTN: SDSTO-AH, Bldg 11, Admin Loop, 0800-1630 hrs dly, Comm: 717-894-7647, ATVN: 795-7647. All ranks. DV/VIP, EX-6223.

LOGISTICAL SUPPORT: Complete spt fac avail. Key phone numbers follow.

PX-894-8383	Caf-894-7998	Chap-894-7319
Commiss-894-7628	Fam Svc-894-7970	Gas S-894-7401
Legal-894-7210	Loc-894-7000	Med-894-7282
NCO C-894-7045	Off C-894-7969	Pkg S-894-8400
Theat-894-7650		

HEALTH & WELFARE: Dispensary/Emerg EX-7282. Inpatient, Walson Army Comm Hosp, Fort Dix, NJ, 609-562-2695. Chapel serving all faiths, EX-7319.

RECREATION: MWR Ofc, EX-7584, Arts & Crafts, EX-7584, Auto Hobby, EX-7584, Sports Rental, EX-7092, Library, EX-7316, Youth Act, EX-7584.

SPACE-A: Depot Army Airfield, infrequent flts via Army exec aircraft, EX-7000. Also, McGuire AFB, NJ, Pax Term and Lounge, Bldg 1706, 24 hrs dly, Comm: 609-724-3078/1706, ATVN: 440-3078/1706, flts to CONUS and overseas.

ATTRACTIONS: Scranton and Wilkes-Barre close by, several ski areas, and Pocono Mountains.

WILLOW GROVE NAVAL AIR STATION (PA01R1)

Willow Grove, PA 19090-5010

Comm: 215-443-1000
ATVN: 991-1000

LOCATION: Take PA Turnpike (I-276), exit 27 N to PA-611, 5 mi to NAS. RM: p-85, ES/15. NMC: Philadelphia, 21 mi S.

Willow Grove NAS, Cont'd

GENERAL INFORMATION: Naval Air Reserve Unit training.

TEMPORARY MILITARY LODGING: Biltg Ofc, Bks No. 5 and BOQ, 24 hrs dly, Comm: 215-443-6059/6060/6076, ATVN: 991-6059/6076/6060. All ranks. DV/VIP, EX-6060/6076.

LOGISTICAL SUPPORT: Most spt fac available. Key phone numbers follow.

NX-443-6029/9031	**Caf-443-6282**	**Chap-443-6002**
Chld C-443-6080	**Conv S-443-6026**	**CPO C-443-6082**
EM C-443-6082	**Gas S-443-6025**	**Legal-443-6056**
Loc-443-6854	**Med-443-6373**	**Off C-443-6081**
Pkg S-443-6078	**Pol-443-6067**	**Theat-443-6082**

HEALTH & WELFARE: Dispensary/Emer EX-1600, Appoint, EX-6373/75. Inpatient, Philadelphia Naval Hosp, 215-465-6836. Chapel serving all faiths, EX-6002/3/4.

RECREATION: Arts & Crafts, EX-6082, Gym, EX-6082, Sports Rental, EX-6086, Swimming, EX-6082, Rec Ctr, EX-6082/6084.

SPACE-A: Pax Term/Lounge, Hangar 80, 0800-2300 Sun-Thurs, 0800-2400 Fri, 0700-2400 Sat, Comm: 215-443-6196, ATVN: 991-6196, flts to CONUS and overseas.

ATTRACTIONS: Close to NJ beaches and ski areas and lakes of Pocono Mountain area of PA.

- NOTES -

RHODE ISLAND

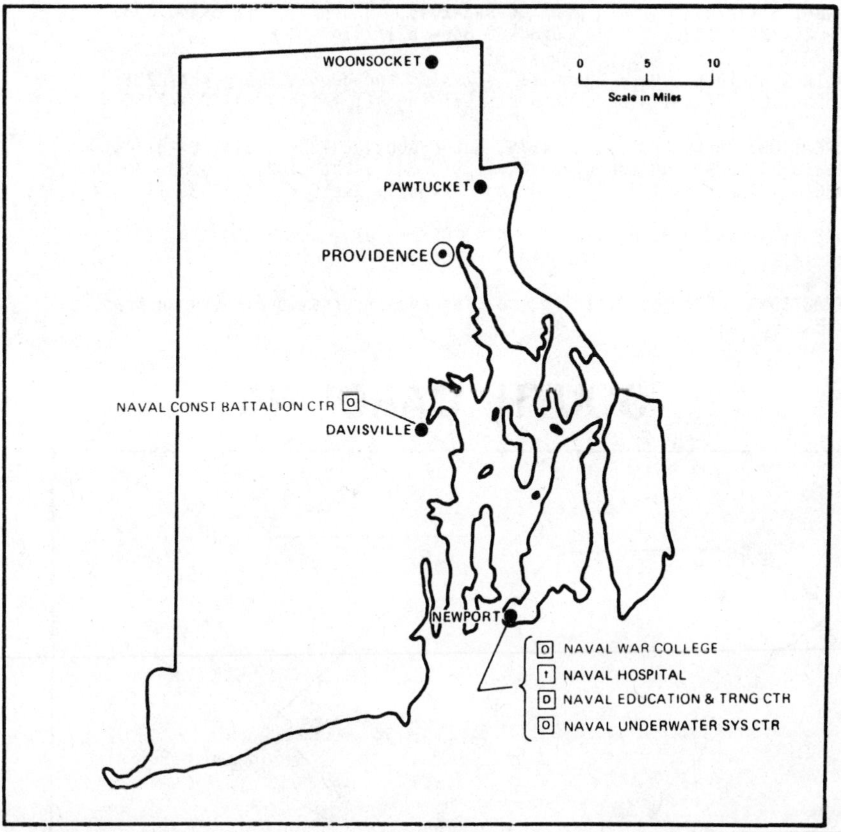

NEWPORT NAVAL EDUCATION & TRAINING CENTER (RI01R1)

Comm: 401-841-2311
ATVN: 948-1110

Newport, RI 02841-5000

LOCATION: From US-1 exit to RI-138 E over Jamestown/Newport bridge (toll) to Newport. Follow signs to Nav Base, gate No 1. RM: p-19, H/24. NMC: Newport, 2 mi S.

GENERAL INFORMATION: Naval War College, Surface Warfare Officers School, Naval Justice School, Naval Hospital, homeport for 7 ships, other training and support activities.

TEMPORARY MILITARY LODGING: Biltg Ofc, Bldg 684 (Officer), Bldg 447 (Enlisted), both 24 hrs dly, Comm: 401-841-3157 (Officer), EX-4410 (Enlisted), ATVN: 948-3157/4410. All ranks. DV/VIP, EX-3456.

LOGISTICAL SUPPORT: Complete spt fac avail. Key phone numbers follow.

NX-846-2555	**Caf-841-3994**	**Chap-841-3456**
Chld C-841-4562	**Commiss-841-2111**	**Conv S-846-2831**

Newport NE & TC, Cont'd

CPO C-841-3877 EM C-841-3994 Fam Svc-841-2283
Gas S-846-8515 Legal-841-3766 Loc-841-4001
Med-841-3111/2222 Off C-846-2515 Pkg S-846-3080
Pol-841-3241/3242 Snack B-846-7671/7814

HEALTH & WELFARE: Naval Hospital, clinics and depts, Emerg, EX-2222, Appoint, EX-3111, 128 bed. Chapels serving all faiths, EX-2234/3456.

RECREATION: Arts/Crafts, EX-4579, Auto Hobby, EX-3026, Gym, EX-3154, Sprts Rental, EX-2568, Rod/Gun Club, EX-1729/3654, Yacht Club, EX-3283, Beach, EX-2314, Library, EX-3044, Swimming, EX-3957, Rec Ctr, EX-4293, ITT, EX-3116.

SPACE-A: Pease AFB, Bldg 237, 0800-1700 hrs dly, Comm: 603-430-2103, ATVN: 852-2103, flts to CONUS and overseas.

ATTRACTIONS: Beaches, fishing, boating and internationally-known Newport Music Festival.

SOUTH CAROLINA

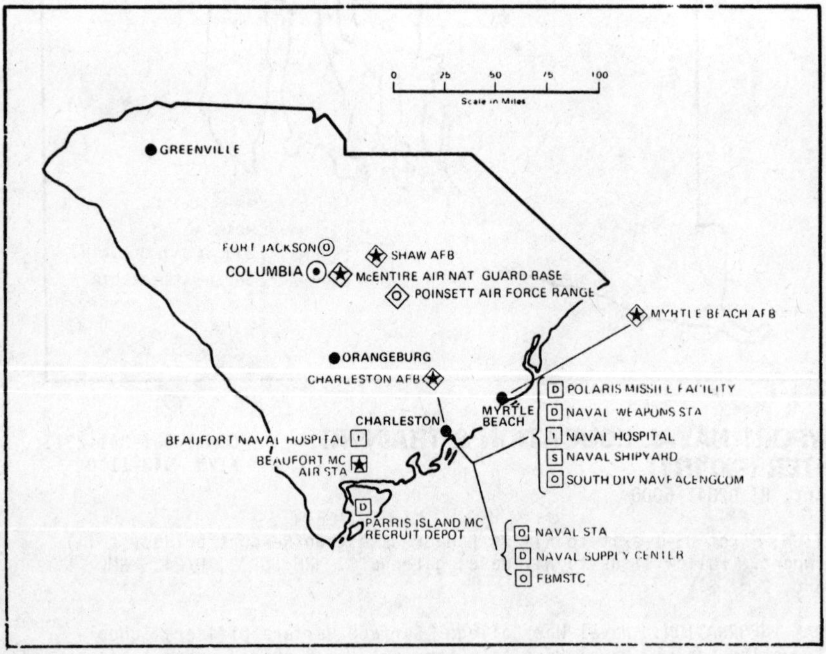

BEAUFORT MARINE CORPS AIR STATION (SC01R1)

Beaufort, SC 29904-5000

Comm: 803-846-2111
ATVN: 832-2111

LOCATION: From I-95 exit at Pocataligo to SC-21, 4 mi to MCAS. Clearly marked. RM: p-88, K/10. NMC: Savannah, GA, 40 mi S.

GENERAL INFORMATION: Marine Aircraft Grp-31, Marine Air Control Squadron.

Beaufort MCAS, Cont'd

TEMPORARY MILITARY LODGING: Biltg Ofc, Bldg 431, 24 hrs dly, Comm: 803-846-7674, ATVN: 832-7674. All ranks. DV/VIP, EX-7121.

LOGISTICAL SUPPORT: Most spt fac available. Key phone numbers follow.

MX-846-2252	Caf-846-2441	Chap-846-7775
Chld C-846-7290	Conv S-846-4372	EM C-846-7592
Fam Svc-846-7353	Gas S-846-8248	Legal-846-7330
Loc-846-7188	Med-846-7311	Off C-846-7541
Pkg S-846-2247	Pol-846-7373	Snack B-846-2497
Staff C-846-7726	Theat-846-7183	

HEALTH & WELFARE: Dispensary/Emer, EX-7311. Inpatient, Naval Hosp, Beaufort, Comm: 803-525-5400/1, ATVN: 832-5400/1. Chapels serving all faiths, EX-7775.

RECREATION: Craft Shops, EX-7340, Bowling, EX-7106, Gym, EX-7192, Spec Svc, EX-7400, Picnic Area, EX-7340, Boat Rental, EX-7570, Library, EX-7682, Swimming, EX-7573, Rec Ctr, EX-7340.

SPACE-A: Pax Term and Lounge, Bldg 860, 0700-1100 dty dys, Comm: 803-846-7143, ATVN: 836-7143, flts to CONUS locations.

ATTRACTIONS: Beaches, seafood, period homes, Savannah easy drive 40 mi S.

CHARLESTON AIR FORCE BASE (SC06R1)

Comm: 803-554-0230
ATVN: 583-0111

Charleston Air Force Base, SC 29404-5415

LOCATION: From I-26 E exit to West Aviation Ave to traffic light, continue through light to 2nd light right, follow road around end of runway to gate 2 (River Gate). RM: p-88, I/12. NMC: Charleston, 5 mi SE.

GENERAL INFORMATION: Military Airlift Command Base. Airlift Wing, AFRES units and other support activities.

TEMPORARY MILITARY LODGING: Biltg Ofc, Bldg 322, intersection of Simpson St and Davis Dr, 24 hrs dly, Comm: 803-552-9900, ATVN: 583-9900. All ranks. DV/VIP, EX-3201.

LOGISTICAL SUPPORT: Complete spt fac avail. Key phone numbers follow.

BX-552-1463	Caf-552-5311	Chap-554-2536
Chld C-554-2588	Commiss-554-3321	Conv S-552-5182
Fam Svc-554-3416	Gas S-552-5161	Legal-554-3581
Loc-554-2669/2671	Med-554-2144	NCO C-554-3514
Off C-554-3587	Pkg S-554-3588	Pol-554-2213
Snack B-552-9415	Theat-554-3532	

HEALTH & WELFARE: Dispensary/Emerg EX-2144. Inpatient, Charleston Naval Hospital, Comm: 803-743-5444, ATVN: 794-5444. Chapels serving all faiths, EX-2536.

RECREATION: Arts & Crafts, EX-3293, Auto Hobby, EX-2973, Bowling, EX-2514, Gym, EX-2560, Saddle Club, EX-2329, Library, EX-3134, Swimming, EX-2213, Rec Ctr, EX-2280, Tennis, EX-2560, Youth Ctr, EX-2789. Fam-Camp, on Base, year round, Comm: 803-554-2280, ATVN: 583-2280, 5 camper spaces w/hookups, 2 camper spaces w/elec, all facilities and activities.

SPACE-A: Pax Term and Lounge, S-164, 24 hrs dly, Comm: 803-554-2347/2610, ATVN: 583-2347-2610, Recording EX-3125, flts to CONUS and overseas.

Charleston AFB, Cont'd

ATTRACTIONS: Beaches, period homes, seafood.

CHARLESTON NAVAL BASE (SC04R1)

Charleston Naval Base, SC 29408-5000

Comm: 803-743-4111
ATVN: 794-4111
FTS: 679-4111

LOCATION: From I-26 take Cosgrove Ave exit to Spruill Ave, right for 3 mi to main gate of Base. RM: p-88, J/16. NMC: Charleston, 5 mi N.

GENERAL INFORMATION: Charleston Naval Shipyard, Naval Supply Center, Cruiser-Destroyer Groups, Submarine Group, Mine Warfare Command, Charleston Naval Hospital, and other support activities.

TEMPORARY MILITARY LODGING: Biltg Ofc, Navy Lodge, Bldg 225, 24 hrs dly, Comm: 803-743-7676/6011, ATVN: 794-7676/6011. All ranks. DV/VIP, EX-4111.

LOGISTICAL SUPPORT: Complete spt fac avail. Key phone numbers follow.

NX-743-3905	Caf-743-6477	Chap-743-6476/5230
Chld C-743-4319	Commiss-743-3080	CPO C-743-7984
EM C-743-5266	Fam Svc-743-6250	Gas S-743-5860
Legal-743-4740	Loc-743-4111	Med-743-5444
Off C-743-6376	Pkg S-743-5287	Pol-743-5555
Snack B-743-5328		

HEALTH & WELFARE: Dispensary/Emer, EX-6179. Inpatient, Charleston Naval Hosp, Comm: 803-743-5130, ATVN: 794-5130, 500 bed. Chapels serving all faiths, EX-5127.

RECREATION: Arts/Crafts, EX-6493/7996, Auto Hobby, EX-6449/7752, Bowling, EX-5416, Golf, EX-5222/7802, Gym, EX-7783/5160, Weight Room, EX-5160/7783, Rod & Gun Club, EX-7621, Library, EX-5399/7900, Rec Ctr, EX-7601. Short Stay Rec Area, on Lake Moultrie, off SC-52, 35 mi from Base, year round, Comm: 803-743-5233, ATVN: 794-5233, 78 camper spaces w/hookups, 150 camper spaces w/o hookups, 60 mobile homes and cabins, full rec area facilities and all activities.

SPACE-A: See Charleston AFB listing.

ATTRACTIONS: Charleston, one of America's most historic cities, features miles of beautiful, sandy beaches, gardens, fresh water lakes, great fishing.

CHARLESTON NAVAL WEAPONS STATION (SC11R1)

Charleston Naval Weapons Station, SC 29408-7000

Comm: 803-743-4111
ATVN: 563-4111

LOCATION: Exit 203 off I-26, US-78, to US-52, to SC-37 (Red Bank Road) to main gate. RM: p-88, J/16. NMC: Charleston, approximately 25 mi.

GENERAL INFORMATION: The Naval Weapons Station occupies 16,331 acres. Established in 1941 the activity provides material and technical support for ammunition and assigned weapons and weapon systems. It operates an explosive ordnance outloading facility. Manages Navy family housing for Charleston area.

TEMPORARY MILITARY LODGING: Navy Lodge at Charleston Naval Base.

LOGISTICAL SUPPORT: Most spt fac avail. Key phone numbers follow.

NX-764-7042	Chld C-764-7065	Commiss-764-7015

Charleston NWS, Cont'd

Fam Svc-764-7294	Gas S-764-7573	Mini Mart-764-7735
Pkg S-743-7680		

HEALTH & WELFARE: Dispensary, 764-7634. Inpatient, Charleston Naval Hosp, 803-743-5130. Chapels serving all faiths, 764-7222.

RECREATION: Morale Welfare Recreation, 764-7601.

SPACE-A: See Charleston AFB listing.

ATTRACTIONS: Historic Charleston nearby.

FORT JACKSON (SC09R1)

Fort Jackson, SC 29207-5000

Comm: 803-751-7511
ATVN: 734-1110

LOCATION: Exit from I-20 N of Fort or from US-76/378 S of Fort, or US-601 E of the Fort. From I-20 take Ft Jackson or Two Notch Rd exit, left on Decker Blvd, right on Percival Rd to Fort main entrance. RM: p-88, E/9. NMC: Columbia, 12 mi SW.

GENERAL INFORMATION: Two Training Brigades for training initial entry soldiers, One Combat Support Training (AIT) Bde, Army Reception Station, and Reserve Command.

TEMPORARY MILITARY LODGING: Biltg Ofc, Bldg 2785, Semmes Rd & Lee Rd, 24 hrs dly, Comm: 803-751-6223/6419, ATVN: 734-6223/6419. All ranks. DV/VIP, EX-5218/5325. Palmetto Lodge Guest House, EX-4779/5205/6134.

LOGISTICAL SUPPORT: Complete spt fac avail. Key phone numbers follow.

PX-787-1950	Chap-751-3121	Chld C-787-3721
Commiss-751-7212	Conv S-787-0590	EM C-751-6179
Gas S-782-1639	Legal-751-6126/4287	Loc-751-7671
Med-751-2128	NCO C-782-1932	Off C-782-3167
Pkg S-782-1601	Pol-751-3113	Snack B-752-8698
Theat-751-4858/7488		

HEALTH & WELFARE: Moncrief Army Community Hospital, clinics and depts, Emer, EX-2168/4444, Appoint, EX-5431, 410 bed. Chapels serving all faiths, EX-3122/3121. Dental, EX-4828/5197/7025.

RECREATION: Arts/Crafts, EX-6359, Auto Hobby, EX-5653/5755, Bowling, EX-4656/4959, ITT, EX-6219, Golf, EX-6367, Gym, EX-5878, Sports Rental, EX-5481, Rifle Range, EX-6954, Skating Rink, EX-6387, Riding Club, EX-6357, Youth Act, EX-5040, Youth Ctr, EX-6552/787-9190, Library, EX-4816/5589, Swimming, EX-4796/7472/7084, Rec Ctr, EX-4829, Outdoor Rec Dir, EX-5481, Picnic Area, EX-6606/4215. Weston Lake Travel Camp, on Post, year round, Comm: 803-751-6013/6070, ATVN: 734-6013/6070, 21 camper spaces w/hookups, 6 cabins, full facilities and all lake activities.

SPACE-A: See Charleston AFB listing.

ATTRACTIONS: Outdoor sports & recreation. Columbia, the state capital. Ernie Pyle Media Center located on Fort Jackson, a memorial to the slain World War II correspondent. Museum traces South Carolina and Fort Jackson history.

MYRTLE BEACH AIR FORCE BASE (SC07R1)

Myrtle Beach Air Force Base, SC 29579-5000

Comm: 803-238-7211
ATVN: 748-7211

Myrtle Beach AFB, Cont'd

LOCATION: From I-95 exit E on US-501, right on US-17 (Business) to AFB. Clearly marked. RM: p-88, F/15. NMC: Myrtle Beach, in city limits.

GENERAL INFORMATION: Tactical Air Command Base. Tactical Fighter Wing and support units.

TEMPORARY MILITARY LODGING: Biltg Ofc, Bldg 126, A Ave & 4th St, 24 hrs dly, Comm: 803-238-7691, ATVN: 748-7691. All ranks. DV/VIP, EX-7673.

LOGISTICAL SUPPORT: Complete spt fac avail. Key phone numbers follow.

BX-238-2221	Caf-238-7027	Chap-238-7697
Chld C-238-7475	Commiss-238-7046	Conv S-238-2512
Fam Svc-238-7857	Gas S-238-2711	Legal-238-7624
Loc-238-7056	Med-238-7333	NCO C-238-5672
Off C-238-7133	Pkg S-238-7843	Pol-238-7659
Snack B-238-5722	Theat-238-7472	

HEALTH & WELFARE: USAF Hospital, clinics and depts, Emer, EX-7333, Appoint, EX-7273, 20 bed. Chapels serving all faiths, EX-7697.

RECREATION: Arts & Crafts, Auto Hobby, Bowling, Golf, Gym, Sports Rental, Picnic Area, Swimming, Rec Ctr, call EX-7477 for all, Library, EX-7086/7195. Fam-Camp, on Base, year round, Comm: 803-238-7708, ATVN: 748-7708, 26 camper spaces w/hookups, 30 camper spaces w/o hookups, full fac and all act, Atlantic Ocean 1000 yds away.

SPACE-A: Pax Term, Lounge, Base Ops, ATC Bldg, Comm: 803-238-7929/7144 and 7159, ATVN: 748-7929/7144/7159, very limited flts to CONUS locations.

ATTRACTIONS: Great beach on the Atlantic Ocean. Minutes away from 36 championship golf courses.

PARRIS ISLAND MARINE CORPS RECRUIT DEPOT (SC08R1)

Comm: 803-525-2111
ATVN: 832-2111

Parris Island, SC 29905-5000

LOCATION: From I-95 S exit at Beaufort to SC-170 or US-21, both E to SC-280 to SC-281 which leads to main gate of Depot. RM: p-88, K/10. NMC: Savannah, 43 mi SW.

GENERAL INFORMATION: Recruit Depot for USMC male recruits east of Mississippi and all female recruits. Naval Hospital.

TEMPORARY MILITARY LODGING: Biltg Ofc, Hostess House, Bldg 200, 24 hrs dly, Comm: 803-525-2976, ATVN: 832-2976. All ranks. DV/VIP, EX-2594/5.

LOGISTICAL SUPPORT: Complete spt fac avail. Key phone numbers follow.

MX-525-3301	Chap-525-2385	Chld C-525-2215
Commiss-525-2385	Conv S-525-2248	EM C-525-3653
Fam Svc-525-3791	Loc-525-3358	Med-525-3351
Off C-525-2905	Pol-525-3444	Snack B-525-2810
SNCO C-525-2452	Theat-525-2841	

HEALTH & WELFARE: Naval Hospital, clinics and depts, Emerg, 525-5400, Appoint, 525-5600, 275 bed. Chapels serving all faiths, EX-2424.

RECREATION: Arts & Crafts, EX-2204, Auto Hobby, EX-3205, Bowling,

Parris Island MCRD, Cont'd

EX-2290, Golf, EX-2240, Boat Basin, EX-3670, Fitness Ctr, EX-3395, Rod & Gun Club, EX-3635, Library, EX-3261, Swimming, EX-3347/3447/3530.

SPACE-A: See Beaufort MCAS listing.

ATTRACTIONS: Great beaches on the Atlantic Ocean. Museum on Base.

SHAW AIR FORCE BASE (SC10R1)

Shaw Air Force Base, SC 29152-5000

Comm: 803-668-8110
ATVN: 965-1110

LOCATION: Off US-76/378, 8 mi W of Sumter. Clearly marked. RM: p-88, E/10. NMC: Columbia, 35 mi W.

GENERAL INFORMATION: Tactical Air Command Base. 9th Air Force Hq, Tactical Fighter Wing, Tactical Air Control Wing and support units.

TEMPORARY MILITARY LODGING: Biltg Ofc, Bldg 929, Myer St, 24 hrs dly, Comm: 803-668-3210/3309, ATVN: 965-3210/3309. All ranks. DV/VIP, EX-8110.

LOGISTICAL SUPPORT: Complete spt fac avail. Key phone numbers follow.

BX-666-3481	**Chap-668-2372**	**Chld C-668-2305**
Commiss-666-2201	**Conv S-666-3467**	**Fam Svc-668-3538**
Gas S-666-3104	**Legal-668-2504**	**Loc-668-8110/1110**
Med-668-3641	**NCO C-666-3651**	**Off C-666-3661**
Pkg S-666-3651	**Pol-668-3626**	**Snack B-666-2460**
Theat-668-2242		

HEALTH & WELFARE: USAF Hospital, clinics and depts, Emerg EX-2571/3641, Appoint, EX-2571, 40 bed. Chapels serving all faiths, EX-2372.

RECREATION: Arts & Crafts, EX-2362, Auto Hobby, EX-2929, Bowling, EX-3553, Aero Club, EX-2057, Golf, EX-3950, Gym, EX-2941/3531, Sports Rental, EX-2204, Rod & Gun Club, EX-2492, Library, EX-3084, Rec Ctr, EX-2205, Swimming, EX-3124/3349, Youth Ctr, EX-2120. Fam-Camp (Lake Wateree), off Base, at Lake Wateree, 24 mi from AFB, Mar-Dec, Comm: 803-668-2205, ATVN: 965-2205, 12 camper spaces w/hookups, 19 mobile homes, full fac and all lake activities.

SPACE-A: Pax Term and Lounge, Bldg 615, 24 hrs dly, Comm: 803-668-2916/3468, ATVN: 965-2916/3468, infrequent flts to CONUS locations.

ATTRACTIONS: Sumter Iris Gardens. Columbia is easy drive. Outdoor sports and recreation.

- NOTES -

SOUTH DAKOTA

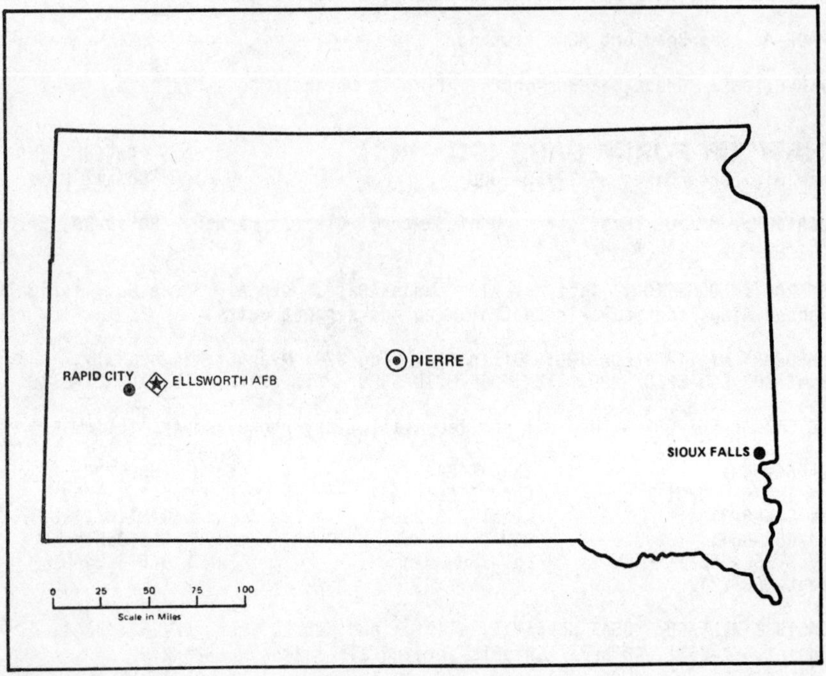

ELLSWORTH AIR FORCE BASE (SD01R3)

Ellsworth Air Force Base, SD 57706-5000

Comm: 605-342-2400
ATVN: 747-1110

LOCATION: Off I-90, 10 mi E of Rapid City. Clearly marked. RM: p-89, F/4. NMC: Rapid City, 10 mi W.

GENERAL INFORMATION: Strategic Air Command base. Missile Wing, Bomb Wing, and Postattack Command and Control System Squadron. B-1B base, Nov '86.

TEMPORARY MILITARY LODGING: Biltg Offc, bldg 1103, 6 St, 24 hrs dly, Comm: 605-923-5861 or 605-399-7091/2762/2763, ATVN: 747-7091/2729/2294. All ranks. DV/VIP, 7513.

LOGISTICAL SUPPORT: Complete spt fac avail. Key phone numbers follow.

BX-923-5821	Caf-923-1455	Chap-342-7282
Chld C-399-2121	Commiss-923-2314	Conv S-923-5231
Fam Svc-399-2438	Gas S-923-1489	Legal-399-2341
Loc-342-2400/1110	Med-399-2361	NCO C-923-1442
Off C-923-7714	Pkg S-342-2068	Pol-399-2001
Snack B-923-1474	Theat-923-1437	

HEALTH & WELFARE: USAF Hospital, clinics and depts, Emerg, EX-3430, Appoint, EX-399-7631, 30 bed. Chapels serving all faiths, EX-7281/2475.

RECREATION: Arts & Crafts, EX-2959, Auto Hobby, EX-2779, Field House, EX-7309, Riding Club, EX-923-1454, Library, EX-7965, Rec Ctr, EX-2358,

Ellsworth AFB, Cont'd

Bowling, EX-2536. Fam-Camp, on Base, May-Oct, Comm: 605-399-2454, ATVN: 747-2454, 14 camper spaces w/hookups, 10 camper spaces w/o hookups, other fac and outdoor activities.

SPACE-A: Pax Term and Lounge, Base Ops bldg, 24 hrs dly, Comm: 605-399-2761/2792, ATVN: 747-2761/2792, flts to CONUS and overseas locations.

ATTRACTIONS: Air & Space Museum on Base, Black Hills/Mt Rushmore nearby, outdoor sports and recreation paradise.

TENNESSEE

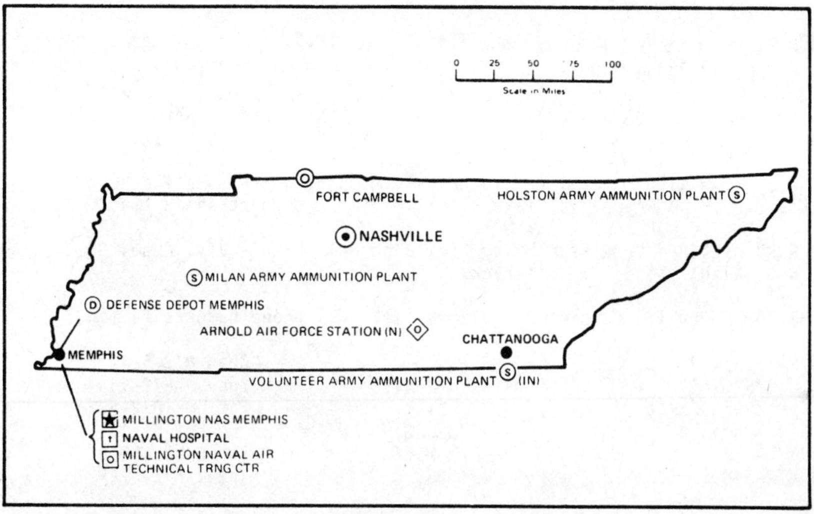

ARNOLD AIR FORCE STATION (TN02R2)

Arnold Air Force Station, TN 37389-5000

Comm: 615-454-3000
ATVN: 340-5011

LOCATION: From Tullahoma, take AEDC access hwy. From I-24 take AEDC exit 117, 4 mi S of Manchester. Clearly marked. RM: p-91, G/17. NMC: Chattanooga, 65 mi SE; Nashville, 65 NW.

GENERAL INFORMATION: Air Force Systems Command. Arnold Engineering Development Center, freeworld's largest complex of wind tunnels, and other R & D test and evaluation facilities.

TEMPORARY MILITARY LODGING: Biltg Offc, bldg 3027, 0600-2200 M-F, 1000-2000 Sat/Sun, Comm: 615-454-5498, ATVN: 340-5498. All ranks. DV/VIP, EX-2611.

LOGISTICAL SUPPORT: Complete spt fac avail. Key phone numbers follow.

BX-455-7153	**Commiss-455-7249**	**Fam Svc-454-7768**
Legal-454-7153/7814	**Loc-454-3000**	**Med-454-5351**
NCO C-455-6113	**Off C-454-7601**	**Pol-454-5222**

HEALTH & WELFARE: Dispensary/Emerg 454-5351, Appoint, 454-5351. Inpatient, Fox Army Comm Hosp, Redstone Arsenal, Huntsville, AL, 205-876-2857. No Chapels

Arnold AFS, Cont'd

on Base.

RECREATION: Arts & Crafts, 454-5471, Tennis, 454-5498, Sailing & Boat Rental, 454-5471, Library, 454-5471. Fam-Camp, on Base, year round, Comm: 615-454-6084, ATVN: 340-6084, 22 camper spaces w/hookups, full fac & lake act.

SPACE-A: Base Ops, Comm: 615-454-7752, ATVN: 340-7752, flts to CONUS locations. Also, McGhee Tyson Airport, Knoxville, ANG, Comm: 615-970-3077, Air Refueling flts, CONUS and overseas, and Nashville Metro Airport, ANG, Comm: 615-361-4600, ATVN: 446-6210, training flts to CONUS locations.

ATTRACTIONS: Chattanooga and Lookout Mountain easy drive SE. Home of Grand Ole Opry, Opryland, Nashville, easy drive NW.

MEMPHIS NAVAL AIR STATION (TN01R2)

Millington, TN 38054-5000 **Comm: 901-872-5111**
ATVN: 966-5111

LOCATION: From US-51 N at Millington, exit to Navy Road, right to first gate on right, main gate. RM: p-90, G/3. NMC: Memphis, 20 mi SW.

GENERAL INFORMATION: Naval Air Technical Training Center, Marine Corps Aviation Training Group, Naval Hospital and other support activities.

TEMPORARY MILITARY LODGING: Biltg Ofc, Bldg S-1, 24 hrs dly, Comm: 901-872-5500, ATVN: 966-5500. All ranks. DV/VIP, EX-5500.

LOGISTICAL SUPPORT: Complete spt fac avail. Key phone numbers follow.

NX-872-7716/0138	Caf-872-1170	Chap-872-5341
Chld C-872-5745	Commiss-872-5122	Conv S-872-1334
CPO C-872-5664/5442	EM C-872-5131	Fam Svc-872-5075
Gas S-872-2241	Legal-872-5201	Loc-872-5111
Med-872-5444	Off C-872-8154	Pkg S-872-5960
PO C-872-5980	Pol-872-5533	Theat-872-5749

HEALTH & WELFARE: Naval Hospital, clinics and depts, Emerg EX-5444, Appoint EX-5801, 230 bed. Chapels serving all faiths, EX-5576.

RECREATION: Auto Hobby, EX-5675, Bowling, EX-5779, Golf, EX-5168, Gym, EX-5383, Rod & Gun Club, EX-5665, Library, EX-5683, Swimming, EX-5163, Rec Ctr, EX-5303, Tennis, EX-5163, Tickets, EX-5303, Riding Stables, EX-5301. Navy Lake Rec Area, on Base, year round, Comm: 901-872-1573, ATVN: 966-5163, 12 camper spaces w/o hookups, other fac and all lake activities.

SPACE-A: Pax Term and Lounge, Bldg N-2, Air Ops, 0700-2300 hrs dly, Comm: 901-872-5332/5245, ATVN: 966-5332/5245, flts to CONUS locations. Also, try Memphis IAP, ANG, Comm: 901-369-4111, ATVN: 966-8210, training flts to CONUS locations.

ATTRACTIONS: Memphis is short drive S. Mississippi River.

TEXAS

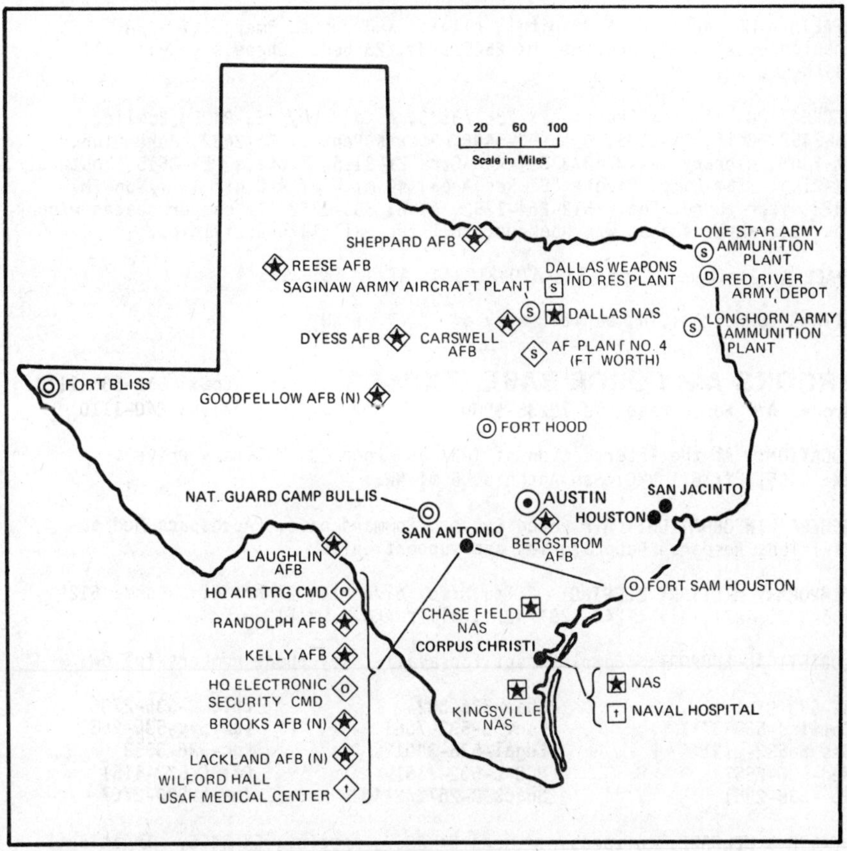

BERGSTROM AIR FORCE BASE (TX27R3)

Bergstrom Air Force Base, TX 78743-5000

Comm: 512-479-4100
ATVN: 685-1110

LOCATION: From US-183 or TX-71. Clearly marked. RM: p-93, WZ/10. NMC: Austin, 7 mi NW.

GENERAL INFORMATION: Tactical Air Command Base. Hq 12th Air Force, Tactical Recon, AFRES units here.

TEMPORARY MILITARY LODGING: Biltg Ofc, Bergstrom Inn, Bldg 2104, 24 hrs dly, Comm: 512-479-2207, ATVN: 685-2207. All ranks. DV/VIP, EX-3710.

LOGISTICAL SUPPORT: Complete spt fac avail. Key phone numbers follow.

BX-479-3311	Chap-479-2111	Chld C-479-2340
Commiss-479-2549	Conv S-385-0012	Fam Svc-479-2244
Gas S-385-0199	Legal-479-3782	Loc-479-2214/2207
Med-479-2333/2368	NCO C-479-3545	Off C-479-2403
Pkg S-385-0171	Pol-479-2604	Snack B-385-8719

Bergstrom AFB, Cont'd

Theat-479-3605

HEALTH & WELFARE: USAF Hospital, clinics and depts, Emerg, EX-2368, Ambulance, EX-2333, Appoint, EX-2545/3903, 25 bed. Chapels serving all faiths, EX-2111.

RECREATION: Arts & Crafts, EX-2381/3630, Auto Hobby, EX-3425, Bowling, EX-2452, Golf, EX-2105, Gym, EX-3496, Sports Rental, EX-2482, Aero Club, EX-2301, Library, EX-3740/3739, Rec Ctr, EX-2125, Tickets, EX-3515, Youth Ctr, EX-2104. Fam-Camp, Circle "B" Rec Area, 40 mi W of AFB off TX-71 on FM-2322, year round, Comm: 512-264-1752, ATVN: 685-1752, 12 camper spaces w/hook-ups, 24 camper spaces w/o hookups, full fac, all lake activities.

SPACE-A: Limited. Comm: 512-479-3123/4, ATVN: 685-3123/4.

ATTRACTIONS: Austin, capital city of TX, 7 mi NW.

BROOKS AIR FORCE BASE (TX26R3)

Brooks Air Force Base, TX 78235-5000 **Comm: 512-536-1110**
ATVN: 240-1110

LOCATION: At the intersection of I-37 and Loop 13 (Military Drive). RM: p-95, EY/16. NMC: San Antonio, 5 mi NW.

GENERAL INFORMATION: Air Force Systems Command base. Aerospace Medical Division, Research Laboratories and support units.

TEMPORARY MILITARY LODGING: Biltg Offc, bldg 214, 24 hrs dly, Comm: 512-536-3291/4271, ATVN: 240-3291/4271, all ranks. DV/VIP.

LOGISTICAL SUPPORT: Complete spt fac avail. Key phone numbers follow.

BX-533-9161	**Chap-536-3824**	**Chld C-536-2736**
Commiss-536-3717	**Conv S-532-7561**	**Fam Svc-536-2531**
Gas S-532-2191	**Legal-536-3301**	**Loc-536-3353**
Med-536-3557	**NCO C-532-7161**	**Off C-532-4161**
Pol-536-2851	**Snack B-2671/2140**	**Theat-536-2767**

HEALTH & WELFARE: Dispensary/Emerg EX-3557, Appoint, EX-3456. Inpatient, USAF Hosp, Lackland AFB, Comm: 512-671-7412, ATVN: 473-7412. Chapels serving all faiths, EX-2128.

RECREATION: Arts/Crafts, EX-2120/2881/2123, Auto Hobby, EX-2624, Bowling, EX-2671, ITT, EX-3691, Golf, EX-2636, Gym, EX-2342, MWR Offc, EX-3691, Library, EX-2634, Picnic Area, EX-3691, Swimming, EX-2703/3477, Rec Ctr, EX-3691, Youth Ctr, EX-2515, Tours, EX-3234.

SPACE-A: No active flight line. Kelly AFB, Pax Term, Lounge, bldg 1614, 0700-1800 hrs dly, Comm: 512-925-5706, ATVN: 945-5706, flts to CONUS locs.

ATTRACTIONS: Close to San Antonio, museums, missions, The Alamo, zoo.

CARSWELL AIR FORCE BASE (TX21R3)

Carswell Air Force Base, TX 76127-5000 **Comm: 817-735-5000**
ATVN: 739-1110

LOCATION: On TX-183. From Fort Worth, W on I-30, exit at Carswell AFB/Horne St. Follow signs to main gate. RM: p-96, D/2. NMC: Fort Worth, 7 mi E.

GENERAL INFORMATION: Strategic Air Command Base. Air Division, Bomb Wing, and AFRES units here.

Carswell AFB, Cont'd

TEMPORARY MILITARY LODGING: Biltg Ofc, ATTN: 7th SVH, Bldg 3140, corner 6th St & Meandering Rd, 24 hrs dly, Comm: 817-735-5274/5449, ATVN: 739-5274/5449. All ranks. DV/VIP, EX-5000/5449.

LOGISTICAL SUPPORT: Complete spt fac avail. Key phone numbers follow.

BX-738-1943	Chap-735-7301	Chld C-735-5753
Commiss-735-7602	Conv S-731-4187	EM C-735-5293
Fam Svc-735-7877	Gas S-738-1002	Legal-735-7594
Loc-735-7082/5000	Med-735-5781	NCO C-735-5293
Off C-735-5631	Pkg S-738-0462	Pol-735-5200
Snack B-738-6221	Theat-735-5298	

HEALTH & WELFARE: USAF Hospital, clinics and depts, Emerg, EX-5781, Appoint, EX-7442, 120 bed. Chapels serving all faiths, EX-7301.

RECREATION: Arts & Crafts, Bowling, Intermural Sports, Golf, Sports Rental, Base Marina, Rec Ctr, Library, Youth Ctr, EX-5000. Fam-Camp, on Base, year round, Comm: 817-735-5274, ATVN: 739-5274, 10 camper spaces w/hookups, some fac and act. Also, Shoreline Rec Area, on Base, year round, Comm: 817-735-7339, ATVN: 739-7339, 10 camper spaces w/o hookups, other fac, water act.

SPACE-A: Pax Term and Lounge, Bldg 1423, 0730-1630 Mon-Fri, Comm: 817-735-5649/7992, ATVN: 739-5649/7992, flts to CONUS and overseas.

ATTRACTIONS: Dallas, Dallas Cowboys, Dallas Mavericks, Texas Rangers, symphony, art, and rodeo.

CHASE FIELD NAVAL AIR STATION (TX23R3)

Comm: 512-354-5377
ATVN: 861-5377

Chase Field Naval Air Station, TX 78103-5000

LOCATION: Off US-181, 5 mi SE of Beeville. Clearly marked. RM: p-95, ER/7. NMC: Corpus Christi, 45 mi SE.

GENERAL INFORMATION: Naval Air Training Wing and support units.

TEMPORARY MILITARY LODGING: Biltg Offc, bldg 2939, Lexington St, 0800-1630 hrs dly, Comm: 512-354-2523, ATVN: 861-2523. All ranks. DV/VIP, EX-5011.

LOGISTICAL SUPPORT: Complete spt fac avail. Key phone numbers follow.

NX-354-5146	Chap-354-4254	Chld C-354-5517
Commiss-354-5457	CPO C-354-5209	EM C-354-5307
Fam Svc-354-5304	Gas S-354-5369	Legal-354-5219
Loc-5011/1110	Med-354-5503/5221	Off C-354-5484
Pol-354-5412	Theat-354-5515	

HEALTH & WELFARE: Dispensary/Emerg, EX-5221, Appoint, EX-5292. Inpatient, Naval Hosp, Corpus Christi NAS, Comm: 512-939-3735, ATVN: 861-3735. Chapels serving all faiths, EX-5255/5254.

RECREATION: Arts & Crafts, EX-5706, Auto Hobby, EX-5466, ITT, EX-5335, Sports Rental, EX-5466, Stables, EX-5515, Gun Club, EX-5515/5162, Hunting, EX-5413, Swimming, EX-5768/5822, Rec Ctr, EX-5462.

SPACE-A: Very Limited, Comm: 512-354-5276/5377, ATVN: 861-5276/5377.

ATTRACTIONS: Corpus Christi, 45 mi SE.

CORPUS CHRISTI NAVAL AIR STATION (TX10R3)

Comm: 512-939-2811
ATVN: 861-1110

Corpus Christi Naval Air Station, TX 78419-5000

LOCATION: On TX-358, SE side of Corpus Christi. The south gate is on NAS Drive. RM: p-95, ET/8. NMC: Corpus Christi, 10 mi W.

GENERAL INFORMATION: Naval Air Training Command, Coast Guard Air Station, and Corpus Christi Army Depot.

TEMPORARY MILITARY LODGING: Biltg Ofc, Bldg 1281, Ocean Dr, Navy Lodge, Bldg 101, 0900-1700 Mon-Fri, Comm: 512-939-6361, ATVN: 861-6361. All ranks. DV/VIP, EX-2380/2389.

LOGISTICAL SUPPORT: Complete spt fac avail. Key phone numbers follow.

NX-939-2033	**Chap**-939-3751	**Nur**-939-2355
Commiss-939-3177	**Conv S**-939-3122	**CPO C**-939-2240
EM C-939-3867	**Fam Svc**-939-3722	**Gas S**-939-2113
Legal-939-3531	**Loc**-939-2383	**Med**-939-3735/3839
Off C-939-2541	**Pkg S**-939-2081	**Pol**-939-3460
Snack B-939-3444	**Theat**-939-3488	

HEALTH & WELFARE: USN Hospital, clinics and depts, Emerg, EX-3735, Appoint, EX-3154, 195 bed. Chapels serving all faiths, EX-3751.

RECREATION: Arts & Crafts, EX-3169, Auto Hobby, EX-3470, ITT, EX-3497, Golf, EX-3250, Gym, EX-3164, Camping Rental, 937-5071, Party House, EX-2444, Skeet Range, EX-2444, Picnic Area, EX-2444, Library, EX-3574. Operates Padre Island Rec Area, off Base, year round, Comm: 512-937-5071, ATVN: 861-3497, 24 camper spaces w/hookups, 4 camper spaces w/water, all fac and water act.

SPACE-A: Pax Term & Lounge, Hangar 58, dty hrs, Comm: 512-939-2505/3385, ATVN: 861-2505/3385, flts to CONUS locations. Also check with CGAS, EX-2052.

ATTRACTIONS: Bay and gulf water sports. Also, symphony, museums, and historical homes.

DALLAS NAVAL AIR STATION (TX12R3)

Comm: 214-266-6111
ATVN: 874-6111

Dallas Naval Air Station, TX 75211-5000

LOCATION: Exit from I-30 at loop 12, W of Dallas, go S on loop 12 to Jefferson Ave exit, NAS on left or S side of Ave. Near Grand Prairie. RM: p-96, E/7. NMC: Dallas, 15 mi NE.

GENERAL INFORMATION: Tactical Naval and Marine Corps Aircraft Units, and ANG and Reserve Air Units here.

TEMPORARY MILITARY LODGING: Biltg Ofc, Bldg 209, Hutchins Hall, Halsey Ave, 24 hrs dly, Comm: 214-266-6468/6134, ATVN: 874-6468/6134. All ranks. DV/VIP, EX-6140.

LOGISTICAL SUPPORT: Complete spt fac avail. Key phone numbers follow.

NX-266-6400	**Caf**-266-6405	**Chap**-266-6132
Chld C-266-6275	**Conv S**-266-6409	**CPO C**-266-6413
EM C-266-6413	**Fam Svc**-266-6137	**Gas S**-266-6443
Legal-266-6105	**Loc**-266-6111	**Med**-266-6283
Off C-266-6134	**Pkg S**-266-6425	**Pol**-266-6105
Snack B-266-6405	**Theat**-266-6427	

Dallas NAS, Cont'd

HEALTH & WELFARE: Dispensary/Emerg EX-6283, Appoint, EX-6283. Inpatient, USAF Hosp, Carswell AFB, Comm: 817-735-5781, ATVN: 739-5781. Chapels serving all faiths, EX-6132.

RECREATION: Arts & Crafts, Auto Hobby, Aero Club, Golf, Sports Rental, Sail Club, Swimming, Rec Ctr, EX-6111. Mountain Creek Lake and Marina, on Base, year round, Comm: 214-266-6427, ATVN: 874-6427, limited fac and lake act.

SPACE-A: Pax Term and Lounge, Bldg 20, 24 hrs dly, Comm: 214-266-6376 and 266-6374/6375, ATVN: 874-6376/6374/6375, flts to CONUS and overseas.

ATTRACTIONS: Great city of Dallas nearby.

DYESS AIR FORCE BASE (TX14R3)

Dyess Air Force Base, TX 79607-5000

Comm: 915-696-0212
ATVN: 461-1110

LOCATION: Six mi SW of Abilene. Main gate is 3 mi E of I-20. Also from I-20 & US-277. RM: p-94, EG/2. NMC: Abilene, 6 mi NE.

GENERAL INFORMATION: Strategic Air Command Base. Air Division, Bomb Wing, Tactical Airlift Wing, and support units.

TEMPORARY MILITARY LODGING: Biltg Ofc, Bldg 7409, Dyess Inn, 5th St, 24 hrs dly, Comm: 915-696-2467, ATVN: 461-2467. All ranks. DV/VIP, EX-4275.

LOGISTICAL SUPPORT: Complete spt fac avail. Key phone numbers follow.

BX-692-5656	**Caf-698-1720**	**Chap-696-4224**
Chld C-696-4337	**Commiss-696-2511**	**Conv S-692-4771**
EM C-696-3220	**Fam Svc-696-2409**	**Gas S-692-6721**
Legal-696-3305	**Loc-696-3113**	**Med-696-2333/4**
NCO C-692-9540	**Off C-692-9577**	**Pkg S-692-1421**
Pol-696-2131	**Theat-696-4320**	

HEALTH & WELFARE: USAF Hospital, clinics and depts, Emerg, EX-2333/4, Appoint, EX-4742/3/4/5, 40 bed. Chapels serving all faiths, EX-4224/5.

RECREATION: Arts & Crafts, EX-4175, Auto Hobby, EX-4179, Bowling, EX-4166, Golf, EX-4384, Gym, EX-4306, Sports Rental, EX-2402, Library, EX-2274, Swimming, EX-3346, Rec Ctr, EX-4305, Tennis, EX-4305, Youth Ctr, 692-7363, Riding Stables, EX-3300.

SPACE-A: Pax Term and Lounge, Bldg 9001, dty hrs, Comm: 915-696-3108, ATVN: 461-2237, flts to CONUS and overseas.

ATTRACTIONS: Abilene and outdoor sports.

FORT BLISS (TX06R3)

Fort Bliss, TX 79916-6200

Comm: 915-568-2121
ATVN: 978-0831

LOCATION: Accessible from I-10 or US-54 N or S. RM: p-93, WW/4. NMC: El Paso, within NE section of the city limits.

GENERAL INFORMATION: Air Defense Artillery Center and School, Sergeant Majors Academy, Air Defense Artillery Brigade, Armored Cavalry Regiment, William Beaumont Army Medical Center, and other support units.

TEMPORARY MILITARY LODGING: Biltg Ofc, ATTN: ATZC-DEH-H, Bldg 504A, Rm 103, Pershing St, 24 hrs dly, Comm: 915-568-4888/2703, ATVN: 978-4888/2703.

Fort Bliss, Cont'd

All ranks. DV/VIP, EX-4192.

LOGISTICAL SUPPORT: Complete spt fac avail. Key phone numbers follow.

PX-562-7600	**Caf-562-2551**	**Chap-568-5992**
Chld C-562-3843	**Commiss-568-6719**	**Conv S-562-3300**
EM C-568-2715	**Fam Svc-568-3680**	**Gas S-562-1545**
Legal-568-6040	**Loc-568-2281**	**Med-569-2331**
NCO C-568-2715	**Off C-568-2738**	**Pkg S-562-4179**
Pol-568-3830	**Snack B-562-5442**	**Theat-568-2233/2132**

HEALTH & WELFARE: William Beaumont Army Medical Center Hospital, clinics and depts, Emerg, EX-2331/2209, Appoint, EX-2211, 463 bed. Chapels serving all faiths, EX-3194/5992 and EX-5106/1519.

RECREATION: Arts/Crafts, EX-4035, Auto Hobby, EX-4035, Bowling, EX-7493/6792/8263/3189, Golf, EX-1273/2066/7255, Gym, EX-6281/8693, Sports Rental, EX-7705, Music/Theater, EX-2595/7427, Picnic Area, EX-6097, Saddle Club, EX-8648, Rod/Gun Club, EX-8956/8753, Library, EX-2489, Swimming, EX-4508, Tickets, EX-6498/5210, Youth Ctr, EX-5495, Rec Ctr, EX-3881/8276. Family Campgrounds, on Post, year round, Comm: 915-568-4693, ATVN: 978-4693, 20 camper spaces w/hook-ups, 50 camper spaces w/o hookups, most other facilities and activities.

SPACE-A: Biggs Army Airfield has flts to CONUS locations via exec aircraft, EX-2121/0831.

ATTRACTIONS: Four museums (Replica, Air Defense Artillery, 30 Armored Cavalry Regiment, and Noncommissioned Officer) on Post, Carlsbad Caverns Natl Park easy drive E, White Sands Natl Monument two hrs drive N, Ciudad Juarez - Mexico's largest city on Mexican border short dist across Rio Grande River.

FORT HOOD (TX02R3)

Fort Hood, TX 76544-5000

Comm: 817-287-1110
288-1110
ATVN: 737/738-1110

LOCATION: From I-35 N exit to US-190 W, 9 mi to Killeen. Main gate is clearly marked. RM: p-94, EK/7. NMC: Killeen, at main entrance.

GENERAL INFORMATION: Army Corps Hq, and two Armored Divisions, Cavalry Brigade, Support Command, Signal Brigade, Military Police Brigade, Military Intelligence Group, Medical Activity, Dental Activity, and Testing Activity.

TEMPORARY MILITARY LODGING: Biltg Ofc, Bldg 36006, Wratten Ave, 24 hrs dly, Comm: 817-287-2818/5290, ATVN: 737-2818/5290. DV/VIP, 287-5001.

LOGISTICAL SUPPORT: Complete spt fac avail. Key phone numbers follow.

PX-532-7200	**Chap-287-5711**	**Commiss-287-6648**
Fam Svc-287-4936	**Gas S-532-5930**	**Legal-287-2993**
Loc-287-2137	**Med-288-8113**	**NCO C-287-5825**
Off C-532-5073	**Pkg S-532-5239**	**Pol-287-2176**
Theat-287-3066		

HEALTH & WELFARE: Darnell Army Community Hospital, clinics/depts, Emer, 288-8111, Appoint, 288-8888, 225 bed. Chapels serving all faiths, 287-5711/4533.

RECREATION: Arts/Crafts, 287-4070/2211, Auto Hobby, 287-2725, Intermural Sports, 287-4250/1435/5739, ITT, 287-INFO/7310 (ATVN: 737-4636), Sports Rntal, 287-4126, Skating Center, 287-5623, Hunting & Fishing, 287-5841, Archery, 287-

Fort Hood, Cont'd

4907, Library, 287-5202/2739/2250 & 288-9520, Swimming/Bch, 287-6644, Rec Ctr, EX-5690/2927/4142 & 288-9828, Theater, 287-4690/3350, Youth Activ's, 287-0335/2330, Boat Rntl, 287-5526, Music Ctr, 287-6337, Dirt Riders Assn, 287-6734, Hunt/Saddle Club, 287-6307, Riding Club, 288-9524, Rod/Gun Club, 532-4552, Sport Parachute Club, 287-4005. West Fort Hood Travel Camp, 64 camper spaces w/hookups, year round, Comm: 817-288-9926, ATVN: 738-9926. Storage for 200 RV. Also operates, Belton Lake Rec Area, on Post, year round, Comm: 817-287-8303, ATVN: 737-8303, 10 cottages, 36 RV campers, 64 camper spaces w/hookups, 20 camper spaces w/o hookups.

SPACE-A: Robert Gray Army Airfield, occasional flts available within US, 288-9718 after 6 pm.

ATTRACTIONS: State capital, Austin, 60 mi S. Great outdoor sports.

FORT SAM HOUSTON (TX18R3)

Fort Sam Houston, TX 78234-5000

Comm: 512-221-1211
ATVN: 471-1211

LOCATION: Accessible from I-410 or I-35/US-81. RM: p-95, EW/16. NMC: San Antonio, NE section of city.

GENERAL INFORMATION: Hq 5th Army, Army Health Services Command, Brooke Army Medical Center, and support activities.

TEMPORARY MILITARY LODGING: Biltg Offc, bldg 367, Stanley Rd, 0730-1600 hrs dly, Comm: 512-221-6125/5946, ATVN: 471-6125/5946. All ranks. DV/VIP, EX-1211.

LOGISTICAL SUPPORT: Complete spt fac avail. Key phone numbers follow.

PX-225-5566	**Chap-221-6688**	**Chld C-225-4683**
Commiss-221-4514	**Conv S-225-3589**	**EM C-224-2721**
Fam Svc-221-2418/4357	**Gas S-228-9748**	**Loc-221-2948**
Med-221-6466	**NCO C-224-2721**	**Off C-224-4211**
Pkg S-223-3427	**Pol-221-6363**	**Snack B-228-9071**
Theat-221-7891		

HEALTH & WELFARE: Brooke Army Medical Center, clinics and depts, Emerg, EX-6466, Appoint, EX-6771. Chapels serving all faiths, EX-2300/6688.

RECREATION: Arts/Crafts, EX-7125, Auto Hobby, 224-7046, Bowling, EX-3683/4740, Golf, EX-5863/4388, Gym, EX-2593, Outdoor Rec, 224-7162, Riding Stables, 224-7207, Outdoor Rec, EX-3703/3404, Hunting & Fishing, EX-7367, Library, EX-3702, Swimming, EX-3290/5993, Rec Ctr, EX-4829/2892/4743/2333, Theater(Plays), EX-3703/5655, Tickets, EX-3703, Youth Ctr, EX-7882, Music and Theater, EX-2186. Fort Sam Houston Rec Area (Canyon Lake), 50 mi N of Fort, year round, Comm: 512-221-3703, ATVN: 471-3703, 32 camper spaces w/hookups, 28 mobile homes, all facilities and lake activities.

SPACE-A: See Kelly AFB listing. In Houston, USCG Air Station, Base Ops, Comm: 713-481-0025, flts to CONUS locations.

ATTRACTIONS: City of San Antonio, Gulf Coast and Mexico both nearby.

GOODFELLOW AIR FORCE BASE (TX24R3)

Goodfellow Air Force Base, TX 76908-5000

Comm: 915-657-3217
ATVN: 477-3217

LOCATION: Off US-87 or US-277. Clearly marked. RM: p-93, WO/16. NMC: San Angelo, 2 mi NW.

Goodfellow AFB, Cont'd

GENERAL INFORMATION: Air Tng Cmd Base. Tech Tng Ctr. No active runway.

TEMPORARY MILITARY LODGING: Biltg Ofc, Bldg 124, Ft McKavitt Rd, 24 hrs dly, Comm: 915-657-3334/3231, ATVN: 477-3334. All ranks. DV/VIP.

LOGISTICAL SUPPORT: Complete spt fac avail. Key phone numbers follow.

BX-655-5789	**Chap-657-3424**	**Chld C-657-3239**
Com C-655-0553	**Commiss-653-2441**	**Conv S-655-5794**
Fam Svc-657-3425	**Gas S-655-5793**	**Legal-657-3302**
Med-657-3145	**Pkg S-657-3249**	**Pol-657-3511**
Snack B-655-2351	**Theat-657-3206**	

HEALTH & WELFARE: Clinic/Emerg, EX-3130, Appoint, EX-3150. Inpatient, USAF Hosp, Dyess AFB, Abilene, 90 mi NE, 915-696-2333. Chapels serving all faiths, EX-3424.

RECREATION: Auto Hobby Shop, EX-3233, Arts & Crafts, EX-3237, Bowling, EX-3227, Gym, EX-3242, Rod & Gun Club, EX-3246, Library, EX-3045, Swimming, EX-3226, Rec Ctr, EX-3247, Theater, EX-3206. Goodfellow Rec Camp, on Lake Nasworthy, 10 mi SW of Base, year round, 915-657-3211, ATVN: 477-3211, 8 camper spaces w/elec, unlimited w/o hookups.

SPACE-A: See Dyess AFB listing.

ATTRACTIONS: Lakes and city of San Angelo.

KELLY AIR FORCE BASE (TX03R3)

Kelly Air Force Base, TX 78241-5000

Comm: 512-925-1110
ATVN: 945-1110

LOCATION: All of the following, I-10, I-35, I-37, I-410 intersect with US-90. From US-90 take either the Gen Hudnell or Gen McMullen exit and go S to AFB. RM: p-95, EY/13. NMC: San Antonio, 7 mi NE.

GENERAL INFORMATION: Air Force Logistics Command Base. San Antonio Air Logistics Center, Hq Electronics Security Command, Electronic Warfare Center, Information and News Center, AFRES and ANG units here.

TEMPORARY MILITARY LODGING: Biltg Ofc, Bldg 1676, Gilmore Rd, 24 hrs dly, Comm: 512-925-8931, 924-7201, ATVN: 945-8931. All ranks. DV/VIP, EX-7678.

LOGISTICAL SUPPORT: Complete spt fac avail. Key phone numbers follow.

BX-924-9247	**Caf-925-3875**	**Chap-925-7874**
Chld C-925-5747	**Commiss-925-5268**	**Fam Svc-925-4181**
Gas S-924-2960	**Legal-925-3095**	**Loc-925-1841**
Med-925-4544	**NCO C-924-4511**	**Off C-924-7127**
Pol-925-6811	**Snack B-924-9247**	**Theat-925-5858**

HEALTH & WELFARE: Dispensary/Emerg, EX-4544, Appoint, EX-8811. Inpatient, USAF Hosp, Lackland AFB, Comm: 512-671-7412, ATVN: 473-7412. Chapels serving all faiths, EX-7874.

RECREATION: Arts/Crafts, EX-8346, Auto Hobby, EX-3872, Bowling, EX-5933/5480, Golf, EX-4006, Gym, EX-4846, Rec Svc Ofc, EX-4584, Aero Club, EX-3825, Library, EX-4116, Swimming, EX-4846, ITT, EX-5858, Youth Ctr, EX-8100. Fam-Camp, on Base, year round, Comm: 512-925-5725, 32 camper spaces w/hookups, full fac and all activities. Also operates Flying K Rec Ranch, on Lake LBJ at Wirtz Dam, 100 mi N of Base, year round, Comm: 512-925-4585, ATVN: 945-4585, 21 mobile homes, all facilities and complete activities.

Kelly AFB, Cont'd

SPACE-A: Pax Term and Lounge, Bldg 1614, 0700-1800 Mon-Fri, 0800-1600 other times, Comm: 512-925-8714/5, ATVN: 945-8714/5, flts to CONUS locations.

ATTRACTIONS: Alamo, River Walk, Tower of the Americas, Zoological and Botanical Gardens/Center, Institute of Texas Cultures and the McNay Art Museum.

KINGSVILLE NAVAL AIR STATION (TX22R3)

Comm: 512-595-6146
Kingsville Naval Air Station, TX 78363-5000
ATVN: 861-6146

LOCATION: Off US-77 S, exit to TX-425 SE to main gate. RM: p-95, ET/7. NMC: Corpus Christi, 30 mi NE.

GENERAL INFORMATION: Tng Air Wing TWO, three Squadrons, & support units.

TEMPORARY MILITARY LODGING: Biltg Ofc, Bldg 3729, dty hrs, Comm: 512-595-6321/6371, ATVN: 861-6321/6371. All ranks. DV/VIP, EX-6136.

LOGISTICAL SUPPORT: Complete spt fac avail. Key phone numbers follow.

NX-595-6184	Chap-595-6331	Chld C-595-6176
Commiss-595-6241/42	Conv S-595-6106	CPO C-595-6261
EM C-595-6413	Fam Svc-595-6333	Gas S-595-6106
Legal-595-6838	Med-595-6305	Off C-595-6121
Pkg S-595-6473	Pol-595-6217	Theat-595-6717

HEALTH & WELFARE: Dispensary/Emerg, EX-6286, Appoint, EX-6319/6342. Inpatient, Naval Hosp, Corpus Christi NAS, Comm: 512-939-3735/3839, ATVN: 861-3735/3839. Chapels serving all faiths, EX-6331/5254.

RECREATION: Arts & Crafts, Auto Hobby, Bowling, Gym, Golf, Swimming, Picnic Area, all EX-6171/2.

SPACE-A: Limited, EX-6136. See Corpus Christi NAS listing.

ATTRACTIONS: Gulf Coast, Corpus Christi, King's Ranch, and fishing in Baffin Bay.

LACKLAND AIR FORCE BASE (TX25R3)

Comm: 512-671-1110
Lackland Air Force Base, TX 78236-5000
ATVN: 473-1110

LOCATION: Off US-90 S. Loop 13 (Military Drive) bisects Lackland AFB. RM: p-95, EY/12. NMC: San Antonio, 6 mi NE.

GENERAL INFORMATION: Air Training Command Installation. Basic and technical training. Wilford Hall Medical Center. Officers' Training School.

TEMPORARY MILITARY LODGING: Biltg Ofc, Bldg 10062, Dimstead St, east side of Base, 24 hrs dly, Comm: 512-671-3281. All ranks. DV/VIP, 671-3622.

LOGISTICAL SUPPORT: Complete spt fac avail. Key phone numbers follow.

BX-674-7832	Caf-674-8962	Chap-671-2911
Chld C-671-3675	Commiss-671-2760	Conv S-674-0848
EM C-671-2347	Fam Svc-671-3609	Fam Spt-671-3069
Gas S-673-6961	Legal-671-3361	Loc-671-2655
Med-670-7412	NCO C-673-6071	Off C-673-5881
Pkg S-671-3603	Pol-671-2018	Snack B-674-8912
Theat-673-8077		

Lackland AFB, Cont'd

HEALTH & WELFARE: Wilford Hall, USAF Medical Center and Regional Hospital, 671-7412, Emerg, 670-7177, 1000 bed. Chapels serving all faiths, 671-4208.

RECREATION: Arts & Crafts, 671-2515, Auto Hobby, 671-3549, Bowling, 671-2271, Golf, 671-3466, Gym, 671-2751, Equip Rental, 673-6470, Rod & Gun Club, 674-7831, Library, 671-3610, Swimming, 671-4267, Rec Ctr, 671-2763, ITT, 671-3133. Lake Medina Recreation Camp, on Lake Medina near dam, year round, Comm: 512-671-4267, ATVN: 473-4267, tent camping, no hookups, boat rentals, other facilities and all lake activities.

SPACE-A: No active runway. See Kelly AFB or Randolph AFB listings.

ATTRACTIONS: City of San Antonio, Zoo, River Walk, The Alamo, El Mercado La Villita, Inst of Texan Cultures, Spanish missions, McNay Art Inst.

LAUGHLIN AIR FORCE BASE (TX05R3)

Comm: 512-298-3511
ATVN: 732-1110

Laughlin Air Force Base, TX 78843-5000

LOCATION: Take US-90 W from San Antonio, 150 mi or US-277 S from San Angelo, 150 mi to Del Rio area. The AFB is clearly marked off US-90. RM: p-93, WU/15. NMC: Del Rio, 6 mi NW.

GENERAL INFORMATION: Air Training Command Base. Undergrad Pilot Training, Flying Training Wing.

TEMPORARY MILITARY LODGING: Biltg Ofc, Bldg 470, 7th St, 24 hrs dly, Comm: 512-298-5731, ATVN: 732-5731. All ranks. DV/VIP, EX-5041.

LOGISTICAL SUPPORT: Complete spt fac avail. Key phone numbers follow.

BX-298-3627	Amn Dining-298-5295	Chap-298-5111
Chld C-298-5419	Commiss-298-5815	Fam Svc-298-5324
Gas S-298-3867	Legal-298-5172	Loc-298-5195
Med-298-5225/3416	NCO C-298-3573	Off C-298-3675
Pkg S-298-3449	Pol-298-5102	Rec Ctr-298-5474
Snack B-298-2711	Theat-298-5144	

HEALTH & WELFARE: USAF Hospital, clinics and depts, Emerg, 298-2333, Appoint, EX-5562. Chapels serving all faiths, EX-5111.

RECREATION: Arts & Crafts, Bowling, EX-5526, Golf, EX-5451, Gym, EX-5251/5326, Marina, EX-5474. Fam-Camp, on Base, year round, Comm: 512-298-5224, ATVN: 732-5224, 15 camper spaces w/hookups, 20 camper spaces w/o hookups, other fac & acts. Also operates Lake Amistad Rec Area, 23 mi NW of Base near Amistad Dam, Comm: 512-775-5971, ATVN: 732-5971, 8 camper spaces w/hookups, complete facilities and lake activities.

SPACE-A: Base Ops, Comm: 512-298-5308, ATVN: 732-5308, occasional flts to CONUS locs.

ATTRACTIONS: On the Mexican border at Ciudad Acuna, Mexico. Outdoor recreation sports, fishing, and boating on Lake Amistad.

RANDOLPH AIR FORCE BASE (TX19R3)

Comm: 512-652-1110
ATVN: 487-1110

Randolph Air Force Base, TX 78150-5000

LOCATION: Take exit 172 off I-35, Pat Booker Rd. Off I-10 take exit 587, TX FM-1604. RM: p-95, EO/6. NMC: San Antonio, 20 mi SW.

Randolph AFB, Cont'd

GENERAL INFORMATION: Air Training Command Base. 12th Flying Training Wing, Hq Air Training Command, Air Force Military Personnel Center, Hq, Air Force Recruiting Service, and support units.

TEMPORARY MILITARY LODGING: Biltg Ofc, ATTN: 12/SVH, Bldg 118, 24 hrs dly, Comm: 512-652-1844, ATVN: 487-1844. All ranks. DV/VIP, 658-4621

LOGISTICAL SUPPORT: Complete spt fac avail. Key phone numbers follow.

BX-652-2681	**Caf-658-1338**	**Chap-652-6121/4755**
Chld C-652-4946	**Commiss-652-6545**	**Fam Svc-652-3060**
Gas S-658-1515	**Legal-652-6781**	**Loc-652-1841**
Med-652-2734	**NCO C-658-3557**	**Off C-658-7445**
Pkg S-658-4544	**Pol-652-5700**	**Shoppette-658-1717**
Snack B-658-1440	**Theat-652-3278 (recording)**	

HEALTH & WELFARE: Primary Care, EX-2666/5588, Appoint, 652-2061/2063. Inpatient, USAF Hosp, Lackland AFB, San Antonio, 671-7412. Chapels serving all faiths, EX-5225/6121.

RECREATION: Art & Crafts, EX-2788, Auto Hobby, EX-2952, Bowling, EX-6271, Gym, 652-2955/5955, Golf, EX-4570, MWR Supply, EX-3702, Hunt & Saddle, EX-2346, MWR Ofc, EX-3012/3125, Library, 652-2617, Rec Supply, 652-5268, Rec Ctr, 652-2301, Youth Act, EX-3298, Skeet Range, EX-2064. Rec Area, Canyon Lake in Jacobs Creek Park, year round (closed M & T), Comm: 512-964-3804, ATVN: 487-3804, 10 camper spaces w/hookups, 45 camper spaces w/o hookups, many pop-up campers w/hookups, resv req, full fac and lake acts.

SPACE-A: Pax Term and Lounge, Hangar 7, east side, 0600-1800 M-F, other days as required, Comm: 512-652-1854, ATVN: 487-1854, flts to CONUS locations.

ATTRACTIONS: San Antonio, New Braunfels and Canyon Lake. Rec areas are nearby.

RED RIVER ARMY DEPOT (TX09R3)

Texarkana, TX 75507-5000

Comm: 214-838-2141
ATVN: 829-2141
FTS: 730-2141

LOCATION: Off I-30 E or W, take Red River Army Depot exit. Route clearly marked. RM: p-94, ED/14. NMC: Texarkana, 20 mi E.

GENERAL INFORMATION: Army Depot Activity for ammunition, missiles and general supplies.

TEMPORARY MILITARY LODGING: Biltg Ofc, Bldg 228, Texas Ave, 0745-1645 hrs dly, Comm: 214-838-3227, ATVN: 829-3227. All ranks. DV/VIP, EX-2316.

LOGISTICAL SUPPORT: Most spt fac avail. Key phone numbers follow.

PX-838-2396	**Caf-838-2341**	**Chap-838-2264**
Legal-838-3258	**Loc-838-2141**	**Med-838-2155**
NCO C-838-2854	**Off C-838-2350**	**Pkg S-838-3466**
Pol-838-2911	**Fam Svc-838-3506**	

HEALTH & WELFARE: Dispensary/Emerg, EX-2650, Appoint, EX-2155. Inpatient, USAF Hosp, Barksdale AFB, Shreveport, LA, 80 mi SE, 318-456-2365. Chapel serving all faiths, EX-2264.

RECREATION: Arts & Crafts, EX-2694, Bowling, EX-2733, Golf, EX-2359, Gym,

Red River Army Depot, Cont'd

EX-2733, Rod & Gun Club, EX-2900, Library, EX-3506, Rec Svc, EX-2694/3506. Elliott Lake Rec Area, on Post, year round, Comm: 214-838-2694, ATVN: 829-3506, 16 camper spaces w/hookups, 4 mobile homes, 5 wooded screen shelters, 2 cabins, all facilities and lake activities.

SPACE-A: Little Rock AFB, AR, Bldg 120, 0500-2100 hrs dly, Comm: 501-988-3684, ATVN: 731-3684, flts to CONUS and overseas.

ATTRACTIONS: Near Texarkana, TX/AR, outdoor sports and recreation.

REESE AIR FORCE BASE (TX20R3)

Comm: 806-885-4511
Reese Air Force Base, TX 79489-5000
ATVN: 838-4511

LOCATION: From I-289 (Loop) take 4th Street, 6 mi W. The road terminates at AFB, main gate is one block N. RM: p-92, WJ/12. NMC: Lubbock, 6 mi E.

GENERAL INFORMATION: Air Training Command Base. Flying Training Wing, Undergraduate Pilot Training, and support activities.

TEMPORARY MILITARY LODGING: Biltg Ofc, Bldg 1142, K St, 24 hrs dly, Comm: 806-885-3155, ATVN: 838-3155. All ranks. DV/VIP, EX-3409.

LOGISTICAL SUPPORT: Complete spt fac avail. Key phone numbers follow.

BX-885-4581	**Caf-885-3226**	**Chap-885-3237**
Chld C-885-3541	**Commiss-885-3312**	**Conv S-885-2427**
EM C-885-3156	**Fam Svc-885-3306**	**Gas S-885-4965**
Legal-885-3506	**Loc-885-3678**	**Med-885-3373**
NCO C-885-3156	**Off C-885-3466**	**Pkg S-885-3412**
Pol-885-3332	**Snack B-885-2120**	**Theat-885-3144**

HEALTH & WELFARE: USAF Hospital, clinics and depts, Emerg, EX-3373, Appoint, EX-3245, 20 bed. Chapels serving all faiths, EX-3237.

RECREATION: Arts & Crafts, EX-3241, Auto Hobby, EX-3142, Bowling, EX-3116, Golf, EX-3819, Gym, EX-3207, Sports Rental, EX-3815, Picnic Area, EX-3207, Rod & Gun Club, EX-3241, Library, EX-3344, Swimming, EX-3371/3667, Rec Ctr, EX-3787, Youth Ctr, EX-3820.

SPACE-A: Pax Term and Lounge, Comm: 806-885-3105/6, ATVN: 838-3105/6, used primarily for student training sortie preparation. Open only to accommodate student training operations. Closed when not supporting student training. Virtually no Space-A flights available to other CONUS locations. Commercial air and bus transportation from Lubbock is poor. No rail connections.

ATTRACTIONS: Adjacent to Lubbock on the E with Texas Tech University and Lubbock Christian College, Ranching Heritage Center, planetarium, museum. Agriculture and ranching. Outdoor recreation and hunting in season.

SHEPPARD AIR FORCE BASE (TX37R3)

Comm: 817-851-2511
Sheppard Air Force Base, TX 76311-5000
ATVN: 736-1001

LOCATION: Take US-281 N from Wichita Falls, exit to TX-325 which leads to main gate. Clearly marked. RM: p-92, WA/6. NMC: Wichita Falls, 5 mi SW.

GENERAL INFORMATION: Air Training Command Base. Sheppard Tech Tng Ctr, Flying Tng Wing & spt units. Sch of Health Care Sciences. 3700 Tech Tng Wing.

TEMPORARY MILITARY LODGING: Biltg Ofc, Bldg 776, Ave H, 24 hrs dly, Comm:

Sheppard AFB, Cont'd

817-851-6655, ATVN: 736-6655. All ranks. DV/VIP, EX-2123.

LOGISTICAL SUPPORT: Complete spt fac avail. Key phone numbers follow.

BX-855-4122	Chap-851-4370	Chld C-851-2038
Commiss-851-4362	Conv S-855-4341	EM C-851-2783
Fam Svc-851-4270	Gas S-855-4071	Legal-851-4262
Loc-851-4403	Med-851-2333	NCO C-851-6427
Off C-851-6460	Pkg S-855-6502	Pol-851-2981/2
Theat-851-4427		

HEALTH & WELFARE: USAF Hospital, clinics and depts, Emer, EX-2333, Appoint, EX-4585, 160 bed. Chapels serving all faiths, EX-4370.

RECREATION: Arts/Crafts, EX-6667, Auto Hobby, EX-4110, Bowling, EX-2677/2170, Golf, EX-6369, Gym, EX-2022, Picnic Area, EX-6210, Library, EX-6152, Swimming, EX-6494, Rec Ctr, EX-6200, Skeet & Trap Range, Gun Club, EX-6852. Operates Lake Texoma Recreational Annex, 120 mi E of AFB, on TX side of lake, year round, Comm: 817-851-2876, ATVN: 736-2876, 12 camper spaces w/hookups, 45 cabins sleeping 4-6 each, all rec area facilities and comp activities on lake. Possum Kingdom Rec Area, 75 mi W of Fort Worth, managed by Fort Sill, OK, Comm: 405-351-2025, ATVN: 639-2025, 4 camper spaces w/hookups, 45 camper spaces w/o hookups, 20 family units, 16 bachelor units, all fac and act.

SPACE-A: Pax Term and Lounge, Bldg 1360, dty hrs, Comm: 817-851-2180/2622, ATVN: 736-2180/2622, flts to CONUS locations.

ATTRACTIONS: Wichita Falls and outdoor sports.

UTAH

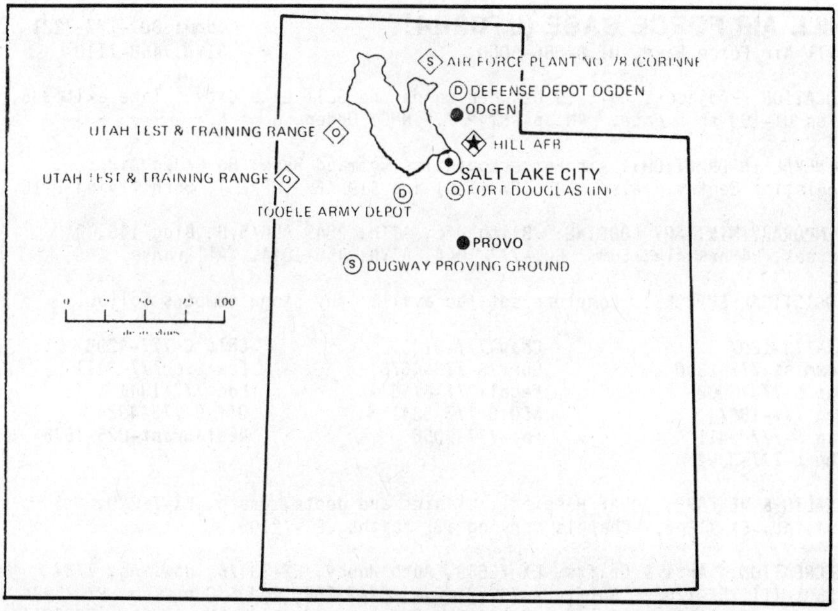

DUGWAY PROVING GROUND (UT04R4)

Dugway Proving Ground, UT 84022-5000

Comm: 801-831-3545
ATVN: 789-1110

LOCATION: Isolated but can be reached from I-80. Take Skull Valley Road (exit 77) for 40 mi S. RM: p-97, E/3. NMC: Salt Lake City, 80 mi NE.

GENERAL INFORMATION: Major Army Test and Evaluation Facility. Range and evaluation units.

TEMPORARY MILITARY LODGING: Biltg Ofc, Bldg 5228, 0730-2000 hrs dty dys, (Mon-Thurs) Comm: 801-831-2333, ATVN: 789-2333. All ranks. DV/VIP, EX-2399/2333.

LOGISTICAL SUPPORT: Complete spt fac avail. Key phone numbers follow.

PX-831-4491	ACS-831-2278	Chap-831-2431
Chld C-831-3345	Commiss-831-3464	Comm Club-831-2901
Conv S-831-4491	Gas S-831-4773	Legal-831-3333
Loc-831-3545/1110	Med-831-2222	Pkg S-831-2898
Pol-831-2933		

When calling from Salt Lake City the prefix is 522 instead of 831.

HEALTH & WELFARE: Dispensary/Emerg, EX-2222/2211, Appoint, EX-2222. Inpatient, USAF Hosp, Hill AFB, Ogden, 801-777-5285. Chapels serving all faiths, EX-2431.

RECREATION: Arts & Crafts, EX-2709, Auto Hobby, EX-2654, Bowling, EX-2687, Gym, 2705, Youth Ctr, EX-2177, Lib, EX-2178, Outdoor Rec Ctr, EX-2318.

SPACE-A: Salt Lake City IAP, AGN, Comm: 801-595-2272, ATVN: 790-9292, flts to CONUS locations via KC-135/10 aircraft.

ATTRACTIONS: The Great Salt Lake and Salt Lake City nearby.

HILL AIR FORCE BASE (UT02R4)

Hill Air Force Base, UT 84056-5000

Comm: 801-777-7221
ATVN: 458-1110

LOCATION: Adjacent to I-15 between Ogden and Salt Lake City. Take exit 336, E on UT-193 to S gate. RM: p-96, P/7. NMC: Ogden, 8 mi N.

GENERAL INFORMATION: Air Force Logistics Command Base, Hq Ogden Air Logistics Center. Also, 388 TFW (TAC) and 419 TFW (AFRES), both flying F-16.

TEMPORARY MILITARY LODGING: Biltg Ofc, ATTN: 2849 ABG/SVH, Bldg 146, D Street, 24 hrs dly, Comm: 801-777-1844, ATVN: 458-1844. All ranks.

LOGISTICAL SUPPORT: Complete spt fac avail. Key phone numbers follow.

BX-773-1207	Chap-777-2106	Chld C-777-4298
Commiss-777-2300	Conv S-773-4673	Fam Svc-777-3411
Gas S-773-3600	Legal-777-6756	Loc-777-1841
Med-777-1847	NCO C-777-3841	Off C-773-4924
Pkg S-777-3411	Pol-777-3056	Restaurant-825-1675
Theat-777-3394		

HEALTH & WELFARE: USAF Hospital, clinics and depts, Emerg, EX-75285, Appoint, EX-71847. Chapels serving all faiths, EX-72106.

RECREATION: Arts & Crafts, EX-72649, Auto Hobby, EX-73476, Bowling, 773-4250, ITT, EX-72892, Golf, EX-73272, Gym, EX-72762, Rod & Gun Club, EX-75984, Library, EX-73833, Swimming, EX-73975, Rec Ctr, EX-73924, Youth Ctr, EX-72419.

Hill AFB, Cont'd

Fam-Camp, on Base, Apr-Oct, Comm: 801-777-3661/2601, ATVN: 458-3661/2601, 14 camper spaces w/hookups, 10 w/elec only, other fac and activities. Operates Carter Creek Rec Area, 100 mi E of AFB, Uintas National Forest, Jul-Oct, Comm: 801-777-2892, ATVN: 458-7892, 6 cabins, 4 camper spaces w/hookups, 2 camper spaces w/o hookups. Also operates Hillhaus Lodge, off Base in Wasatch National Forest, 27 mi E of AFB near Snowbasin ski area, Comm: 801-777-2892, ATVN: 458-2892, 2 suites, male and female sleeping lofts, snow skiing, full fac and act.

SPACE-A: Bldg 1, N side, 0600-2200 hrs dly, Comm: 801-777-2887/1854, ATVN: 458-2887/1854.

ATTRACTIONS: Salt Lake City, snow skiing, boating, hiking, fishing, camping, and hunting. Park City, Temple Square.

TOOELE ARMY DEPOT (UT05R4)

Tooele, UT 84074-5008

Comm: 801-833-3211
ATVN: 790-1110

LOCATION: From W I-80, exit 99, to UT-36 south for about 15 mi to main entrance. RM: p-97, E/5. NMC: Salt Lake City, 40 mi NE.

GENERAL INFORMATION: Army Depot for maintenance, general supplies and ammunition.

TEMPORARY MILITARY LODGING: Biltg Ofc, Bldg S-1, 0730-1600 hrs dly, Comm: 801-833-3209/2124, ATVN: 790-3209/2124. All ranks. DV/VIP, EX-2693.

LOGISTICAL SUPPORT: Most spt fac avail. Key phone numbers follow.

PX-882-4235	**Chap-833-2244**	**Comm C-833-3548**
Legal-833-2536	**Loc-833-3211**	**Med-833-2572**
Pkg S-833-3548/3422	**Pol-833-2314/2559**	

HEALTH & WELFARE: Dispensary/Emerg, EX-2666, Appoint, EX-2572. Inpatient, USAF Hosp, Hill AFB, Ogden, 801-777-5285. Chapels serving all faiths, EX-2244.

RECREATION: Arts/Crafts, EX-2940, Auto Hobby, EX-2873, Bowling, EX-2849/882-9922, Sports, EX-2005, ITT, EX-3219, Gym, EX-3189, Sports Rental, EX-2107, Swimming, EX-3189, Library, N/A, Riding Stables, EX-3345. Oquirrh Hills Travel Camp, on Post, May-Oct, Comm: 801-833-3129, ATVN: 790-3129, 14 camper spaces w/o hookups, fac and act at Camp and on Post. No military dining fac or restaurants available on Post, but Community Club serves lunch Mon-Fri 11:00-12:30, sandwiches available Tues & Thurs 5:30-8:30 pm.

SPACE-A: See Hill AFB and Dugway Proving Ground listings.

ATTRACTIONS: Salt Lake City nearby, Tooele has bowling alley, theater in town and drive-in, some restaurants. The Utah desert is available for exploration, camping, motor cycling, hiking, and/or to enjoy sight seeing in nearby canyons and mountains. View largest open pit copper mine from top of Settlement Canyon. Fishing available 1 mile from Tooele city center at Settlement Canyon Dam and 35 miles SW of Tooele at Vernon Dam. Rock hunting in nearby deserts and mountains. Overnight camping, hiking, motor cycling, fishing, rock hunting, deer hunting (during season), available in local canyons.

VERMONT

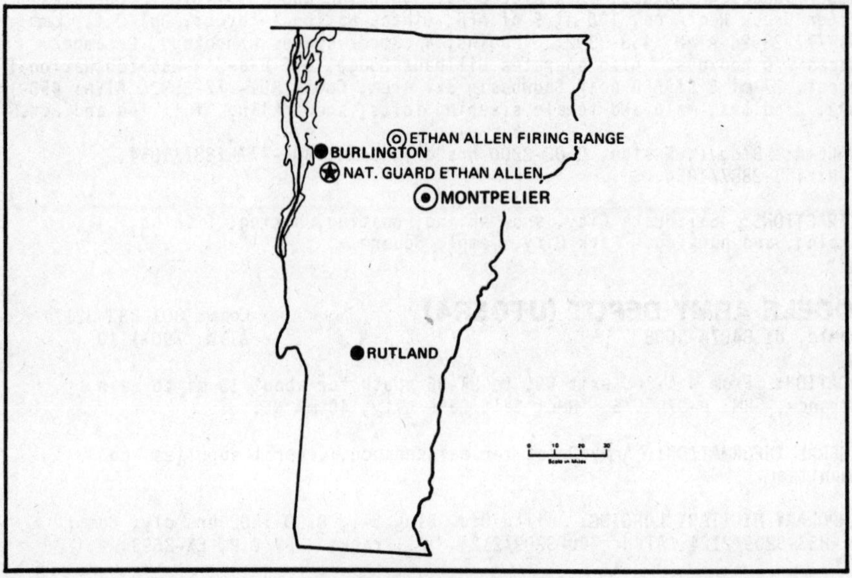

- None -

VIRGINIA

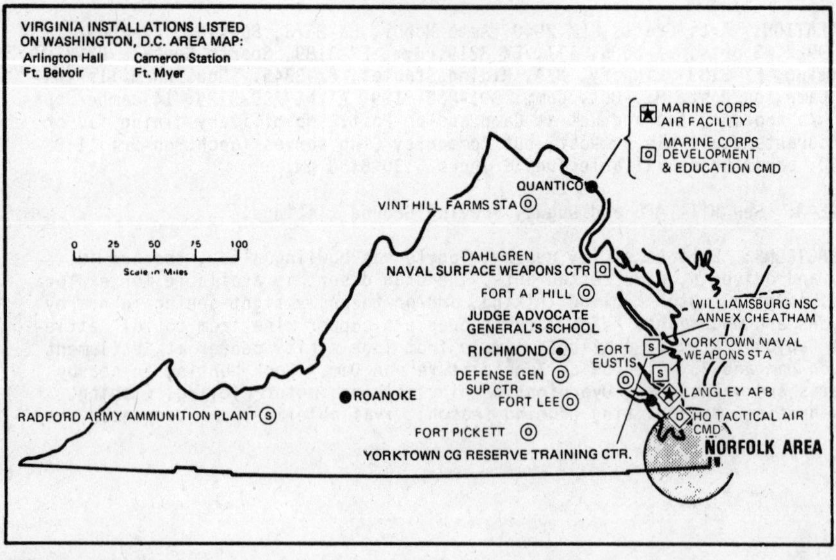

ARLINGTON HALL STATION (VA20R1)

Arlington, VA 22212-5042

Comm: 202-692-5145
ATVN: 222-5061

LOCATION: On US-50 W, 4000 Arlington Blvd. Main ent is 100 yds from George Mason Dr overpass to US-50 W. RM: p-42, I/7. NMC: Washington, DC, 3 mi E.

GENERAL INFORMATION: Hq US Army Intelligence and Security Command, Defense Communications Agency, and support units.

TEMPORARY MILITARY LODGING: None. See Fort Myer listing.

LOGISTICAL SUPPORT: Most spt fac avail. Key phone numbers follow.

PX-920-9115	**Chap-692-6323**	**Legal-692-6397/6883**
Med-692-6276	**NCO C-920-9106**	**Off C-920-7300/1**
Pol-692-6276		

HEALTH & WELFARE: Clinic/Emerg, 692-6276. Inpatient, Walter Reed Army Med Ctr, Washington, DC, 202-576-3501. Chapels serving all faiths, 692-6323.

RECREATION: Auto Hobby, 692-6262, Bowling, 692-5598, Physical Fitness Ctr, 692-5737, ITT, 692-3343, Library, 692-5236, Multi-Crafts, 692-6262, Rec Ctr, 692-6330, Rec Equip, 692-5484, Swimming, 692-6347, Tennis, 692-5737, Ath Field, 692-5737.

SPACE-A: See Andrews AFB, MD, listing.

ATTRACTIONS: Washington, DC, National Zoo, Smithsonian Museums, Kennedy Center for the Performing Arts and Mt Vernon.

CAMERON STATION (VA29R1)

Alexandria, VA 22304-5050

Comm: 202-545-6700
ATVN: 284-7199

LOCATION: From I-395 exit to VA-236 (Duke Street) E for 2 mi, entrance to Station is on right. RM: p-42, J/6. NMC: Washington, DC, 5 mi N.

GENERAL INFORMATION: Hq Defense Logistics Agency, Joint Personal Property Shipping Office, Hq Capital Region AAFES.

TEMPORARY MILITARY LODGING: None.

LOGISTICAL SUPPORT: Most spt fac avail. Key phone numbers follow.

PX-370-4700	**Commiss-274-6531/6527**	**Conv S-370-1950**
Gas S-370-4835	**Hot Deli-370-4700**	**Loc-545-6700**
Med-274-6545	**NCO C-274-7687**	**Off C-751-4292**
Pol-274-6516		

HEALTH & WELFARE: Clinic/Emerg, EX-7168, Appoint, EX-6545. Inpatient, Dewitt Army Hosp, Fort Belvoir, 703-664-4656. Chapels serving all faiths at Fort Myer, 696-3685.

RECREATION: Picnic Area, EX-7199, Rec Equip, EX-7127, Tennis, (no reservations). Trailer Park, on Base, May-Sep, Comm: 202-274-7199, ATVN: 284-7199, 12 camper spaces w/o hookups, limited fac and activities. Depending upon availability, a 7-10 day limit may be imposed.

$3.00/day

SPACE-A: See Andrews AFB, MD, listing.

ATTRACTIONS: Lovely lake with water fowl on Post (great for picnicking). Old Towne Alexandria, Washington, DC, and Mt Vernon nearby.

CHEATHAM ANNEX NAVAL SUPPLY CENTER (VA02R1)

Comm: 804-887-7000
ATVN: 953-7010

Williamsburg, VA 23185-5000

LOCATION: From I-64 take exit 57 to US-199 E to main gate of Cheatham Annex. RM: p-99, B/19. NMC: Williamsburg, 6 mi W.

GENERAL INFORMATION: Annex of Norfolk Naval Supply Center, Depot activity for Naval supplies and other special programs.

TEMPORARY MILITARY LODGING: Biltg Ofc, Spec Svc, Bldg 235, 0800-1700 hrs wkdys, Comm: 804-887-7224/7101/2, ATVN: 953-7224/7101/2. All ranks. DV/VIP, EX-7108.

LOGISTICAL SUPPORT: Most spt fac avail. Key phone numbers follow.

BX-887-3582	**Conv S-887-3582**	**Gas S-887-3582**
Pol-887-7222	**Med-887-7222**	**Theat-887-7154**

HEALTH & WELFARE: Med Emerg-887-7222. Inpatient, USAF Hosp, Langley AFB, Hampton, Comm: 804-764-6800, ATVN: 432-6800. No chapel.

RECREATION: Auto Hobby, EX-7101/2, Bowling, EX-7614, ITT, EX-7101, Golf, EX-7102, Swimming, EX-7102, Spec Svc Ofc, EX-7101/2. Rec Cabins and RV Park, on Base, year round, Comm: 804-887-7224, ATVN: 953-7224, 10 camper spaces w/ hookups, 6 rec cabins, 3 mobile homes, full fac and activities on York River.

SPACE-A: Langley AFB, Base Ops Bldg, 24 hrs, Comm: 804-764-5807 and 764-4698/4311, ATVN: 432-5807/4698/4311, flts to CONUS locations.

ATTRACTIONS: Historic Williamsburg and Yorktown both nearby, Water Country, and Busch Gardens.

DAHLGREN NAVAL SURFACE WEAPONS CENTER (VA06R1)

Comm: 703-663-8291
ATVN: 249-8291

Dahlgren, VA 22448-5000

LOCATION: From I-95 in Fredericksburg, E on VA-3 to VA-206 (17 mi), left at Arnolds Corner, E to Dahlgren (11 mi). Also, US-301 S to VA-206, E to main gate of Center. RM: p-99, J/21. NMC: Washington, DC, 38 mi N.

GENERAL INFORMATION: Naval Surface Weapons Research and Development Ctr.

TEMPORARY MILITARY LODGING: Biltg Ofc, Bldg 217, Welsh Rd, 0700-1530 hrs dly, Comm: 703-663-7118/8846, ATVN: 249-7118/8846. All ranks. DV/VIP, EX-8846.

LOGISTICAL SUPPORT: Complete spt fac avail. Key phone numbers follow.

BX-663-2121	**Caf-663-3841**	**Chap-663-8297**
Chld C-663-8820	**Commiss-663-7317**	**EM C-663-8785**
Gas S-663-2050	**Legal-663-7121**	**Loc-663-8216/8701**
Med-663-8241	**Off C-663-2610**	**Pol-663-8500**
Snack B-663-7261	**Theat-663-7336**	

HEALTH & WELFARE: Dispensary/Emerg, EX-8241. Inpatient, Dewitt Army Hosp, Fort Belvoir, 30 mi N, 703-664-1938. Chapel serving all faiths, EX-8241.

RECREATION: Sports Rental, EX-8585, Library, EX-7474, Tennis, EX-8580, Bowling, EX-7327, Golf, EX-3002, Gym, EX-8580, Swimming, EX-8088.

Dahlgren NSWC, Cont'd

SPACE-A: Andrews AFB, MD, Pax Term and Lounge, Bldg 1245, 0600-2200 hrs dly, Comm: 301-981-3527/8, ATVN: 858-3527/8, flts to CONUS and overseas.

ATTRACTIONS: Center is on the Potomac River. Washington, DC, nearby.

DAM NECK FLEET COMBAT TRAINING CENTER ATLANTIC (VA25R1)

Comm: 804-433-2000
ATVN: 433-2000

Virginia Beach, VA 23461-5000

LOCATION: On ocean front, 2 mi SE of Oceana NAS, off Oceana Blvd. RM: p-99, P/24. NMC: Virginia Beach, 2 mi NE.

GENERAL INFORMATION: Fleet Combat Tng Ctr with many tenant commands, incl Naval Guided Missiles School and Fleet Combat Directional Support Systems.

TEMPORARY MILITARY LODGING: BOQ-Biltg Ofc, Off, Bldg 241 and 225, Regulus Ave, BEQ-Bldg 550, Terrier Ave, both dty hrs, Comm: 804-433-6366, ATVN: 433-6366 (Off), Comm: EX-6691, ATVN: 433-6691 (Enl). All ranks. DV/VIP, EX-2000.

LOGISTICAL SUPPORT: Most spt fac avail. Key phone numbers follow.

BX-433-6146	Chap-433-6602	CDO-433-6234
Conv S-433-7754	Gas S-433-7794	Loc-433-6000
Med-433-6327	MMO-433-6211	Off C-433-6141
Pol-433-6186	Theat-433-6337	

HEALTH & WELFARE: Dispensary/Emerg, EX-6327. Inpatient, Portsmouth Naval Hosp, Comm: 804-398-5300/5330, ATVN: 564-5300/5330. Chapels serving all faiths, EX-6602.

RECREATION: Bowling, EX-6341, Swimming, EX-6551, Golf, EX-2866, Gym, EX-6101, Library, EX-6565, Off/CPO/Enl Beach. Stewart Campground, off Base, 35 mi S of Norfolk, in Chesapeake, operated by and located on Naval Security Group Activity near Atlantic beaches, Apr-Oct (Nov-Mar, self-contained only), Comm: 804-421-8262, ATVN: 564-1336, EX 262, 18 camper spaces w/elec, 13 camper spaces w/hookups. Horseback riding (Oceana NAS), EX-3266.

SPACE-A: Oceana NAS, Pax Term and Lounge, Bldg 100, 24 hrs dly, Comm: 804-433-2260/2162, ATVN: 433-2260/2162, flts to CONUS and overseas. Also w/in 20 miles, NAS Norfolk, w/in 35 miles Langley AFB.

ATTRACTIONS: Virginia Beach, water sports, close to Norfolk, Williamsburg.

DEFENSE GENERAL SUPPLY CENTER (VA30R1)

Comm: 804-275-3861
ATVN: 695-1110

Richmond, VA 23297-5000

LOCATION: From I-95 (Richmond-Petersburg Turnpike) exit 6 or 7 to US-1/301. Clearly marked on US-1/301. RM: p-99, N/20. NMC: Richmond, 8 mi N

GENERAL INFORMATION: General Supply Ctr of the Defense Logistics Agency.

TEMPORARY MILITARY LODGING: Limited. Try VOQ, EX-3861.

LOGISTICAL SUPPORT: Limited spt fac avail. Key phone numbers follow.

PX-275-2654	Caf-271-7254	Commiss-275-3904

Defense GSC, Cont'd

Gas S-275-1478	Loc-275-3861	Med-275-3821
NCO C-275-3707	Off C-275-4258	Pkg S-275-1498
Pol-275-3861	Snack B-275-3572	

HEALTH & WELFARE: Clinic, 275-3821. Inpatient, Kenner Army Hosp, Fort Lee, Petersburg, 25 mi S, 804-734-3443.

RECREATION: None. Canadian elk roam the 20-acre pasture of the Center.

SPACE-A: None.

ATTRACTIONS: Historic Richmond is short drive N.

FORT A.P. HILL (VA17R1)

Bowling Green, VA 22427-5000

long drive

Comm: 804-633-5041
ATVN: 934-8110

LOCATION: From I-95, take Bowling Green/Fort A. P. Hill exit, US-17 (bypass) E to VA-2 S to Bowling Green, take VA-301 NE to main gate. Also, exit I-95 to VA-207 N to VA-301 N and to main gate. RM: p-99, K/20. NMC: Fredericksburg, 14 mi NW.

GENERAL INFORMATION: Army Active, Reserve and ANG Training Center.

TEMPORARY MILITARY LODGING: Biltg Ofc, Bldg TT-0114, 4th St, 0800-1630 hrs dty dys, Comm: 804-633-8459, ATVN: 934-8459. All ranks. DV/VIP, EX-8459.

LOGISTICAL SUPPORT: Limited spt fac avail. Key phone numbers follow.

PX-633-5420	Med-633-8216	Aud/Theat-633-8375/8219
Comm C-633-8398		

HEALTH & WELFARE: Dispensary/Emerg, EX-8216. Inpatient, Dewitt Army Hosp, Fort Belvoir, Comm: 703-781-8670, ATVN: 354-8670. No chapel.

RECREATION: Rec Ctr, Pool, Hunting, Skeet Range, Fishing, Recreation Facilities, on Post, 48 camper spaces w/hookups, 3 log cabins, and "The Lodge" (9 bedrooms), Comm: 804-633-8219, ATVN: 934-8219.

SPACE-A: See Andrews AFB, MD, listing.

ATTRACTIONS: Nearby historic Fredericksburg and Civil War battlefields.

FORT BELVOIR (VA12R1)

Fort Belvoir, VA 22060-5000

Comm: 703-664-6071
ATVN: 354-6071

LOCATION: From I-95 S or US-1 S take Fort Belvoir exits. Signs clearly mark support facilities. RM: p-99, I-21. NMC: Washington, 10 mi NE.

GENERAL INFORMATION: Army Engineer Center and School, Defense Management College, Defense Mapping School and support units.

TEMPORARY MILITARY LODGING: Biltg Ofc, Bldg 470, Snow Loop, 24 hrs dly, Comm: 703-664-2005, ATVN: 354-2005. All ranks. DV/VIP, 664-1244.

LOGISTICAL SUPPORT: Complete spt fac avail. Key phone numbers follow.

PX-781-7070	Chap-664-6211	Chld C-664-3875
Commiss-664-4892/6704	Conv S-780-4976	EM C-780-0962
Fam Svc-664-3965	Gas S-781-7643	Legal-664-4018

Fort Belvoir, Cont'd

Loc 664-3096	Med-664-1938	NCO C-780-0962
Off C-780-0930	Pkg S-780-0954	Pol-664-1251
Snack B-781-7080/8295	Theat-664-2213/664-3211	

HEALTH & WELFARE: Dewitt Army Hospital, clinics and depts, Emerg, 664-1938, Appoint, 781-8670, 120 beds. Chapels serving all faiths, 664-3391/6211/1218.

RECREATION: Arts & Crafts, 664-5931, Auto Hobby, 664-3889, Bowling, 664-6094, Equip Rental, 664-2714, Golf, 664-6016/1978, Gym, 664-2996/2988, Library, 664-6255/2387/2589, Marina, 664-4835, Picnic Area, 664-2714, Rec Ctr, 664-1847, Skeet Range, 664-2242, Stables, 664-5896, Swimming, 664-3612, Tennis, 664-3455/3571, Wildlife Refuge, 664-4131, Youth Ctr, 664-6571.

SPACE-A: Davison Army Airfield, Base Ops, 0730-1630 hrs dly, Comm: 703-664-6565, ATVN: 354-6565, flts to CONUS locations via exec aircraft.

ATTRACTIONS: Washington, DC. Eighteenth century homes in the area.

FORT EUSTIS (VA10R1)

Fort Eustis, VA 23604-5015

Comm: 804-878-5251
ATVN: 927-5251

LOCATION: From I-64, exit 60A to VA-105, W to Fort. RM: p-99, N/22. NMC: Newport News, 13 mi SW.

GENERAL INFORMATION: Army Transportation School and Ctr and support units.

TEMPORARY MILITARY LODGING: Biltg Ofc, Bldg 2110, Pershing Ave, 24 hrs dly, Comm: 804-878-2337/5807, ATVN: 927-2337/5807. All ranks. DV/VIP, EX-3475.

LOGISTICAL SUPPORT: Complete spt fac avail. Key phone numbers follow.

PX-887-3564	Caf-887-0494	Chap-878-3989
Chld C-878-3294	Commiss-878-5966	Conv S-887-5966
EM C-887-0986	Fam Svc-878-3638	Gas S-887-0392
Legal-878-5286	Loc-878-5251	Med-878-4555
NCO C-887-0787	Off C-887-0191	Pkg S-887-0882
Pol-878-4555	Snack B-887-0494	Theat-878-2744

HEALTH & WELFARE: McDonald Army Hospital, clinics and depts, Emerg, EX-4555. Chapels serving all faiths, EX-3989.

RECREATION: Arts/Crafts, EX-3285, Auto Hobby, EX-5440, Bowling, EX-5482, ITT, EX-3694, Golf, EX-2965, Gym, EX-2328, Sprts Rntal, EX-4448, Lib, EX-5017, Swim, EX-5544, Rec Ctr, EX-3717, Tennis, EX-2565, Flying Club, EX-3722, Saddle Club, EX-2835, Skeet Range, EX-4412, Go-Cart Track, EX-2028, Marina, EX-4416.

SPACE-A: Felker Army Airfield, Base Ops, flts via exec aircraft, EX-5251. Also, see Norfolk NAS listing.

ATTRACTIONS: On the James River, historic Williamsburg nearby, US Transportation Museum (free), Busch Gardens, Mariners' Museum.

FORT LEE (VA15R1)

Fort Lee, VA 23801-5000

Comm: 804-734-1011
ATVN: 687-0111
FTS: 927-1011

LOCATION: From 1-95 take Fort Lee/Hopewell exit, and follow VA-36 to main gate. RM: p-99, N/20. NMC: Petersburg, 3 mi W.

Fort Lee, Cont'd

GENERAL INFORMATION: Quartermaster Center and School, Troop Support Agency, Army Logistics Management Center, Army Information Systems Software Development Center, Army Logistics Center, other logistics and reserve units.

TEMPORARY MILITARY LODGING: Biltg Ofc, Bldg P-8025, Mahoe Ave, 24 hrs dly, Comm: 804-734-1877/4023, ATVN: 687-1877/4023. All ranks. DV/VIP, EX-3475.

LOGISTICAL SUPPORT: Complete spt fac avail. Key phone numbers follow.

PX-861-6580	**Caf-861-6480**	**Chap-734-2937/4437**
Chld C-861-3373	**Commiss-734-3663**	**Conv S-861-6241/5540**
EM C-861-8797	**Fam Svc-734-3707**	**Gas S-861-6794**
Legal-734-4058	**Loc-734-2021**	**Med-734-3637**
NCO C-861-6280	**Off C-861-5480**	**Pkg S-861-3297**
Pol-734-2072	**Snack B-861-6378**	**Theat-734-1118**

HEALTH & WELFARE: Kenner Army Hospital, clinics and depts, Emerg, EX-3637, Appoint, EX-3443, 120 beds. Chapels serving all faiths, EX-2937/4437.

RECREATION: Arts & Crafts, EX-1825, Auto Hobby, EX-3376, Bowling, EX-2265, Golf, EX-5717, Gym, EX-2282, Sports Rental, EX-2882, Riding Club, 733-9838, Rod & Gun Club, 733-0145, Flying Club, EX-4147, Boats, EX-2882, Music Ctr, EX-4919, Library, EX-2322, Swimming, EX-2975, Rec Ctr, EX-1432, Live Theater, EX-1679, Youth Ctr, EX-1861. Freedom Star Recreation Area, on Post, off Ave A & 38 St, Mar-Nov, Comm: 804-734-2882/5038, ATVN: 687-2882/5038, 7 camper spaces w/hookups, 6 camper spaces w/o hookups, full spt facilities and all activities.

SPACE-A: Army Airfld, Base Ops, flts via exec aircrft, CONUS loc, EX-1011.

ATTRACTIONS: Quartermaster Museum, Petersburg National Battlefield Park, Colonial Williamsburg, Richmond's Virginia and Poe Museums + Monument Avenue; Petersburg's Blandford Church and Cemetery and Trapezium House, Colonial Heights' Lee Headquarters (Violet Bank), and Hopewell's Appomattox Manor.

FORT MONROE (VA13R1)

Fort Monroe, VA 23651-5000

Comm: 804-727-2111
ATVN: 680-2111

LOCATION: From I-64 exit at Hampton and follow tour signs through Phoebus to Fort. RM: p-98, B/3. NMC: Hampton, 1 mi N.

GENERAL INFORMATION: Hq Army Training & Doctrine Command, Naval Surface Weapons Center, Continental Army Band, and support units.

TEMPORARY MILITARY LODGING: Biltg Ofc, ATTN: ATZG-EHH, Bldg 28, Murray St, 0800-1645 dty dys, Comm: 804-727-2128, ATVN: 680-2128. All ranks. DV/VIP, EX-4401.

LOGISTICAL SUPPORT: Complete spt fac avail. Key phone numbers follow.

PX-722-0794	**Chap-727-2611**	**Chld C-727-3595**
Commiss-727-3326	**Conv S-722-0794**	**EM C-727-2796**
Fam Svc-727-3878	**Gas S-722-4843**	**Legal-727-2156**
Loc-727-2317	**Med-727-2219**	**NCO C-727-2796**
Off C-727-2406/2432	**Pkg S-727-2487**	**Pol-727-2238**
Snack B-722-3864	**Theat-727-2793-2004**	

HEALTH & WELFARE: Dispensary/Emerg, EX-2219, Appoint, EX-2219. Inpatient, USAF Hosp, Langley AFB, 764-6800. Chapels serving all faiths, EX-2611/2730.

RECREATION: Arts/Crafts, EX-2924, Auto Hobby, EX-2311, Moral Support Ofc,

Fort Monroe, Cont'd

EX-2783, ITT, EX-2685, Golf, Langley AFB, 764-4547, Historic and Arch Society, EX-2526/2504, Fishing/Boating, EX-4308, Library, EX-2909, Swimming, EX-2696 (enlisted), 4181 (officer), Rec Ctr, EX-3957, Youth Ctr, EX-2685/6/3957, YMCA, 723-1852, Bowling, 727-2939.

SPACE-A: See Norfolk NAS listing.

ATTRACTIONS: Williamsburg, Jamestown, Norfolk, Casemate Museum (727-3973).

FORT MYER (VA24R1)

Fort Myer, VA 22211-5000

Comm: 202-545-6700
ATVN: 222-6700

LOCATION: Adjacent to Arlington National Cemetery. Take Fort Myer exit from Washington Blvd, at 2nd St, or enter from US-50 (Arlington Blvd) first gate. Also, exit from Boundary Drive to 12th St North entrance near the Iwo Jima Memorial. RM: p-42, I/7. NMC: Washington, DC, 1 mi NE.

GENERAL INFORMATION: Hq 3rd Infantry (Old Guard), Hq US Army Band, and support units.

TEMPORARY MILITARY LODGING: Biltg Ofc, Bldg T-49, Jackson St, 24 hrs dly, Comm: 202-692-3576, ATVN: 222-3576. All ranks. DV/VIP, EX-7051.

LOGISTICAL SUPPORT: Complete spt fac avail. Key phone numbers follow.

PX-522-4575	**Chap-696-3685**	**Chld C-527-1300**
Commiss-692-3685	**EM C-527-1300**	**Fam Svc-696-3520**
Gas S-522-2584	**Legal-696-3026**	**Loc-545-6700**
Med-696-3628	**NCO C-527-1300**	**Off C-524-7000**
Pol-692-9477	**Snack B-528-1039/527-5838**	**Theat-696-3049**

HEALTH & WELFARE: Clinic/Emerg, EX-9700, Appoint, EX-1530. Inpatient, Dewitt Army Hosp, Fort Belvoir, 703-664-1938. Chapels serving all faiths, Post, EX-2506, Memorial, EX-9490.

RECREATION: Arts & Crafts, EX-6783, Auto Hobby, EX-6787, Bowling, EX-9590, Gym, EX-9655, Library, EX-9574, Multi-Crafts Shops, EX-0957/6783, Museum, EX-9721, Rec Ctr, EX-9339/9343, Rec Equip, EX-9375, Swimming, EX-2940/7000, Tennis/Racquet Ball, EX-9649, Youth Ctr, EX-3858.

SPACE-A: See Andrews AFB, MD, listing.

ATTRACTIONS: Arlington National Cemetery, barracks and activities of the Old Guard, Old Guard Museum, great nation's capital, Washington, DC, is nearby.

FORT PICKETT (VA16R1)

Blackstone, VA 23824-5000

Comm: 804-292-8261
ATVN: 438-8261

LOCATION: On US-460, 1 mi from Blackstone. Clearly marked. RM: p-99, O/18. NMC: Richmond, 60 NE.

GENERAL INFORMATION: National Guard and Reserve Training Post for all branches of services.

TEMPORARY MILITARY LODGING: Biltg Ofc, ATTN: ATZM-FPE-H, Bldg T-469, Military Rd, 0730-1600 dty dys, Comm: 804-292-8309/8320, ATVN: 438-8309/8320. All ranks. DV/VIP, EX-8303.

LOGISTICAL SUPPORT: Most spt fac avail. Key phone numbers follow.

Fort Pickett, Cont'd

PX-292-5489	Chap-292-2601	EM C-292-2336
Gas S-292-5527	Legal-292-2168	Loc-292-2419
Med-292-2528	NCO C-292-2336	Off C-292-2336
Pol-292-8444/5	Snack B-292-5489	Theat-292-8466

HEALTH & WELFARE: Dispensary/Emer, EX-2211. Inpatient, Kenner Army Hosp, Ft Lee, Comm: 804-734-3637, ATVN: 687-3637. Chapel serving all faiths, EX-2601.

RECREATION: Bowling, EX-8518, Swimming, EX-8538, Tennis, EX-2233, Gym, EX-2233, Field House, EX-2233, Craft Shop, EX-2616, Racquetball Court. Travel Camp, on Post, year round, Comm: 804-292-8309, ATVN: 438-8309, 27 camper spaces w/hookups, 8 tent camp sites, full fac and all act, great hunting in season.

SPACE-A: Blackstone Army Airfield, very limited via exec aircraft to CONUS locations, EX-8261.

ATTRACTIONS: Petersburg Civil War Battlefields, 35 mi; Appomattox Court House Natl Hist Pk, 60 mi; Buggs Is Lake/Lake Gaston, 60 mi.

FORT STORY (VA08R1)

Fort Story, VA 23459-5000

Comm: 804-422-7210
ATVN: 927-9210

LOCATION: From S exit of Chesapeake Bay Bridge Tunnel (US-13), E on US-60 (Atlantic Ave) to Fort Story. Clearly marked. From I-64 take US-60 E. From VA-44 (Norfolk-VA Beach Expway) exit US-58, left turn to N on Atlantic Av (US-60) to 89th St to Ft Story. RM: p-98, D/6. NMC: Virginia Beach, 3 mi S.

GENERAL INFORMATION: Sub-Installation of Fort Eustis. Army Transportation Units, Test Activity, Army Logistics, and Over-the-Shore Training & Test Site.

TEMPORARY MILITARY LODGING: Biltg Ofc, Bldg 300, Rm 206, Guadalcanal Rd, 0730-2330 hrs dly, Comm: 804-422-7321, ATVN: 927-9321. All ranks. DV/VIP, Fort Eustis, Comm: 804-878-5206, ATVN: 927-5206/5207.

LOGISTICAL SUPPORT: Complete spt fac avail. Key phone numbers follow.

PX-422-7858/425-0001	Chap-422-7552/7665	Commiss-422-7646/7636
Conv S-428-0188	Fam Svc-422-7311	Gas S-428-0188
Legal-422-7311	Loc-422-7095/7682	Med-422-7822/7802
NCO C-428-2418/422-7405	Off C-425-6631/6632	Pkg S-422-7863
Pol-422-7141	SDO-422-7454	Theat-422-7472

HEALTH & WELFARE: All care, Portsmorth Naval Hospital, Comm: 804-398-5330. Chapels serving all faiths, EX-7552/7665.

RECREATION: Arts & Crafts, EX-7736, Auto Hobby, EX-7713, Bowling, EX-7084, Int'l Sports, EX-7975, Rec Ctr & Sports Rental, EX-7472, Tennis, EX-7975, Fishing Charters, EX-7140, Lib, EX-7525, Tickets, EX-7472. Travel Camp, on Post, Apr-Oct, Comm: 804-422-7601, ATVN: 927-9601, 27 camper spaces w/hookups, 9 mobile homes. Golf (464-7849) and Sailing (464-7126) avail at Little Creek NAB.

SPACE-A: See Oceana NAS, Norfolk NB (NAS), and Langley AFB listings.

ATTRACTIONS: Virginia Beach City and resort beaches nearby. See the Cross at Cape Henry (located on Fort Story), site of the first stop of English settlers in the USA.

HENDERSON HALL USMC (VA21R1)

Arlington, VA 22214-5000

Comm: 703-694-2200
ATVN: 224-2013

Henderson Hall, Cont'd

LOCATION: From Columbia Pike, VA-244, exit to South Orme St, Sheraton National Hotel at corner, W to dead end, right to 2nd gate left. Also, from Memorial Drive (VA-27) exit to Columbia Pike, first traffic light right onto Southgate Rd, enter first gate on right. Across Southgate Road from Navy Annex. RM: p-42, I/7. NMC: Washington, DC, 1 mi NE.

GENERAL INFORMATION: Hq Battalion USMC, HQMC and support units.

TEMPORARY MILITARY LODGING: None.

LOGISTICAL SUPPORT: Most spt fac avail. Key phone numbers follow.

MCX-979-8420	**Chap-694-2779**	**EM C-694-2295**
Fam Svc-694-5064	**HQMC-694-1235**	**Legal-694-3800**
Loc-694-2013	**Pol-694-2200**	**SNCO C-694-2125**

HEALTH & WELFARE: Pentagon Dispensary/Emerg, 695-3581. Inpatient, Bethesda Naval Hosp, MD, 202-295-4810.

RECREATION: Gym, 694-2415, Rec Equip, 694-2415, Racquetball, EX-2415, Swimming, EX-2415, Tickets, 697-2703.

SPACE-A: See Andrews AFB listing.

ATTRACTIONS: Adjacent to Arlington National Cemetery and Pentagon Building. Provides support to personnel at HQMC (Navy Annex across street). New exchange opened July 30, 1986 featuring underground parking, Burger King and Baskin Robbins. A new gym complex is currently under construction.

JUDGE ADVOCATE GENERAL'S SCHOOL (VA01R1)

Comm: 804-293-6885
ATVN: 274-7110

Charlottesville, VA 22901-5000

LOCATION: On the grounds of the University of Virginia at Charlottesville. RM: p-99, K/17. NMC: Charlottesville, within city limits.

GENERAL INFORMATION: Army Judge Advocate General's School.

TEMPORARY MILITARY LODGING: Biltg Ofc, Rm B017, 0800-1630 dty dys, Comm: 804-293-2402, ATVN: 274-2402. Officers only. DV/VIP, EX-6885.

LOGISTICAL SUPPORT: Limited spt fac avail. Key phone numbers follow.

PX-293-3383	**Legal-293-4731**	**Off C-293-4590**

HEALTH & WELFARE: UVA Emerg Rm, 924-2231. Inpatient, Kenner Army Hosp, Fort Lee, Petersburg, 100 mi SE, 804-734-3637. Religious services on campus.

RECREATION: All the facilities of UVA.

SPACE-A: None.

ATTRACTIONS: Historic UVA, and city of Charlottesville.

LANGLEY AIR FORCE BASE (VA07R1)

Comm: 804-764-9990
ATVN: 432-1110

Langley Air Force Base, VA 23665-5000

LOCATION: From I-64 E in Hampton take Armistead Ave exit, R to stop light; R onto LaSalle Ave and enter AFB. RM: p-98, A/2. NMC: Hampton, 1 mi W.

Langley AFB, Cont'd

GENERAL INFORMATION: Tactical Air Command Base. Hq Tactical Air Command, HQ Air Division Tactical Air Command, 1st Tactical Fighter Wing, 5th Weather Wing and support units.

TEMPORARY MILITARY LODGING: Biltg Ofc, 1 CSG/SVH, Bldg 75, Nealy Ave, 24 hrs dly, Comm: 804-764-4051, ATVN: 432-4051. All ranks. DV/VIP, EX-3467.

LOGISTICAL SUPPORT: Complete spt fac avail. Key phone numbers follow.

BX-766-1253	**Caf-766-1255**	**Chap-764-7847**
Chld C-764-3585	**Commiss-766-1304**	**Conv S-766-1249**
Fam Svc-764-4875	**Gas S-766-1286**	**Legal-764-3276**
Loc-764-5615	**Med-865-6800**	**NCO C-766-1220**
Off C-766-1361	**Pkg S-766-4330**	**Rec Ctr 764-7175**
Snack B-766-1237	**Theat-764-2580**	

HEALTH & WELFARE: USAF Hospital, clinics and depts, Emerg EX-6800, Appoint, EX-6400/6010/6020, 105 bed. Chapels serving all faiths, EX-7847.

RECREATION: Arts/Crafts, EX-4647, Wood Shop, EX-3978, Auto Hobby, EX-4607, Bowling, EX-2433, Golf, EX-4547, Gym, EX-5791, Sports Rental, EX-4616, Aero Club, EX-7486, Marina, EX-7220, Scuba Club, EX-2312, Rifle Club, EX-2312, Skeet Club, EX-3769, Eagle Park, EX-5792, Library, EX-2906, Swimming, EX-5791 Rec Ctr, EX-7175, Youth Ctr, EX-2300. Big Bethel Rec Area, 7 mi from Base at Big Bethel Reservoir, Apr-Nov, Comm: 804-764-2470, ATVN: 432-2470, no camper spaces, other fac and all activities are complete.

SPACE-A: Base Ops, 24 hrs dly, Comm: 804-764-5807/4698/4311, ATVN: 432-5807/4698/4311, flts to CONUS locations.

ATTRACTIONS: Virginia Tidewater Area, Norfolk nearby. Great fishing, beaches, and water sports.

LITTLE CREEK NAVAL AMPHIBIOUS BASE (VA19R1)

Comm: 804-464-7000
ATVN: 680-0111

Norfolk, VA 23521-5000

LOCATION: From I-64 S through Hampton Roads Bridge Tunnel take Northampton Blvd exit, 5 mi to Base, exit VA-225. From Chesapeake Bay Bridge Tunnel proceed W on US-60 to Base. RM: p-98, D/5. NMC: Norfolk, 11 mi SW.

GENERAL INFORMATION: Naval Amphibious School, Landing Force Training Command, Special Warfare Group, Atlantic homeport for approximately 30 ships.

TEMPORARY MILITARY LODGING: Biltg Ofc, ATTN: Code N73, Bldg 3408, 24 hrs dly, Comm: 804-464-7522 (Off), EX-8728 (Enl), ATVN: 680-7522/8728. All ranks. DV/VIP, EX-7000. Navy Lodge, Comm: 804-464-6215.

LOGISTICAL SUPPORT: Complete spt fac avail. Key phone numbers follow.

NX-460-1941	**Chap-464-7427**	**Chld C-464-7901**
Commiss-464-3561	**Conv S-464-3090**	**CPO C-464-7789/7710**
EM C-464-7949/3581	**Fam Svc-464-7563**	**Galley-464-7546**
Gas S-460-1941	**Legal-464-7331**	**Loc-464-7226**
Med-464-7879	**Off C-460-1111**	**Pkg S-464-7466**
Pol-464-7621	**Snack B-464-2663**	**Theat-464-7534**

HEALTH & WELFARE: Clinic/Emerg, EX-7879/7858, Appoint, EX-7869. Inpatient, Portsmouth Naval Hosp, Comm: 804-398-5330, ATVN: 564-5330. Chapels serving all faiths, EX-7427/7429.

Little Creek NAB, Cont'd

RECREATION: MWR Dir, EX-7905, Auto Hobby, EX-7339, Ceramics Hobby, EX-7998, Woodworking Hobby, EX-7605, Bowling, EX-7953, Golf, EX-8526, Gym, EX-7478, Sports Rental, EX-7735, Ballet and Slimnastic Classes, EX-7665, Picnic Area, EX-7369, Tennis, EX-8118, Fishing Boat (Gator III), EX-7140, Boats/Campers, EX-7516, Library, EX-7691, Swimming, EX-8139, Tickets, EX-7665.

SPACE-A: Norfolk NAS, Bldg LP-84, 24 hrs dly, Comm: 804-444-4118/4148, ATVN: 564-4118/4148, flts to CONUS and overseas.

ATTRACTIONS: Water sports, beaches, fishing, Norfolk.

NORFOLK AREA

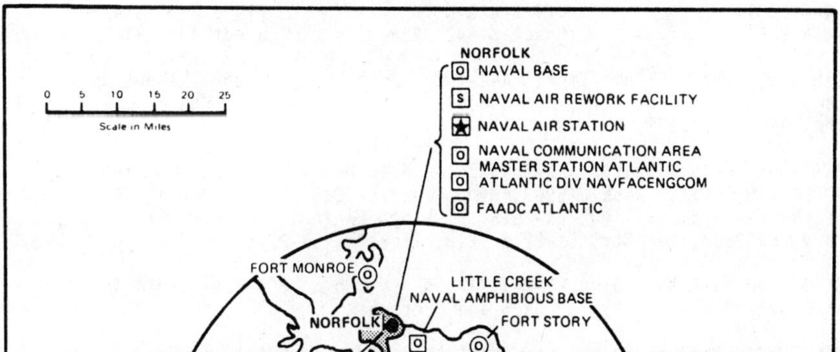

NORFOLK NAVAL BASE (VA18R1)

Norfolk, VA 23511-5000

Comm: 804-444-0000
ATVN: 564-0111
FTS: 954-0111

LOCATION: From north take I-64 W, take Naval Base exit, follow signs. From south take I-64 exit to I-564 into Gate 3. RM: p-98, C/3. NMC: Norfolk, in city limits.

GENERAL INFORMATION: CINC Atlantic Fleet, SACLANT (NATO), FMF Atlantic,

Norfolk NB, Cont'd

Surface and Submarine Forces Atlantic, Air Forces Atlantic, Armed Forces Staff College, and the world's largest Naval Base including the Norfolk NAS, and Norfolk Naval Station.

TEMPORARY MILITARY LODGING: Biltg Ofc, ATTN: Code 06, Bldg J-53, Carter Hall, 0730-1615 hrs dty dys, Comm: 804-444-2839/4294, ATVN: 564-2839/4294. Navy Lodge, 489-2656. All ranks. BOQ, 444-3250.

LOGISTICAL SUPPORT: Complete spt fac avail. Key phone numbers follow.

NX-440-2000	**Caf-440-2241**	**Chap-444-7156**
Chld C-444-4672	**Commiss-423-3188**	**Conv S-423-3751**
CPO C-444-4033	**EM C-444-4576**	**Fam Svc-444-2102/6289**
Gas S-444-2266	**Legal-444-4672**	**Loc-444-0000**
Med-444-1531	**Off C-423-8450**	**Pkg S-489-2684/5**
Pol-444-2361	**Snack B-440-2282**	**Theat-444-2216**

HEALTH & WELFARE: Clinic/Emerg, 444-1531, Appoint, 398-5330. Inpatient, Portsmouth Naval Hosp, Comm: 804-398-5413, ATVN: 564-5413. Chapels serving all faiths, 444-7156/7157.

RECREATION: Arts & Crafts, EX-3846/2614, Auto Hobby, EX-8159, Bowling, EX-7908/2178, Golf, EX-5572/5267, Gym, EX-2276/7218, Sports Rental, EX-2536, Rec Park, EX-4184, Sailing, EX-2918, Fishing, EX-2631, Rec Ofc, EX-7218/2276, Library, EX-2888, Rec Ctr, EX-2276/7218, Tickets, EX-2225.

SPACE-A: Norfolk NAS, Bldg LP-84, 24 hrs dly, Comm: 804-444-4118/4148, ATVN: 690-4118/4148, flts to CONUS and overseas.

ATTRACTIONS: Gardens by the Sea, MacArthur Memorial, Williamsburg VA nearby, Norfolk NB tour which includes a stop at the Hampton Roads Naval Museum, city of Norfolk VA, Hampton Roads Museum.

NORFOLK NAVAL SHIPYARD (VA26R1)

Portsmouth, VA 23709-5000

Comm: 804-396-3000
ATVN: 961-0111

LOCATION: From I-264 in Portsmouth exit on Effingham St S to enter the Shipyard. RM: p-98, E/4. NMC: Portsmouth, within city limits.

GENERAL INFORMATION: World's largest shipyard devoted exclusively to ship repair work.

TEMPORARY MILITARY LODGING: Biltg Ofc, 24 hrs, Comm: 804-396-5979 (Off), EX-3119 (Enl), ATVN: 961-5979/3119. All ranks. DV/VIP, EX-3000.

LOGISTICAL SUPPORT: Complete spt fac avail. Key phone numbers follow.

NX-396-7068	**Chap-396-5021/2**	**Chld C-396-5863**
Commiss-396-6309	**EM C-396-5054/3660**	**Fam Svc-444-6289**
Gas S-397-3511	**Legal-444-4496**	**Loc-396-3000**
Med-396-3678/3268	**Off C-397-1328/4281**	**Pkg S-399-6673**
Pol-396-7266	**Snack B-396-7068**	

HEALTH & WELFARE: Clinic/Emerg, EX-3678, Appoint, EX-3278. Inpatient, Portsmouth Naval Hosp, Comm: 804-398-5330, ATVN: 564-5330. Chapels serving all faiths, 444-7156 (Duty Chaplain), EX-5021/2.

RECREATION: MWR Ofc, EX-3835, Bowling, EX-3808/7051, Gym, EX-3845, Athl Fac, EX-3845/3835, Pool Tables, EX-3835, Library, 444-1571, Swimming, EX-3835.

Norfolk Nav Shipyd, Cont'd

SPACE-A: See Norfolk Naval Base listing for Norfolk NAS.

ATTRACTIONS: Virginia Beach nearby. Tidewater outdoor sports. Naval Shipyard Museum, Mariners' Museum, college sports, Triple-A baseball.

OCEANA NAVAL AIR STATION (VA09R1)

Virginia Beach, VA 23460-5120

Comm: 804-433-2000
ATVN: 433-2000

LOCATION: From I-64 exit to Norfolk-Virginia Beach Expressway (VA-44 E), east on Virginia Beach Blvd. Bordered by Oceana Blvd (VA-615) and London Bridge Rd. Also, bordered by Potters and Harpers Rds. RM: p-98, F/7. NMC: Virginia Beach, in city limits.

GENERAL INFORMATION: Tactical Wings Atlantic, Fighter Wing One, Medium Attack Wing One, and fleet and training squadrons.

TEMPORARY MILITARY LODGING: Biltg Ofc, Bldg 460, G St, 24 hrs dly, Comm: 804-433-3293/4, ATVN: 433-3293/4. All ranks, space available only.

LOGISTICAL SUPPORT: Complete spt fac avail. Key phone numbers follow.

NX-425-6710	Caf-425-6710	Chap-433-2871
Chld C-433-3164	Commiss-428-2931	Conv S-425-6710
EM C-433-2453	Fam Svc-433-2912	Gas S-425-6710
Legal-433-2946	Loc-433-2366	Med-433-2221
NCO C-433-2637	Off C-428-0036	Pkg S-433-2840
Pol-433-3123		

HEALTH & WELFARE: Clinic/Emerg, EX-2221, Appoint, EX-2336/2151. Inpatient, Portsmouth Naval Hosp, 804-398-5330. Chapels serving all faiths, EX-2871.

RECREATION: Auto Hobby, EX-3403, Bowling, EX-2167, Golf, EX-2866, Gym, EX-2695, Rifle and Pistol, Skeet & Trap, EX-2875, Fishing Boat & Campers, EX-3215, Picnic Area, EX-3381, Stables, EX-3266, Athletic and Rec, EX-2695, Spec Svc, EX-3381, Library, 433-2401, Swimming - Off, EX-3285, Enl, EX-2825.

SPACE-A: Pax Term and Lounge, Bldg 100, 24 hrs dly, Comm: 804-433-2162, ATVN: 433-2162, flts to CONUS and overseas.

ATTRACTIONS: Virginia Beach, resort beaches, and entertainment nearby.

PORTSMOUTH NAVAL HOSPITAL (VA27R1)

Portsmouth, VA 23708-5000

Comm: 804-398-5001
ATVN: 961-0111

LOCATION: Off I-264 in Portsmouth, take Effingham St exit to hospital. RM: p-98, E/3. NMC: Norfolk, 2 mi N.

GENERAL INFORMATION: One of the largest and the oldest medical centers in the Navy, supports the entire Tidewater area.

TEMPORARY MILITARY LODGING: Biltg Ofc, Enl only, Comm: 804-398-5703, ATVN: 961-5703. DV/VIP, none.

LOGISTICAL SUPPORT: Limited spt fac avail. Key phone numbers follow.

NX-397-7069	Chap-398-5550	CPO C-397-2431
EM C-398-5427	Fam Svc-444-6289	Legal-398-5453
Loc-398-5001	Med-398-5190	Off C-399-2376
Pkg S-399-2376	Pol-444-2361	Theat-398-5094

Portsmouth Naval Hosp, Cont'd

HEALTH & WELFARE: Hospital, clinics and depts, Emerg, EX-5190, Appoint, EX-5330, 500 bed. Chapels serving all faiths, EX-5550.

RECREATION: Spec Svc Ofc, EX-5101/5094, Athletic Fields/Courts, EX-5094, Bowling, EX-5700, Gym, EX-5097, Swimming, EX-5946, Picnic Area, EX-5094, Library, EX-5858.

SPACE-A: See Norfolk Naval Base for Norfolk NAS listing.

ATTRACTIONS: Tours: Portsmouth Harbor, MacArthur Memorial, Naval Museums.

QUANTICO MARINE CORPS COMBAT DEVELOPMENT COMMAND (VA11R1)

Quantico, VA 22134-5000

Comm: 703-640-2121
ATVN: 278-2121

LOCATION: From I-95 north or south take Quantico/Triangle VA exit. US-1 is N-S adjacent to the Base. Directions to Base clearly marked. RM: p-99, I/20. NMC: Washington, DC, 30 mi N.

GENERAL INFORMATION: Marine Corps Education, Development and Air Facility.

TEMPORARY MILITARY LODGING: Biltg Ofc, Bldg 2034, Little Hall, Barnett Ave, 24 hrs dly, Comm: 703-640-2711/2983, ATVN: 278-2711/2983. All ranks. DV/VIP, EX-2786.

LOGISTICAL SUPPORT: Complete spt fac avail. Key phone numbers follow.

MX-640-7171	Chap-640-2131	Chld C-640-2615
Commiss-640-2476	Conv S-640-7595	EM C-640-2780
Fam Svc-640-2659/2650	Gas S-640-6334	Legal-640-3261/2776
Loc-640-2121	Med-640-2525	Off C-640-6115
Pkg S-640-2587	Pol-640-2251/Info-2253	Snack B-640-7171
SNCO C-640-3166/7	Theat-640-2279/2638	

HEALTH & WELFARE: Clinic/Emerg, EX-911, Appoint, EX-2161. Inpatient, Dewitt Army Hosp, Fort Belvoir, 703-664-1938. Chapels serving all faiths, EX-2131.

RECREATION: Auto Hobby, EX-2729, Bowling, EX-2210, Special Interest Clubs, EX-3196, Golf, EX-2424, Gym/Racquet Sports, EX-2003, Library, EX-2414, Marina, EX-5270, Museum, EX-2607, Rec Ofc, EX-2002, Rifle/Pistol Club, 640-6336, Stables, EX-2288, Swimming, EX-2973, Tennis, EX-2003, Youth Ctr, EX-2249. Lunga Reservoir, on Base, year round, Comm: 703-640-2014, ATVN: 278-2014, 20 camper spaces w/elec & sewer, 6 camper spaces w/o hookups, comp fac/all act.

SPACE-A: See Andrews AFB, MD, and Fort Belvoir listings.

ATTRACTIONS: Borders Potomac River. Washington, DC, easy drive, 30 mi N.

VINT HILL FARMS STATION (VA03R1)

Warrenton, VA 22186-5013

Comm: 703-347-6000
ATVN: 249-0111

LOCATION: From Washington, DC, area, take I-66 to VA-29/211 S, proceed 4 mi to VA-215 (Mitchell Road), left for 2 mi to Station main gate. RM: p-99, I/19. NMC: Washington, DC, 35 mi NE.

GENERAL INFORMATION: Signal Warfare Laboratory, Electronic Materiel Readiness Activity, Intelligence Units.

Vint Hill Farms Sta, Cont'd

TEMPORARY MILITARY LODGING: Biltg Ofc, Bldg 250, Helms Rd, 0730-1615 hrs dly, Comm: 703-347-6325/6670, ATVN: 249-6325/6670. All ranks. DV/VIP, EX-6670.

LOGISTICAL SUPPORT: Complete spt fac avail. Key phone numbers follow.

PX-349-1877	Chap-347-6361/6717	Commiss-347-6512
Conv S-349-1870	EM C-347-5172	Fam Svc-347-6489
Gas S-349-1898	Legal-347-6357	Loc-347-6000
Med-347-6680	NCO C-347-5172	Off C-347-6322
Pkg S-349-1880	Pol-347-6236	Theat-349-6225

HEALTH & WELFARE: Clinic/Emerg, EX-6680. Inpatient, Walter Reed Med Ctr, Washington, DC, 202-576-3501. Chapel serving all faiths, EX-6361/6717.

RECREATION: Auto Hobby, EX-6419, Bowling, EX-6427, Gym, EX-6384, Library, EX-6466, Multi-Craft, EX-6784, Rec Equip, EX-6383, Swimming, EX-6774, Tennis, EX-6334, Youth Ctr, EX-6205. Rec Area, on Post, year round, Comm: 703-347-6334, ATVN: 249-6334, 15 camper spaces w/o hookups, limited fac and act.

SPACE-A: See Andrews AFB, MD, listing.

ATTRACTIONS: Beautiful Northern Virginia horse farms, historic Middleburg and Warrenton nearby.

YORKTOWN COAST GUARD RESERVE TRAINING CENTER (VA28R1)

Comm: 804-898-3500
FTS: 827-2222

Yorktown, VA 23690-5000

LOCATION: From I-64 exit to US-238 E and follow to CG Base entrance. RM: p-99, D/19. NMC: Newport News, 15 mi SE.

GENERAL INFORMATION: Coast Guard Training Schools.

TEMPORARY MILITARY LODGING: None.

LOGISTICAL SUPPORT: Limited spt fac avail. Key phone numbers follow.

CGX-898-2153	Chap-898-2390	CPO C-898-2294
EM C-898-2325	Loc-898-3500	Med-898-2230
Off C-898-2286	Pol-898-3500	

HEALTH & WELFARE: Clinic, 898-2230. Inpatient, McDonald Army Hosp, Fort Eustis, Newport News, 804-878-8005. Chaplain serving all faiths, 898-2390.

RECREATION: Auto Hobby, 898-2279, Gym, 898-2128, Library, 898-2396, Bowling, 898-2100. Campground, on Base, year round, Comm: 804-898-2222, FTS: 827-2222, 10 camper spaces w/hookups, complete fac and activities on bay.

SPACE-A: See Langley AFB listing.

ATTRACTIONS: Historic Yorktown, Jamestown, and Williamsburg nearby, Busch Gardens, Water Country USA.

YORKTOWN NAVAL WEAPONS STATION (VA14R1)

Comm: 804-887-4000
ATVN: 953-0111

Yorktown, VA 23691-5000

LOCATION: From I-64 exit to US-143 W, .5 mi to US-238 left, .5 mi to gate 3,

Yorktown NWS, Cont'd

Skiffes Creek. RM: p-99, D/19. NMC: Newport News, 15 mi SE.

GENERAL INFORMATION: Naval Mine Warfare Engineering Activity, Naval Ophthalmic Support and Training Activity, and support units.

TEMPORARY MILITARY LODGING: Biltg Ofc, Bldg 704/706, 0730-1600 dty dys, Comm: 804-887-7621, ATVN: 953-7621. All ranks. DV/VIP, EX-4444.

LOGISTICAL SUPPORT: Complete spt fac avail. Key phone numbers follow.

NX-887-2307	Caf-887-2220	Chap-887-4711
Chld C-887-4824	Commiss-887-4947	Conv S-887-3402
EM C-887-4555	Gas S-887-2363	Legal-887-7187
Loc-887-4000	Med-887-7404	Off C-887-4272
Pkg S-887-4429	Pol-887-4676	Theat-887-4601

HEALTH & WELFARE: Dispensary/Emerg, EX-7255, Appoint, EX-7255. Inpatient, McDonald Army Hosp, Fort Eustis, Comm: 804-878-8005, ATVN: 927-8005. Chapels serving all faiths, EX-4711.

RECREATION: Auto Hobby, EX-7294, Bowling, EX-4601, Golf, EX-4323, Gym, EX-4601, Handball/Racquetball, EX-4278, Library, EX-4720/4726, Boats & Motors, EX-4601, Picnic Area, EX-4601, Swimming, EX-4601, Tickets, EX-4601, Youth Ctr, EX-4380. Campsite, on Base, Apr-Oct, Comm: 804-887-4601/4741, ATVN: 953-4601/4741, 10 camper spaces w/hookups, full fac and act on Chesapeake Bay.

SPACE-A: See Fort Eustis and Langley AFB listings.

ATTRACTIONS: Historic Yorktown and Williamsburg are both nearby.

WASHINGTON

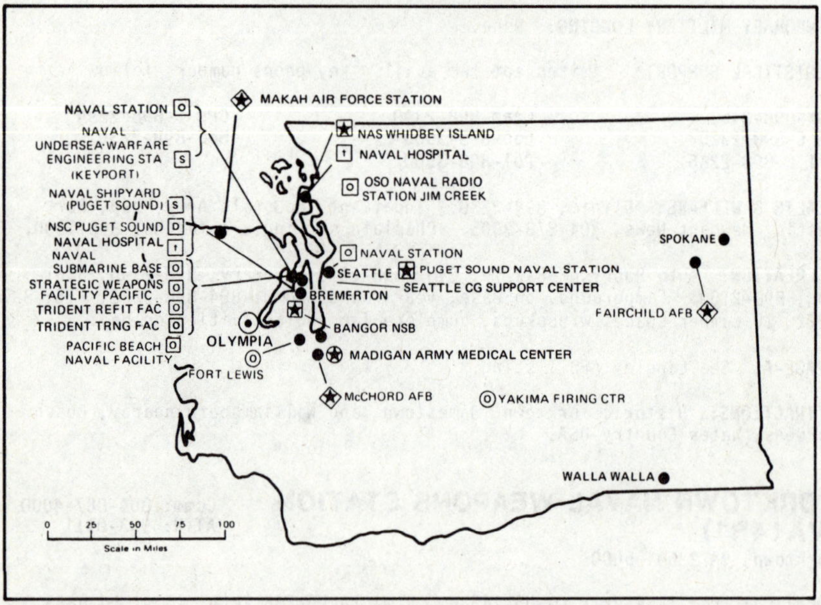

BANGOR NAVAL SUBMARINE BASE (WA08R4)

Bremerton, WA 98315-5000

Comm: 206-396-4864/4800
ATVN: 744-4864/4800

LOCATION: Approximately 15 mi N of Bremerton, on WA-3, exits to Base clearly marked. RM: p-100, G/10. NMC: Tacoma, 40 mi SE.

GENERAL INFORMATION: Trident (Ohio-class) Submarines, Commander Submarine Group 9, Submarine Squadron 17, Trident Training Facility, Strategic Weapons Facility (Pacific), Trident Refit Facility, Marine Barracks.

TEMPORARY MILITARY LODGING: None.

LOGISTICAL SUPPORT: Complete spt fac avail. Key phone numbers follow.

NX-779-3365	**Caf-779-3365**	**Chap-396-6005**
Chld C-779-4066	**Commiss-396-6025**	**Conv S-779-3365**
CPO C-779-9907	**EM C-779-9907**	**Fam Svc-396-4115**
Gas S-779-3365	**Legal-396-6157**	**Loc-396-4800**
Med-396-4391	**Off C-779-9907**	**Pkg S-779-7034**
Pol-396-4312	**Snack B-779-9907**	**Theat-779-3668/4440**

HEALTH & WELFARE: Clinic/Emerg, 479-9321, Appoint, 779-9792. Inpatient, Bremerton Naval Hosp, Comm: 206-479-9344, ATVN: 439-9344. Chapels serving all faiths, 396-6005/4864.

RECREATION: Arts & Crafts, 779-7815, Auto Hobby, 779-3596, Woodworking, 396-2246, Bowling, 779-2838, Gym, 779-4852, Library, 779-9724, Swimming, 779-4817, Rec Ctr, 779-3815, Keyport Lagoon Rec Area, 396-2169.

SPACE-A: McChord AFB, Bldg 1179, 24 hrs, Comm: 296-984-2657, ATVN: 976-2657/5944, flts to CONUS and overseas.

ATTRACTIONS: Kitsap and Olympic Peninsulas, Puget Sound and the Seattle/Tacoma area.

FAIRCHILD AIR FORCE BASE (WA02R4)

Fairchild Air Force Base, WA 99011-5000

Comm: 509-247-1212
ATVN: 352-1110

LOCATION: Take US-2 exit from I-90 W of Spokane. Follow US-2 through Airway Heights, after 2 mi left to Base main gate and vistors' control center. RM: p-101, G/23. NMC: Spokane, 12 mi E.

GENERAL INFORMATION: Strategic Air Command Base. Air Division, Bomb Wing, Air Refueling Wing, and support units.

TEMPORARY MILITARY LODGING: Biltg Ofc, ATTN: 92-CSG/SVH, Bldg 2392, Short St, 24 hrs dly, Comm: 509-247-5737/5519, ATVN: 352-5737/5519. All ranks. DV/VIP, EX-2127.

LOGISTICAL SUPPORT: Complete spt fac avail. Key phone numbers follow.

BX-244-2832	**Caf-244-2022**	**Chap-247-2264**
Chld C-247-2403	**Commiss-244-5591**	**Fam Svc-247-2577**
Gas S-244-2692	**Legal-247-5231**	**Loc-247-5875**
Med-247-5661	**NCO C-244-3622**	**Off C-244-3644**
Pkg S-244-2601	**Pol-247-5493**	**Snack B-244-3061**
Theat-247-5600		

HEALTH & WELFARE: USAF Hospital, clinics and depts, Emer, EX-5661, Appoint, EX-2361, 45 bed. Chapels serving all faiths, EX-2264.

Fairchild AFB, Cont'd

RECREATION: Arts & Crafts, EX-2810/5189, Auto Hobby, EX-2310, Bowling, EX-2422, Gym, EX-5432, Library, EX-5556, Swimming, EX-2242, Rec Ctr, EX-5649, Teen Club, EX-5603. Clear Lake Resort, 12 mi from AFB, W of Spokane off I-90, Apr-Sep (limited hrs), Comm: 509-299-5129, ATVN: 352-5129, 10 camper spaces w/hookups, 4 mobile homes, 3 cabins, complete facilities, all lake activities.

SPACE-A: Pax Term and Lounge, Bldg 1, 24 hrs dly, Comm: 509-247-5435 and 247-5439, ATVN: 352-5435/5439, flts to CONUS and overseas.

ATTRACTIONS: Spokane, parks, EXPO site, theater, nightlife.

FORT LEWIS (WA09R4)

Fort Lewis, WA 98433-5000

Comm: 206-967-5151
ATVN: 357-5151

LOCATION: On I-5 in Puget Sound area, 14 mi N of Olympia. Exits clearly marked. RM: p-100, J/10. NMC: Tacoma, 12 mi N.

GENERAL INFORMATION: Corps HQ, Division, ROTC Region HQ.

TEMPORARY MILITARY LODGING: Biltg Ofc, ATTN: AFZH-EHH, Bldg 2110, Utah & Pendleton Sts, 24 hrs at lodge, Comm: 206-967-6754/7862/2815, ATVN: 357-6754/7862/2815. All ranks. DV/VIP, EX-7385.

LOGISTICAL SUPPORT: Complete spt fac avail. Key phone numbers follow.

PX-964-3161	Caf-964-4424	Chap-967-4849
Chld C-967-7562	Commiss-967-3141	Conv S-964-2141
EM C-964-0450	Fam Svc-967-7166	Gas S-964-5656
Legal-967-3111	Loc-967-3101	Med-967-6972
NCO C-964-2555	Off C-964-0331	Pkg S-964-4311
Pol-967-3107	Snack B-2141	Theat-4329

HEALTH & WELFARE: Madigan Army Medical Center, Tacoma, Emerg, EX-6931/6972, Appoint, EX-2251, 396 bed. Chapels serving all faiths, EX-3126.

RECREATION: Craft Shops, EX-5001, Bowling, EX-4661, ITT, EX-5009, Golf, EX-6522, Gym, EX-5896/4467/5975/7471, Sports Rental, EX-5415/4314, Stadium, EX-2912, Handball, EX-3694, Fitness Ctr, EX-3508, Parachute Club, EX-3906, Flying Club, EX-4932, Riding Club, EX-2318, Hunting/Fishing, EX-6263, Marina, EX-5415/4314, Archery, EX-7788, Library, EX-5015/2703/2824, Swimming, EX-5026, Rec Ctr, EX-2539/2802/6549, Art/Theat, EX-5636, Youth Ctr, EX-5850/5776. Travel Camp, on Post, year round, Comm: 206-967-7744, ATVN: 357-7744, 24 camper spaces w/hookups, 4 camper spaces w/o hookups, 7 mobile homes, tent camping, complete facilities and all activities on lake. Also, Westport Rec Park, operated by and located on USCG Station, Grays Harbor, Westport, Mar-Oct, Comm: 206-268-0121, FTS: 420-9306, 6 camper spaces w/hookups, 6 tent camping spaces, complete facilities and all activities on Pacific Ocean.

SPACE-A: No Space-A. See McChord AFB listing.

ATTRACTIONS: Military Museum, Tacoma, Olympia nearby.

MADIGAN ARMY MEDICAL CENTER (WA15R4)

Tacoma, WA 98431-5000

Comm: 206-967-7986/7082
ATVN: 357-7986/7082

LOCATION: From I-5 N or S take the Madigan exit. Route clearly marked. RM: p-100, I/10. NMC: Tacoma, 12 mi N.

Madigan AMC, Cont'd

GENERAL INFORMATION: An Army Regional Medical Center.

TEMPORARY MILITARY LODGING: Biltg Ofc, Bldg 9900, Lincoln St, 24 hrs dly, Comm: 206-967-6180, ATVN: 357-6180. All ranks. DV/VIP, EX-3101.

LOGISTICAL SUPPORT: Complete spt fac avail at Center or Fort Lewis. Key phone numbers follow.

PX-964-3161	Caf-964-3456	Chap-967-6872
Chld C-967-7561	Commiss-967-5808	Conv S-964-2112
Fam Svc-967-7166	Gas S-964-5211	Legal-967-6737/6148
Loc-967-3101	Med-967-6972/6931	Off C-964-4054
Pkg S-964-4311	Pol-967-3107/3108	Snack B-964-3456
Theat-967-6207		

HEALTH & WELFARE: Army Hospital, clinics and depts, Emerg, EX-6972/6931, Appoint, EX-2251, 396 bed. Chapels serving all faiths, EX-6872.

RECREATION: See Fort Lewis listing.

SPACE-A: See McChord AFB listing.

ATTRACTIONS: Puget Sound area, Tacoma and Olympia.

MAKAH AIR FORCE STATION (WA04R4)

Neah Bay, WA 98357-5000

Comm: 206-645-2231
ATVN: 490-6343

LOCATION: From Port Angeles, take WA-112 W to Neah Bay and the Makah AFS. RM: p-100, D/5. NMC: Seattle, 170 mi SE.

GENERAL INFORMATION: NORAD Station.

TEMPORARY MILITARY LODGING: Biltg Ofc, 0730-1630 hrs, NCO Open Mess, 0900-2000 hrs dly (except Sun), Comm: 206-645-2375-EX-304, NTAS: 490-6304. All ranks. DV/VIP, EX-2231.

LOGISTICAL SUPPORT: Limited spt fac avail. Key phone numbers follow.

BX-EX-315	Commiss-EX-360	EM C-EX-304
Gas S-EX-315	Loc-EX-370	Med-EX-352
NCO C-EX-304	Off C-EX-304	Pol-EX-305
Theat-EX-301		

HEALTH & WELFARE: Dispensary, EX-352. Inpatient, Bremerton Naval Hosp, 206-478-9232. Chapel but no Chaplain.

RECREATION: Local outdoor recreation.

ATTRACTIONS: The most northwestern tip in the continental United States.

McCHORD AIR FORCE BASE (WA05R4)

McChord Air Force Base, WA 98438-5000

Comm: 206-984-1910
ATVN: 976-1110

LOCATION: From I-5 take exit 125. Clearly marked. RM: p-100, I/10. NMC: Tacoma, 8 mi N.

GENERAL INFORMATION: Military Airlift Command Base. Military Airlift Wing, Air Division, AFRES units, and other support activities.

McChord AFB, Cont'd

TEMPORARY MILITARY LODGING: Biltg Ofc, Bldg 166, Main St, 24 hrs dly, Comm: 206-584-1471, ATVN: 976-5613. All ranks. DV/VIP, EX-2621.

LOGISTICAL SUPPORT: Complete spt fac avail. Key phone numbers follow.

BX-582-9450	Chap-984-5556	Chld C-984-2958
Commiss-984-3285	Conv S-582-0788	EM C-984-2918
Fam Svc-984-2475	Gas S-584-2534	Legal-984-5511
Loc-984-2474	Med-984-5601	NCO C-584-1371/5134
Off C-582-4222/5581	Pkg S-588-2527	Pol-984-5624
Snack B-582-2092	Theat-984-2965	

HEALTH & WELFARE: Clinic/Emerg, EX-5601, Appoint, EX-5601. Inpatient, Madigan Army Med Ctr, Tacoma, 206-967-6972. Chapels serving all faiths, EX-5556.

RECREATION: Arts & Crafts, EX-2989, Auto Hobby, EX-2407, Bowling, EX-2007, Golf, EX-2109, Gym, EX-3335, Sports Rntl, EX-3237, Library, EX-5322, Swimming, EX-6808, Rec Ctr, EX-2918, Tickets, EX-2918, Youth Ctr, EX-2203. Fam-Camp, on Base, year round, Comm: 206-984-3281, ATVN: 976-3281, 18 camper spaces w/hookups, some fac and activities.

SPACE-A: Pax Term and Lounge, Bldg 1179, 24 hrs dly, Comm: 206-984-2657, ATVN: 976-5944, flts to CONUS and overseas.

ATTRACTIONS: Tacoma and Puget Sound area.

PUGET SOUND NAVAL SHIPYARD (WA11R4)

Comm: 206-476-3466
ATVN: 439-2011

Bremerton, WA 98314-5000

LOCATION: Off WA-3/16 in Bremerton. The 24 hr gate is at Montgomery and Decatur Ave. Also, can be reached via a 1 hr ferry ride from Seattle. RM: p-100, G/10. NMC: Bremerton, in the city.

GENERAL INFORMATION: Homeport for ships undergoing repair and other ships.

TEMPORARY MILITARY LODGING: Biltg Ofc, Bldg 865, 0730-1600 dty hrs, Comm: 206-476-7627, ATVN: 439-7627. BEQ only.

LOGISTICAL SUPPORT: Complete spt fac avail. Key phone numbers follow.

NX-478-5500	Caf-476-5529	Chap-476-2184
Chld C-478-4012	Commiss-377-3723	CPO C-478-2635
EM C-478-3341	Fam Svc-476-2747	Gas S-478-5506
Legal-476-2120	Loc-476-3466/2011	Med-476-2660
Off C-373-5014	Pkg S-377-8434	Pol-476-3393
Theat-476-2223		

HEALTH & WELFARE: All care at Bremerton Naval Hospital, Comm: 206-479-6600, ATVN: 439-6600. Chapels serving all faiths, EX-2184.

RECREATION: Bowling, Hobby Shops, Swimming, Tennis, Picnic Area, Gym, call 476-3466.

SPACE-A: See McChord AFB listing.

ATTRACTIONS: Puget Sound and Seattle/Tacoma area.

PUGET SOUND NAVAL STATION (WA10R4)

Comm: 206-526-3444
ATVN: 941-3444

Seattle, WA 98115-5001

Puget Sound NS, Cont'd

LOCATION: At 7500 Sand Point Way (WA-513), NE. Accessible from I-5 N and WA-520. RM: p-100, J/4. NMC: Seattle, in the city.

GENERAL INFORMATION: Support Activity for the Navy in the Seattle area.

TEMPORARY MILITARY LODGING: Biltg Ofc, Bldg 224, Eagles Nest Inn, 24 hrs dly, Comm: 206-526-3883, ATVN: 941-3883. All ranks. DV/VIP, EX-3211.

LOGISTICAL SUPPORT: Complete spt fac avail. Key phone numbers follow.

NX-527-7804	**Commiss-523-6935**	**CPO C-526-3539**
EM C-526-3539	**Gas S-527-7829**	**Legal-526-3835**
Loc-526-3211	**Med-526-3331**	**Off C-522-3445**
Pol-526-3211		

HEALTH & WELFARE: Dispensary/Emerg, EX-3331. Inpatient, Uniformed Services Med Treat Fac, Seattle, 206-234-7650. Chapel serving all faiths, 526-3865.

RECREATION: Fishing and camping equipment, gymnasium, sauna, basketball court, swimming pool, racquetball, tennis courts, and weight room; Special Services, Comm: 206-526-3531, ATVN: 941-3531.

ATTRACTIONS: Seattle and the Puget Sound area.

WHIDBEY ISLAND NAVAL AIR STATION (WA06R4)

Comm: 206-257-2211
ATVN: 820-0111

Oak Harbor, WA 98278-5000

LOCATION: Take WA-20 to Whidbey Island, 3 mi W of WA-20 on Ault Field Road. RM: p-100, E/10. NMC: Seattle, 60 mi SE.

GENERAL INFORMATION: Tactical Electronic Warfare Wing, Naval Air Reserve Units.

TEMPORARY MILITARY LODGING: Biltg Ofc, Bldg 973 and 11, dty hrs, Comm: 206-257-2529/2076, ATVN: 820-2529/2076. All ranks. DV/VIP, EX-2529.

LOGISTICAL SUPPORT: Complete spt fac avail. Key phone numbers follow.

NX-679-5200	**Caf-679-5205**	**Chap-257-2414**
Chld C-257-3302	**Commiss-257-3318**	**Conv S-679-5204**
CPO C-257-2505	**EM C-257-3308**	**Fam Svc-257-2901**
Gas S-679-5216	**Legal-257-2011**	**Loc-257-2351**
Med-679-5353	**Off C-257-2521**	**Pkg S-675-4231**
Pol-257-3122	**SDO-257-2631**	**Snack B-679-5237**
Theat-257-3122		

HEALTH & WELFARE: Naval Hospital, clinics and depts, Emerg, EX-2303/5353, Appoint, EX-2901, 35 bed. Chapels serving all faiths, EX-2414.

RECREATION: Arts/Crafts, EX-2173, Auto Hobby, EX-2295, Bowling, EX-2432, Flying Club, EX-5200, Int'l Sports, EX-2432, Golf, EX-2178, Gym, EX-2420, Rec Areas, EX-2431, Sailing Club, EX-2432, Rod/Gun Club, EX-3183, Karate/Judo/Backpacking, EX-2432, Archery Club, EX-9723/0151, Library, EX-2702. Operates Jim Creek Rec Area, 60 mi SE of Base, near Darrington, Apr-Sep, Comm: 206-257-2432, ATVN: 820-2432, 23 camper spaces w/o hookups, 4 camp trailers in place, most fac and act. Also operates Cliffside Rec Vehicle Park, on Base, 20 Mar-1 Dec, Comm: 206-257-2432, ATVN: 820-2432, 20 camper spaces w/hookups, 3 mobile homes, some fac and water sport act.

Whidbey Island NAS, Cont'd

SPACE-A: Pax Term and Lounge, Hangar 2, Bldg 139, 0730-1630 hrs dly, Comm: 206-257-2604/2681, ATVN: 820-2604/2681, flts to CONUS locations.

ATTRACTIONS: Outdoor sports (mostly water), fishing, great seafood.

WEST VIRGINIA

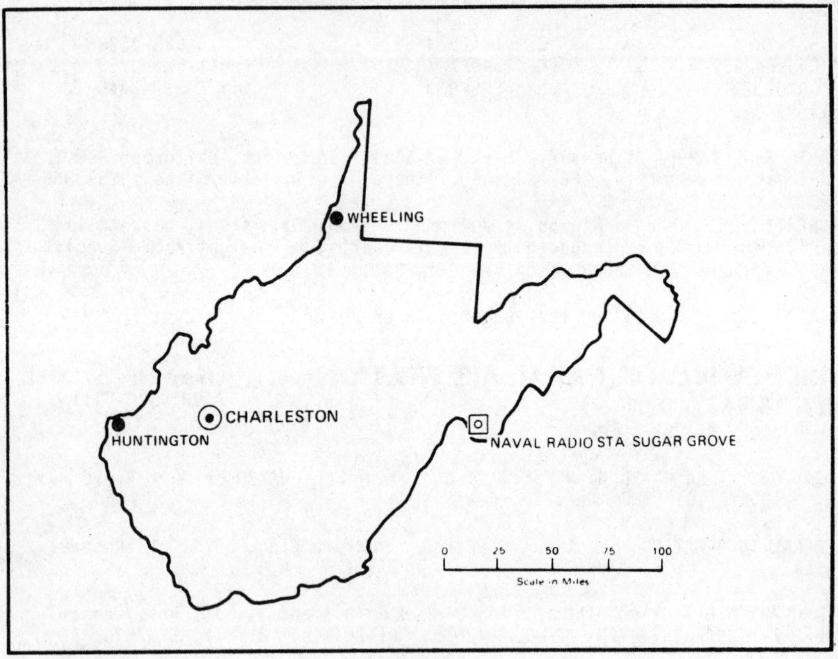

- None -

WISCONSIN

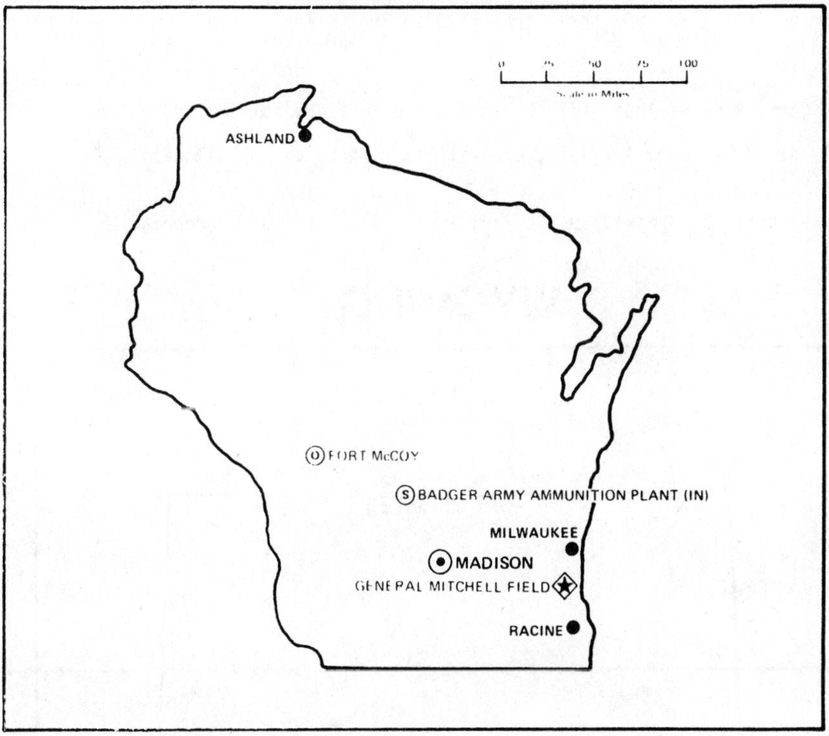

FORT McCOY (WI02R2)

Sparta, WI 54656-5000

Comm: 608-388-2222
ATVN: 280-2222

LOCATION: From W on I-90 to N on WI-27 to NE on WI-21 to Fort. From E on I-90, to W on WI-21 to Fort. RM: p-103, O/7. NMC: La Crosse, 35 mi SW.

GENERAL INFORMATION: Training Fort for Army National Guard and Reserve Units and individuals.

TEMPORARY MILITARY LODGING: Biltg Ofc, Bldg 2168, 8th St, 24 hrs, Comm: 608-388-2107, ATVN: 280-2107. All ranks. DV/VIP, EX-3713/4142.

LOGISTICAL SUPPORT: Most spt fac avail. Key phone numbers follow.

PX-388-2243	Post Rest-388-2342	Chap-388-4203
*EM C-388-3792	Gas S-388-3343	Legal-388-2304
Loc-388-3116	*NCO C-388-3792	*Off C-388-3792
Pol-388-4202	SDO-388-2216	*Snack B-388-3542
*Theat-388-2952		

*** Some activities are seasonal: Bldg 101, Sep-Apr; May-Aug: Off Club-Bldg 905, EM Club-Bldg 101, NCO Club-Bldg 1849.**

HEALTH & WELFARE: Clinic/EX-2414. Inpatient, Great Lakes Naval Hosp, Chicago, IL, 200 mi SE, 312-688-2492. Chapel serving all faiths, EX-4203.

Fort McCoy, Cont'd

RECREATION: Arts & Crafts, EX-4353, Auto Hobby, EX-4353, Gym, EX-2290, Sports Rental, EX-3202, Fishing & Hunting, EX-2102, Library, EX-2410, Swimming, EX-3202/2690, Serv Club, EX-3213/2971, Tennis, EX-3202. Squaw Lake Rec Area, 33 mi W of Post, Apr-Sep, Comm: 608-388-3517/2808, ATVN: 280-3517/2808, 116 camper spaces w/hookups, 5 camper spaces w/o hookups, comp fac and act. The USCG operates Sherwood Point Rec Fac, Sturgeon Bay, 35 mi NE of Green Bay, Comm: 414-291-1886, FTS: 362-1886, rec cottage, spt fac and act.

SPACE-A: McCoy Army Airfield, exec aircraft, flts to CONUS locations, EX-2222.

ATTRACTIONS: La Crosse, Sparta, and Tomah. Outdoor sports/recreation.

WYOMING

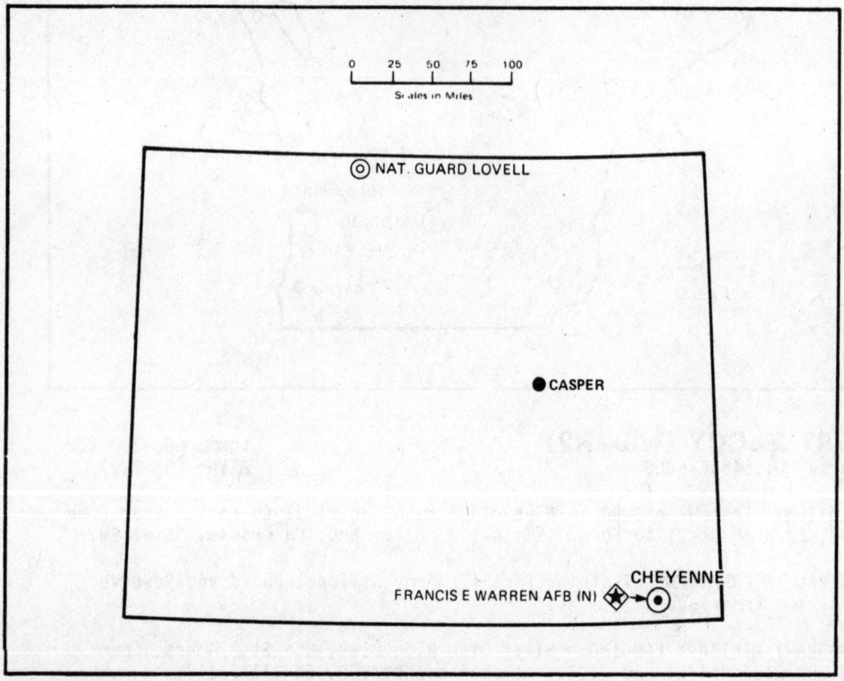

FRANCIS E. WARREN AIR FORCE BASE (WYO1R4)

Comm: 307-775-1110
ATVN: 481-1110

Francis E. Warren Air Force Base, WY 82005-5000

LOCATION: Off I-25, 2 mi N of I-80. Clearly marked. RM: p-104, J/2. NMC: Cheyenne, adjacent to city.

GENERAL INFORMATION: Strategic Air Command base, Air Division, Strategic Missile Wing, and support units.

TEMPORARY MILITARY LODGING: Biltg Offc, Bldg 211, Randall St, 24 hrs dly, Comm: 307-775-3077/2360, ATVN: 481-3077/2360. All ranks. DV/VIP, EX-3610/3750.

Francis E Warren AFB, Cont'd

LOGISTICAL SUPPORT: Complete spt fac avail. Key phone numbers follow.

BX-634-1593	Chap-775-2451	Chld C-775-2639
Commiss-775-3937	Conv S-632-6177	EM C-775-2564
Fam Svc-775-3739	Gas S-634-7432	Legal-775-2256
Loc-775-1110	Med-775-2277	NCO C-638-8993
Off C-632-9286	Pkg S-775-3008	Pol-775-3501
Snack B-634-1595	Theat-775-2345	

HEALTH & WELFARE: USAF Hospital, clinics and depts, Emerg, EX-3460, 25 bed. Chapels serving all faiths, EX-2451/3715.

RECREATION: Bowling, EX-2210, Gym, EX-2304, Library, EX-3460. Fam-Camp, on Base, year round, Comm: 307-775-3077, ATVN: 481-3077, 24 camper spaces w/ hookups (Oct-Apr, elec only), other facilities and water sports activities.

SPACE-A: Limited. Check Base Ops. Also, Cheyenne Munic Arprt, Bldg 115, main hangar, 0730-1630 M-F, Comm: 307-772-6347, ATVN: 943-6347, ANG flts to CONUS locations.

ATTRACTIONS: Cheyenne, outdoor sports, and recreation.

Please, see Central Order Coupons Pages 259 & 261

OUTSIDE CONTINENTAL UNITED STATES

ALASKA

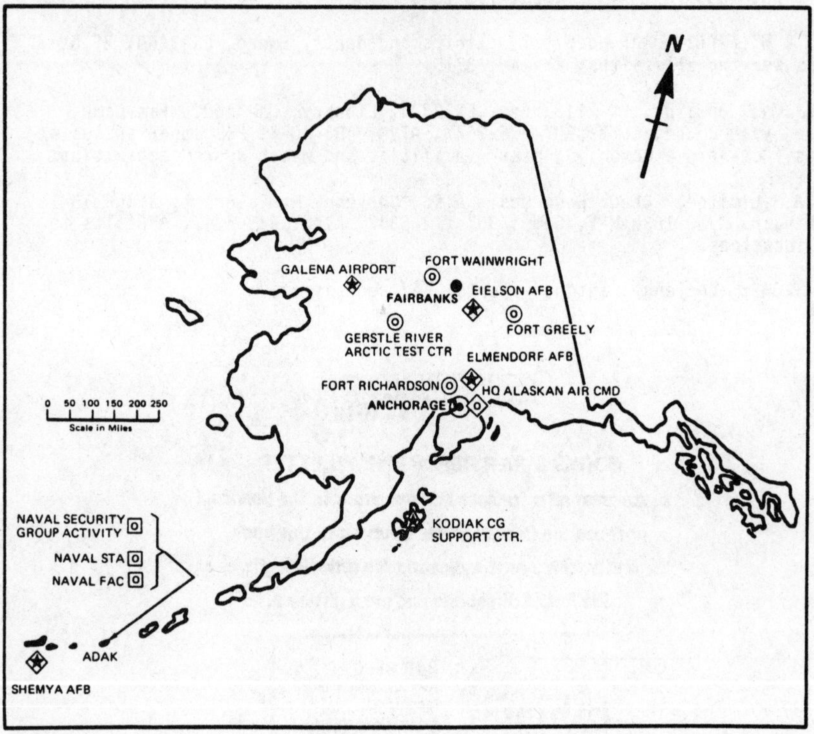

ADAK NAVAL STATION (AK04R5)

FPO Seattle 98791-5000

Comm: 907-592-8001
ATVN: 317-592-8001

LOCATION: On Adak Island of the Aleutian Island chain, accessible only by air or ship. RM: p-6, K/7. NMC: Anchorage, 1200 air mi NE. Note: Closed Station, only assigned personnel and cleared, sponsored, guest are allowed on Base. Write to Commander for clearance.

GENERAL INFORMATION: Naval Security Group Activity, Naval Facility, Patrol Wings, and support units.

TEMPORARY MILITARY LODGING: Biltg Ofc, Housing Ofc, 0800-1700 hrs dly, Comm: 907-592-4226/7, ATVN: 317-592-4226/7. Navy Lodge, EX-8076. All ranks. DV/VIP, EX-4226/8001.

LOGISTICAL SUPPORT: Complete spt fac avail. Key phone numbers follow.

NX-592-4154	**Caf-592-8161**	**Chap-592-4169/8203**

Adak NS, Cont'd

Commiss-592-8368	**Conv S-592-8466**	**EM C-592-8104**
Fam Svc-592-4291	**Gas S-592-8242**	**Legal-592-8028**
Med-592-4222	**NCO C-592-8393**	**Off C-592-8468**
Pkg S-592-8474	**Pol-592-8051**	**Snack B-592-8280**
Theat-592-8347		

HEALTH & WELFARE: Naval Branch Hospital, Emerg, EX-4222, 15 bed. Chapel serving all faiths, EX-4169/8203.

RECREATION: Arts & Crafts, Auto Hobby, Woodworking, Photo, Boat Basin, Skeet Range, Gym, Roller Rink, Swimming, Racquetball/Handball, Tennis, Hunting & Fishing. Call 592-8348 or 592-8344 for all recreation.

SPACE-A: Pax Term and Lounge, Term Bldg, 0800-1700 hrs dly, Comm: 907-592-8089, ATVN: 317-592-4196/8089, flts to AK locations.

ATTRACTIONS: Great outdoors.

EIELSON AIR FORCE BASE (AK15R5)

Eielson Air Force Base, AK 99702-5000

Comm: 907-377-1110
ATVN: 317-377-1110

LOCATION: On the Richardson Highway (AK-2), AFB is clearly marked. RM: p-6, E-7. NMC: Fairbanks, 26 mi NW.

GENERAL INFORMATION: Alaskan Air Command Base. Tactical Fighter Wing, Tactical Air Units, and support activities.

TEMPORARY MILITARY LODGING: Biltg Ofc, ATTN: 343 CSG/SVH, Bldg 5219, Coman Dr, 24 hrs dly, Comm: 907-377-1844, ATVN: 317-377-1844. All ranks. DV/VIP, EX-6101.

LOGISTICAL SUPPORT: Complete spt fac avail. Key phone numbers follow.

BX-372-2139	**BX Caf-377-1341**	**Chap-377-2130**
Chld C-377-3237	**Commiss-377-4255**	**Conv S-377-2210**
Fam Svc-377-2242	**Gas S-372-5391**	**Legal-377-2284**
Loc-377-1841	**Med-377-2259**	**NCO C-372-2162**
Off C-377-1121	**Pkg S-377-1134**	**Pol-377-5130**
Rec C-372-3114	**Snack B/Bowl A-377-5154**	**Theat-377-1172**

HEALTH & WELFARE: Clinic/Emer, 377-2296. Inpatient, Bassett Army Comm Hosp, Fort Wainwright, Fairbanks, Comm: 907-353-5143, ATVN: 317-353-5143. Chapels serving all faiths, 377-2130.

RECREATION: Arts & Crafts, EX-1167, Auto Hobby, EX-3190, Bowling, EX-1129, Flying Club, EX-1223, Gym, EX-1231, Sports Rental, EX-4214, Boat Shop, EX-1232, Library, EX-3174, Swimming, EX-1231, Rec Ctr, EX-3114, Youth Ctr, 372-1234. Fam-Camp, on Base, May-Sep, Comm: 907-377-1232, ATVN: 317-377-1232, 24 camper spaces w/hookups. Ravenwood Ski Lodge, on Base, Oct-Apr, Comm: 907-377-1232, day use only, snow skiing, rec room and spt fac. Also operates Birch Lake Rec Area, 38 mi S of AFB, May-Sep, Comm: 977-488-6161, ATVN: 317-377-5113, 39 camper spaces w/hookups, 2 deluxe cabins, 16 family cabins, 2 smaller cabins, all facilities and all activities on the lake.

SPACE-A: Pax Term and Lounge, Bldg 2216, 0730-1600 hrs, M-F, or as required, Comm: 907-377-1854, ATVN: 317-377-1854, flts to CONUS and in AK.

ATTRACTIONS: Fairbanks and North Pole, outdoor sports, and recreation.

ELMENDORF AIR FORCE BASE (AK09R5)

Elmendorf Air Force Base, AK 99506-5000

Comm: 907-552-1110
ATVN: 317-552-1110

LOCATION: Off Glenn Highway. Take Muldoon, Boniface, Post Rd, or Government Hill exits. The AFB is adjacent to Fort Richardson. RM: p-6, A/11. NMC Anchorage, 2 mi SW.

GENERAL INFORMATION: Alaskan Air Command Base. Hq AAC, 21st Tactical Fighter Wing, NORAD Regional Operations Center, and support units.

TEMPORARY MILITARY LODGING: Biltg Ofc, Bldg 31-250, Acacia Dr, 24 hrs dly, Comm: 907-552-2454/4701, ATVN: 317-552-2454/4701. All ranks. DV/VIP, EX-4860.

LOGISTICAL SUPPORT: Complete spt fac avail. Key phone numbers follow.

BX-753-4260	**Caf-753-6146**	**Chap-552-4422**
Chld C-753-5151	**Commiss-552-4877**	**EM C-552-2674**
Fam Svc-552-2694	**Gas S-753-7120**	**Legal-552-3046**
Loc-552-4860/2983	**Med-552-2748**	**NCO C-753-6131**
Off C-753-3131	**Pkg S-753-5151**	**Pol-552-3421**
Snack B-753-2280	**Theat-552-2344**	

HEALTH & WELFARE: USAF Hospital, clinics and depts, Emerg, EX-5555, Appoint, EX-3361, 95 bed. Chapels serving all faiths, EX-4422.

RECREATION: Arts/Crafts, EX-2031/2470, Bowling, EX-3669, Flying Club, 753-4167, Golf, EX-2773, Sports Rental, EX-4088, Racquetball Court, EX-5353/3669, Tennis, EX-2850/5353, Riding Club, EX-4987, Jogging Club, EX-5353/3504, Field House, EX-5353/3504, Library, EX-3787, Swimming, EX-2956, Rec Ctr, EX-4706, MWR Ofc, EX-2468, Youth Ctr, EX-2266. Fam-Camp, on Base, May-Oct, Comm: 907-552-2468, ATVN: 317-552-2468, 39 camper spaces w/hookups, other fac & act. Also operates Seward Rec Area, 130 mi SW of AFB, May-Sep, Comm: 907-552-5526, ATVN: 317-552-5526, 30 camper spaces w/hookups, 33 camper spaces w/o hookups, 11 mobile homes, 12 motel units, 1 dormitory, comp support fac, all act.

SPACE-A: Pax Term/Lounge, Bldg 32-233, 24 hrs dly, Comm: 907-552-5589, ATVN: 317-552-5589, flts to CONUS, AK, & overseas. Also, Anchorage IAP, Rm 202, Intl Term, 24 hrs dly, Comm: 907-552-3111, ATVN: 317-552-3111. (All Space-A seats assigned by Elmendorf AFB Space-A center.) Check ANG units-Anchorage IAP.

ATTRACTIONS: Anchorage, outdoor sports, and recreation.

FORT GREELY (AK10R5)

Fort Greely, AK 98733-5000

Comm: 907-872-4220
ATVN: 317-872-4206

LOCATION: Off AK-4, 6 mi S of junction of AK-2 and AK-4. Five mi S of Delta Junction. RM: p-6, F/6. NMC: Fairbanks, 105 mi NW.

GENERAL INFORMATION: Army Northern Warfare Training Center, Army Cold Region Test Center, and support units.

TEMPORARY MILITARY LODGING: Biltg Ofc, ATTN: FH/HSG Project Manager, Bldg 601, Arctic Ave, 0730-1630 hrs, M-F, Comm: 907-873-3109, ATVN: 317-873-3109. All ranks. DV/VIP, EX-4220.

LOGISTICAL SUPPORT: Complete spt fac avail. Key phone numbers follow.

PX-872-4102	**Caf-895-3110**	**Chap-872-3109**
Chld C-873-4227	**Commiss-872-4116**	**Conv S-895-3110**
EM C-895-3014	**Fam Svc-872-3197**	**Gas S-895-3210**

Fort Greely, Cont'd

Legal-872-2180	**Loc-873-4255**	**Med-872-3198**
NCO C-895-3014	**Off C-895-3015**	**Pkg S-895-3014**
Pol-872-3212	**Snack B-895-3110**	**Theat-873-3283**

HEALTH & WELFARE: Dispensary/Emer, EX-3198. Inpatient, Bassett Army Comm Hosp, Ft Wainwright, Fairbanks, Comm: 907-353-5143, ATVN: 317-353-5143. Chapels serving all faiths, EX-3109.

RECREATION: Arts & Crafts, EX-3181, Auto Hobby, EX-3293, ITT, EX-3285, Boat Shop, EX-4173, Outdoor Sports Ctr, EX-3183, Hunting & Fishing, EX-4188, Rec Ctr, EX-3285, Library, EX-3285.

SPACE-A: Allen Army Airfield, flts via exec aircraft to AK locations, EX-4220.

ATTRACTIONS: Outdoor sports and recreation.

FORT RICHARDSON (AK03R5)

Fort Richardson, AK 99505-5000
Comm: 907-864-0121
ATVN: 317-864-0121

LOCATION: Main gate is on Glenn Highway, 5 mi S of Eagle River. RM: p-6, B/12. NMC: Anchorage, 8 mi SW.

GENERAL INFORMATION: Hq Light Infantry Brigade, Special Troops, and other support units.

TEMPORARY MILITARY LODGING: Biltg Ofc, ATTN: ASFZT-EH-H, Bldg 724, Quartermaster Dr, 0730-1600 hrs dty dys, Comm: 907-863-8216, ATVN: 317-863-8216. All ranks. DV/VIP, 863-7204/1115.

LOGISTICAL SUPPORT: Complete spt fac avail. Key phone numbers follow.

PX-428-1335	**Caf-428-1314**	**Chap-863-1187**
Chld C-428-1240	**Commiss-862-9184**	**Conv S-428-1394**
EM C-428-1360	**Fam Svc-864-1199**	**Gas S-428-1248**
Legal-863-1182	**Loc-862-0192**	**Med-862-0110**
NCO C-428-1360	**Off C-428-1330**	**Pkg S-428-1207**
Pol-864-0127	**Snack B-428-0055/1393**	**Theat-428-1200**

HEALTH & WELFARE: Clinic/Emerg, 911, Appoint, 862-6285/2100. Inpatient, USAF Hosp, Elmendorf AFB, Comm: 907-552-2748, ATVN: 317-552-2748. Chapels serving all faiths, 864-1190.

RECREATION: Arts/Crfts, 863-5188, Auto Hobby, 863-7218, Bowling, 862-4196, ITT, 862-3217, Gym, 863-6191, Sports Rental, 428-1309, Flying Club, 428-1341, Fieldhouse, 863-2282, Outdr Rec Ctr, 863-9199, Swim Team, 862-1281, Ski Slope/Lodge, 428-1309/5288, Snowmachiners Club, 863-0147, Otter Lake Ldge, 863-9191, Outdoor Sports Ctr, 428-1309, Library, 862-9188, Swimming, 862-1281, Rec Ctr, 863-5212, Youth Ctr, 862-4290/863-0280. Seward Army Rec Camp at Resurrection Bay near Seward, 133 mi S of Fort, May-Sep, Comm: 907-862-3217, ATVN: 317-862-3217, 45 camper spaces w/hookups, 22 cabins, 3 pop-up campers, all spt fac and comp act, water & mountains.

SPACE-A: See Elmendorf AFB listing.

ATTRACTIONS: Great outdoors and Anchorage.

FORT WAINWRIGHT (AK07R5)

Fort Wainwright, AK 99703-5320
Comm: 907-353-6070 (Aft hrs/7500)
ATVN: 317-353-6070 (Aft hrs/7500)

Fort Wainwright, Cont'd

LOCATION: From Fairbanks, take Airport Way E which leads to the Main Gate of the Post. RM: p-6, E/12. NMC: Fairbanks, 3.5 mi W.

GENERAL INFORMATION: Infantry Brigade, Aviation Battalion, and other support units.

TEMPORARY MILITARY LODGING: Biltg Ofc, Bldg 1045 Murphy Hall, Gaffney Rd, 0730-1630 hrs, dty dys, Comm: 907-353-6259/6263, ATVN: 907-353-6259/6263. All ranks. DV/VIP, EX-6211/6066.

LOGISTICAL SUPPORT: Complete spt fac avail. Key phone numbers follow.

PX-353-6057	ACS-353-6369	Chap-353-6610
Chld C-356-1365	Commiss-353-7452	Fam Svc-353-7515
Gas S-356-1263	Legal-353-6773	Loc-353-6586
Med-353-5110	NCO C-353-6716	Off C-356-6109
PAO-353-6211	Pol-353-7535	Post Taxi-353-6093
Snack B-353-7300	Theat-353-6792	

HEALTH & WELFARE: Bassett Army Community Hospital, clinics and depts, EX-5143, Appoint, EX-5172/3, 50 beds. Chapels serving all faiths, 353-7380/6610.

RECREATION: Arts/Crafts, EX-6425, Auto Hobby, EX-6335, Bowling, EX-6731, Flying Club, EX-6671, Golf, EX-6749, Gym, EX-7080/6748, Ski Lodge, EX-6325, Boat Shop/Outdr Equipment, EX-6349/6350, Library, EX-6114, Swimming, EX-6466, Rec Ctr, EX-6021. Glass Park, on Post, year round, Comm: 907-353-6350, ATVN: 317-353-7117, 12 camper spaces, some w/elec.

SPACE-A: See Eielson AFB listing or call 353-7402.

ATTRACTIONS: Fairbanks (2d lgst city), North Pole, Eielson AFB w/in 20-mi rad. Summer Weather 80's-90's Jn-Aug, Golden Days (mid-Jul), World Eskimo Indian Olympics, Tanana Valley St Fair, Jesse Owens Games, area tours, hiking/boating/hunting/fishing/camping, Alaskalnd, Denali Natl Pk, MtMcKinley (hi'st pt, NoAmer) by car, bus or train, then bus to Main Tower. Winter NORTHERN LIGHTS, Luge Run (Bobsled), skiing/hunting/ice fishing, Chena Hot Sprgs, Dog Sled Races, Eskimo Winter Olympics, Special Olympics, Nenana Ice Classic, Ice Festival. Winter temps Sep-May/Jn fr +20 to -60 degrees in Interior of AK. Contin daylight, summer; contin night time, winter...LAND OF THE MIDNIGHT SUN.

KODIAK COAST GUARD SUPPORT CENTER (AK08R5)

Comm: 907-487-5265
ATVN: 317-487-5265

Kodiak, AK 99619-5000

LOCATION: From Kodiak City, take main road SW for 7 mi. Base is on left side. RM: p-6, I/5. NMC: Kodiak, 7 mi NE.

GENERAL INFORMATION: Support Center, CG Air Station, LORAN Station, and home port for 4 ships.

TEMPORARY MILITARY LODGING: Guest house, 0745-1630 hrs, dty dys, Comm: 907-487-5446, ATVN: 317-487-5310/5446. All ranks. DV/VIP, EX-5265.

LOCATION: Complete spt fac avail. Key phone numbers follow.

CGX-487-5102	Chap-487-5730/1	Commiss-487-5204
Conv S-487-5359	CPO C-487-5440	EDF-487-5710
EM C-487-5110	Fam Svc-487-5481/5402	Gas S-487-5107
Legal-487-5474	Loc-487-5252	Med-487-5222

Kodiak CG Spt Ctr, Cont'd

Off C-487-5440	**OOD-487-5266**	**Pkg S-487-5475**
Pol-487-5267	**Snack B-487-5748**	**Theat-487-5819/5108**

HEALTH & WELFARE: Coast Guard Support Center Hosp, Emerg, EX-5222, 18 bed. Chapels serving all faiths, EX-5730/1.

RECREATION: Arts & Crafts, EX-5471, Auto Hobby, EX-5844, Bowling, EX-5401, Gym, EX-5272, Boat House, EX-5047, Teen Center, EX-5271. Racquetball, Weight Room, and Outdoor sports, EX-5272.

SPACE-A: Pax Term, Lounge, Hangar #1, flt hrs, Comm: 907-487-5149, ATVN: 317-487-5149, flts to AK and CONUS.

ATTRACTIONS: Hunting and fishing paradise.

SHEMYA AIR FORCE BASE (AK11R5)

APO Seattle 98736-5000

Comm: 907-392-3000
ATVN: 317-392-3000

LOCATION: At the western tip of the Aleutian Islands chain, midway between Anchorage and Tokyo, accessible via air only. RM: p-6, J/5. NMC: Anchorage, 2000 air miles NE. Note: Visits require prior approval by the base commander.

GENERAL INFORMATION: Alaskan Air Command base. Support base.

TEMPORARY MILITARY LODGING: Biltg Offc, 24 hrs dly, Comm: 907-382-3000. Ask for Biltg. 70 units. All ranks. DV/VIP, EX-3000.

LOGISTICAL SUPPORT: Limited spt fac avail. Key phone numbers follow.

BX-382-3364	**Chapel-382-3366**	**Cons Opn Ms-382-3362**
Med-382-3550	**NCO C-382-3362**	**Pol-382-3377**
Snack B-382-3364	**Theat-382-3261**	

HEALTH & WELFARE: Dispensary/Emerg, EX-382-3550. Inpatient via MEDEVAC. Chapel serving all faiths, EX-382-3366.

RECREATION: Arts & Crafts, EX-3211, Bowling, EX-3665, Gym, EX-3265, Sports Rental, EX-3211.

SPACE-A: Pax Term, Base Ops, 24 hrs dly, Comm: 907-382-1200, ATVN: 317-382-3000, flts to AK, CONUS and overseas.

ATTRACTIONS: Summer fishing, fair only. No hunting on these islands.

-NOTES-

GUAM

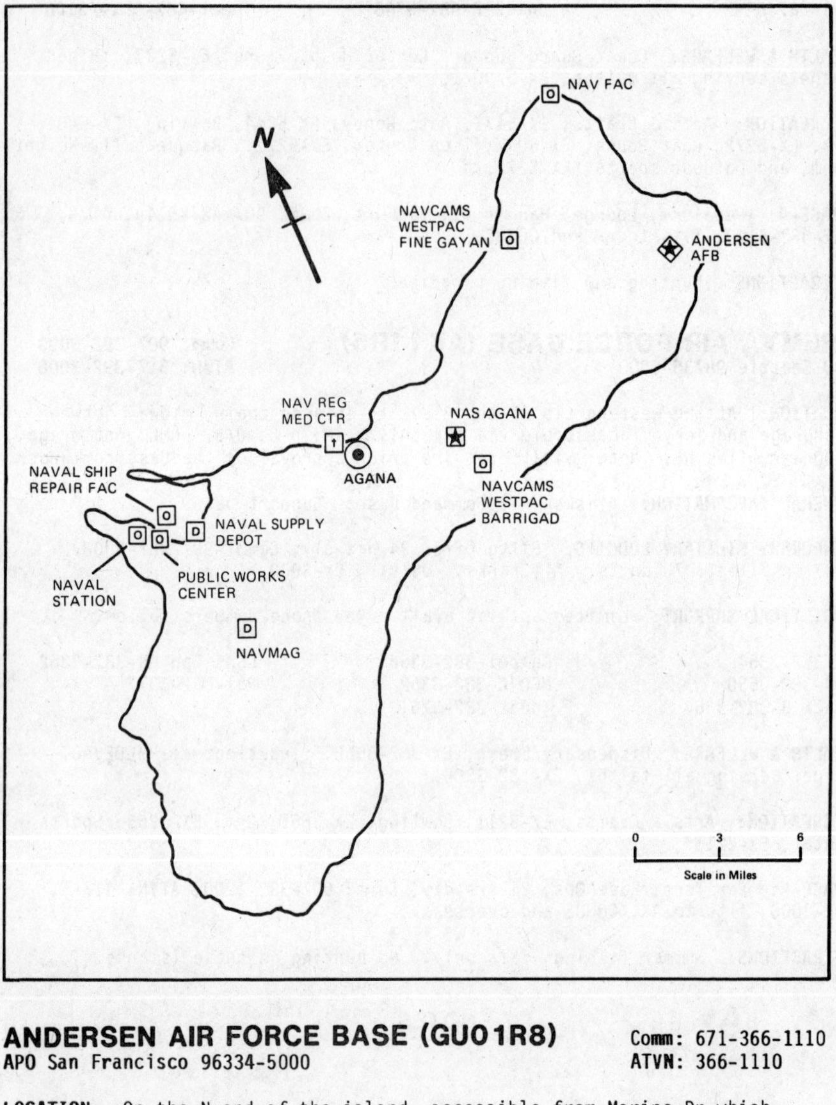

ANDERSEN AIR FORCE BASE (GUO1R8)

APO San Francisco 96334-5000

Comm: 671-366-1110
ATVN: 366-1110

LOCATION: On the N end of the island, accessible from Marine Dr which extends entire length of the island of Guam. NMC: Agana, 15 mi S.

GENERAL INFORMATION: Strategic Air Command Base. Air Division, Strategic Wing, MAC Squadron, and other support units.

TEMPORARY MILITARY LODGING: Biltg Ofc, Bldg 27006, 4th & Marianas Sts, 24 hrs dly, Comm: 671-366-8144, ATVN: 366-8144. All ranks. DV/VIP, 366-4228.

LOGISTICAL SUPPORT: Most spt fac avail. Key phone numbers follow.

Andersen AFB, Cont'd

BX-362-1136	**Caf-362-3247**	**Chap-366-6139**
Commiss-366-2214	**Garage-362-5286**	**Med-363-2978**
Off C-362-4269	**Pol-363-2913**	**Snack B-362-3149**
Theat-366-5102/4180		

HEALTH & WELFARE: Clinic/Emer, 366-4267, Appoint, 363-2978. Inpatient, Naval Hosp, 344-9356. Chapels serving all faiths, 366-6139 (after hrs 366-8144).

RECREATION: Arts & Crafts, 362-7173, Auto Hobby, 362-1258, Bowling, 362-3238, Golf, 362-5143, Gym, 366-8282, Recreation Supply, 366-5197, Library, 366-4191/4294, Youth Ctr, 366-5206.

SPACE-A: Pax Term and Lounge, Bldg 17002, 24 hrs dly, Comm: 671-366-5165, ATVN: 366-5165, flts to CONUS and overseas.

ATTRACTIONS: Great beaches, scuba diving, snorkeling, and many other popular water sports.

GUAM NAVAL AIR STATION (GU03R8)

FPO San Francisco 96637

Comm: 671-344-8103/05
ATVN: 344-8103/05

LOCATION: Naval Air Station is in the middle of the island of Guam, on Marine Drive. Clearly marked. NAS: Agana, 3 mi SW.

GENERAL INFORMATION: The mission of U.S. Naval Air Station, Agana, Guam is to maintain and operate facilities and provide services and material to support operations of aviation activities and units of the operating forces of the U.S. Navy and other activities and units as designated by the Chief of Naval Operations.

TEMPORARY MILITARY LODGING: Biltg Ofc, Bldg 1300/15, 24 hrs dly, Comm: 671-344-8125/5221, ATVN: 344-8125/5221. All ranks. DV/VIP, EX-8103.

LOGISTICAL SUPPORT: Complete spt fac avail. Key phone numbers follow.

NX-342-1118	**Caf-342-1256**	**Car Rntl-342-1288**
Chap-344-6119	**Chld C-344-5285**	**Conv S-342-1132**
CPO C-342-2269	**EM C-342-2274**	**Fam Svc-333-2056/57/58**
Gas S-342-4114	**Legal-344-6189**	**Loc-344-8103**
Med-344-8201	**Off C-342-2124**	**Pkg S-342-5156**
Snack B-342-2116	**Theat-344-5124**	

HEALTH & WELFARE: Clinic/Emerg, 344-6141, Appoint, 344-8201. Inpatient, Naval Hosp, Guam Naval Base, 344-9356. Chapels serving all faiths, 344-6119.

RECREATION: Gym, 344-5136, Hobby Complex, 344-5153, Swimming, 344-7233, Library, 344-6109, Bowling, 344-6157, Rec Off, 344-5137.

SPACE-A: Base Ops, 344-7127, flts CONUS and overseas. Also, see Andersen AFB listing (preferred - more flights).

ATTRACTIONS: Water sports, beaches, and tours of the island. Call the District II Rec Off for current info, 344-5137.

GUAM NAVAL BASE (GU02R8)

FPO San Francisco 96630-0051

Comm: 671-339-4262
ATVN: 321-1110

LOCATION: S on Marine Drive. Clearly marked. NMC: Agana, 10 mi N.

Guam NB, Cont'd

GENERAL INFORMATION: Hq Naval Forces Marianas. Naval Hospital.

TEMPORARY MILITARY LODGING: Biltg Ofc, Bldg 2000, 24 hrs dly, Comm: 671-339-7224/5259, ATVN: 321-7224/5259. All ranks. DV/VIP, EX-5202.

LOGISTICAL SUPPORT: Complete spt fac avail. Key phone numbers follow.

BX-332-1232	**Caf**-332-1217	**Chap**-339-5138
Commiss-339-5237	**Conv S**-332-7181	**EM C**-332-8193
Fam Svc-333-2056	**Gas S**-332-7183	**Legal**-339-3155
Loc-321-1110	**Med**-344-9351	**NCO C**-342-2274
Off C-342-2124	**Pkg S**-332-8191	**Pol**-349-5216
Snack B-332-7182	**Theat**-332-8215	

HEALTH & WELFARE: Naval Hospital, clinics and depts, Emerg, 344-9351, Appoint, 344-9235, 350 bed. Chapels serving all faiths, EX-2126.

RECREATION: Gym, 332-8202/8203, Golf, 344-5277, Beach, 332-8201, Swimming, 332-8200, Deep Sea Fishing, 332-8216, Hobby Complex, 332-8207, Library, 339-6159, Bowling, 332-8213/8205.

SPACE-A: See Guam NAS and Andersen AFB listings.

ATTRACTIONS: Great beaches and water sports.

HAWAII

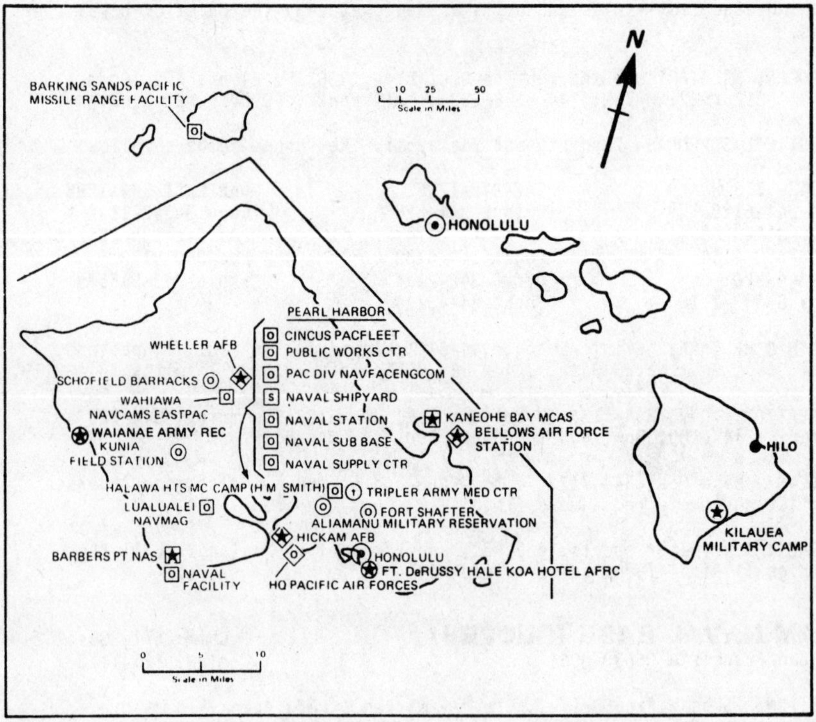

BARBERS POINT NAVAL AIR STATION (HI 01R6)

Comm: 808-684-6266
ATVN: 430-1111

Barbers Point Naval Air Station, HI 96862-5050

LOCATION: Take H-1 W (toward Waianae) to Barbers Point NAS/Makakilo exit, S for 2.5 mi to main gate. RM: p-7, I/3. NMC: Honolulu, 12 mi E.

GENERAL INFORMATION: Tactical Naval Aircraft Squadrons, Coast Guard Air Station, and support units.

TEMPORARY MILITARY LODGING: Biltg Ofc, Bldg 1788, Bealleau Woods, off Enterprise, 24 hrs dly, Comm: 808-684-2248/2290, ATVN: 430-2248/2290. All ranks. DV/VIP, 684-1181.

LOGISTICAL SUPPORT: Complete spt fac avail. Key phone numbers follow.

NX-684-6271	**Chap-684-3111**	**Chld C-684-7192**
Commiss-684-9126	**Conv S-684-6271**	**CPO C-684-5222**
EM C-684-7133	**Fam Svc-684-7198**	**Gas S-684-5112**
Legal-684-4247	**Loc-684-3222**	**Med-684-6241**
Off C-684-4108	**Pkg S-684-7197**	**Pol-684-6222**
SDO-684-6266	**Snack B-684-6271**	**Theat-684-4243**

Iroquois Point has BX-499-1355, Conv S-499-1355, Snack B-499-2873, Gas S-499-1355, Yacht Club, 499-1500. Lualualei has BX-668-3448, Gas S-668-3316, Pkg S-696-7142, Beach Cottages and Rec Ctr, EX-2883. (Lualualei is a 20 min ride from Barbers Point.)

HEALTH & WELFARE: Clinic, 684-6241, Emerg, 684-3133, Appoint, 684-6201. Inpatient, Tripler Army Med Ctr, Comm: 808-433-6661, ATVN: 315-6661. Chapels serving all faiths, 684-3111/3188.

RECREATION: Bowling, 684-9151, Intl Sports, 684-7498, Golf, 684-7123, Gym, 684-2120, Riding, 684-681-3963, Library, 684-5217, Swim, 684-5133, Rec Ctr, 684-8281, Tennis, 684-5170, Tickets, 684-9151, Picnic Area. Rec Area, on Base, year round, Comm: 808-682-5282, ATVN: 682-5282, 20 2-bdrm cottages, 14 camper spaces w/water faucet only, all fac on Base, comp beach act.

SPACE-A: Pax Term and Lounge, Bldg 4, 24 hrs dly, Comm: 808-684-2174/2190, ATVN: 430-2174/2190, flts to CONUS and overseas.

ATTRACTIONS: Honolulu, beaches, island tours.

BELLOWS AIR FORCE STATION (HI18R6)

Comm: 808-259-8841
ATVN: 259-8841

Waimanalo, HI 96795-5000

LOCATION: On the eastern shore of Oahu. From Honolulu, take H-1 E to HI-61, N to HI-72, S to AFS. Clearly marked. RM: p-7, H/7. NMC: Kaneohe, 7 mi NW.

GENERAL INFORMATION: Pacific AF Base, rec ctr-uniformed services personnel and fams, Det 1, 15 Air Base Wing, communications site, Marine tng area.

TEMPORARY MILITARY LODGING: Biltg Ofc, PO Box 1010, 0800-1800 M-F, 0900-1800 hrs other days, Comm: 808-259-8841 (Off), EX-8841 (Enl), ATVN: 259-8841. All ranks. DV/VIP, EX-8841.

LOGISTICAL SUPPORT: Most spt fac avail. Key phone numbers follow.

BX-259-5913	**EM C-259-5915**	**Loc-259-5428**
Off C-259-5915	**Pkg S-259-8419**	**Pol-259-5955**

Bellows AFS, Cont'd

HEALTH & WELFARE: Clinic, Kaneohe Marine Corps AS, 24 hr Emerg, 257-3133. Inpatient, Tripler Army Med Ctr, 433-5811. No chapel.

RECREATION: Complete beachfront Rec Center, 38 camper spaces w/o hookups, many beach cottages, all fac and activities. Golf Driving Range, EX-8841, Marina & Boathouse, EX-5447. Also, Mini-Golf.

SPACE-A: See Hickam AFB listing.

ATTRACTIONS: Great beaches on windward side of Island of Oahu. Watch out for the strong undertow . Blowhole is nearby and beautiful Kaneohe Bay.

CAMP H. M. SMITH MARINE CORPS (HI15R6)

Comm: 808-477-5106/6330
ATVN: 477-5106/6330

Camp H. M. Smith, HI 96861-5001

LOCATION: Off H-1 W in Halawa Heights. Take Halawa Heights Road. Clearly marked. RM: p-7, H/5. NMC: Honolulu, 10 mi SE.

GENERAL INFORMATION: Hq US Pacific Command, Hq Fleet Marine Force Pacific, and support units.

TEMPORARY MILITARY LODGING: None available.

LOGISTICAL SUPPORT: Limited spt fac avail. Key phone numbers follow.

MX-477-6380	Chap-477-6461	EM C-477-6360
Fam Svc-474-2220	Gas S-477-5058	Legal-477-6003
Loc-477-6353	Med-477-6414	Off C-477-6491
Pol-477-6353	SDO-477-5106/6330	Snack B-477-6328
SNCO C-477-5056		

HEALTH & WELFARE: Emer 477-6411. Inpatient, Tripler Army Med Ctr, 433-5811. Chapel serving all faiths, 477-6461.

RECREATION: Bowling, EX-6382, Riding, EX-9646, Fitness Ctr, EX-5197, Racquet Sports, EX-5142, Spec Svc, EX-5142, Swimming, EX-6207, Tennis, EX-5176, Tickets, EX-5142, Camp Hawkins Campground, EX-6092.

SPACE-A: See Barbers Point NAS listing or Hickam AFB listing.

ATTRACTIONS: Great view of Honolulu and Pearl Harbor from Officers' Club and Staff NCO Club.

FORT DeRUSSY, HALE KOA HOTEL AFRC (HI08R6)

Comm: 808-955-0555 (24hrs)
800-367-6027 (0800-1600 hrs dly HI time, except hol)

2055 Kalia Road
Honolulu, HI 96815-1998

LOCATION: At 2055 Kalia Road, Waikiki Beach, Honolulu. Fort DeRussy is on Waikiki Beach, between Ala Moana Blvd, Kalakaua Ave & Saratoga Rd, 6 mi E of Honolulu IAP. RM: p-7, I/5. NMC: Honolulu, in city limits.

GENERAL INFORMATION: Hale Koa Hotel and Armed Forces Recreation Center, HI National Guard and USA Reserve Units.

TEMPORARY MILITARY LODGING: Hale Koa Hotel, 14 stories w/420 A/C guest rms + all mod resort hotel fac. 1st cl beachfront resort built by & run for active

Fort DeRussy Hale Koa, Cont'd

duty & retired uniformed personnel, all services. Rates from $28.00 to $73.00 per rm per night, dbl occup. For complete details on eligibility, rate structure and facilities, see Military Living's publications Temporary Military Lodging Around The World or Military Camping & Rec Areas Around The World.

LOGISTICAL SUPPORT: Most spt fac avail. Key phone numbers follow.

PX-955-0060	**Dining Room-955-0555**	**Coffee House-955-0555**
Lounge-955-0555	**Tour Ofc-946-5204**	**Activity Desk-955-0555**
Flower Shop-942-9833	**Pkg S-955-1033**	**Snack B-533-8260**

HEALTH & WELFARE: All care, Tripler Army Medical Center, 433-5811. Chapel located on Fort DeRussy, 1/2 block from Hale Koa Hotel.

RECREATION: Athl Fld, EX-0555, Racquetball/Handball/Tennis, EX-0555, Swimming, EX-0555, Video Games, EX-0555, Volleyball, EX-0555. Pic area/beach open.

SPACE-A: See Hickam AFB listing.

ATTRACTIONS: Resort hotels, entertainment, beaches, zoo, water sports.

FORT SHAFTER (HI09R6)

Fort Shafter, HI 96858-5000

Comm: 808-471-7411
ATVN: Op 430-0111

LOCATION: Take H-1 W exit at Fort Shafter. Clearly marked. RM: p-7, J/8. NMC: Honolulu, 7 mi E.

GENERAL INFORMATION: US Army Western Command (WESTCOM) and US Army Support Command, HI (USASCH).

TEMPORARY MILITARY LODGING: Biltg Ofc, Bldg 453B, Burr Rd, 0730-1600 hrs dty dys, Comm: 808-438-1102, ATVN: 438-1102. All ranks. DV/VIP, EX-1577.

LOGISTICAL SUPPORT: Complete spt fac avail. Key phone numbers follow.

PX-845-9681	**Caf-845-9636**	**Chld C-847-4940**
Commiss-438-1367	**Gas S-845-9694**	**Loc-438-1904**
NCO C-848-0431	**Off C-847-6468**	**Pkg S-848-0488**
Pol-438-2885	**Snack B-848-0477**	**Theat-438-9979**

Coast Guard Housing Area at Red Hill has BX-839-7306, Gas S-833-2448, some rec fac. The Coast Guard Base, Sand Is has BX-847-6286, Cons C-847-6467, Gas S-847-6286, + some rec fac. On Kauai, Pacific Missile Rnge Fac, Barkng Sands, BX-471-6204. On Maui, ANG, PX-244-3006.

HEALTH & WELFARE: Tripler Army Medical Center, nearby, Emerg, 433-5700/6629, Appoint, 433-6075. Chapels serving all faiths, 438-9745.

RECREATION: Athletic Fields, 438-1152, Bowling, 438-9996, Golf, 847-6075, Hobby Shops, 428-1071, Handball/Racquetball, 438-1152, Spec Svc, 655-9238, Rec Ctr, 438-1522, Swimming, 847-6726/438-1790, Tennis, 438-1152, Wt Rm, 438-1152.

SPACE-A: See Hickam AFB listing.

ATTRACTIONS: Honolulu, USS Arizona National Memorial, National Memorial Cemetery of the Pacific (Punch Bowl), Pearl Harbor.

HICKAM AIR FORÇE BASE (HI11R6)

Hickam Air Force Base, HI 96853-5000

Comm: 808-422-0531
ATVN: 430-0111

Hickam AFB, Cont'd

LOCATION: Adjacent to the Honolulu IAP. Accessible from H-1 or HI-92. Clearly marked. RM: p-7, L/7. NMC: Honolulu, 6 mi E.

GENERAL INFORMATION: Pacific Air Force Base. Hq PACAF, 834th Airlift Division, 15th Air Base Wing Support Units.

TEMPORARY MILITARY LODGING: Biltg Ofc, 15 ABW/SVH, Bldg 1153, 24 hrs dly, Comm: 808-423-5511, ATVN: 430-5511. All ranks.

LOGISTICAL SUPPORT: Complete spt fac avail. Key phone numbers follow.

BX-423-1304	Caf-422-8000	Chap-449-1754
Chld C-449-9845	Commiss-449-9660	Conv S-422-8404
EM C-449-2097/1092	Fam Svc-449-6603	Gas S-422-8666
Legal-449-1737	Loc-422-0531/0111	Med-449-9901/6248
NCO C-449-1092	Off C-449-1998	Pol-449-6372
Snack B-422-8008	Theat-449-9979	

HEALTH & WELFARE: Clinic/Emerg, EX-6333, Appoint, EX-9901. Inpatient, Tripler Army Med Ctr, 488-5811. Chapels serving all faiths, EX-1754.

RECREATION: Arts & Crafts, 449-1568, Auto Hobby, 449-2554, Athletic Fields, 449-6686, Bowling, 449-2702/1044, Golf, 449-6490/2093, Gym, 449-6711, Ticket, 449-0548, Marina, 449-5215, Rec Ctr, 449-6635, Equip Rental, 449-6870, Spc Svc, 449-1334, Swimming, 449-6573/1070/6685, Tennis, 449-2598, Library, 449-2831, Hunting & Fishing, 449-6721, Youth Ctr, 449-1492/6758.

SPACE-A: Pax Term and Lounge, Bldg 2028, 24 hrs dly, Comm: 808-449-2807/1270, ATVN: 430-2807/1270, flts to CONUS and overseas.

ATTRACTIONS: Honolulu, Waikiki Beaches, and Pearl Harbor.

KANEOHE MARINE CORPS AIR STATION (HI12R6)

Comm: 808-257-2747
ATVN: 430-1110

Kaneohe Bay, HI 96863-5001

LOCATION: At end of H-3 on the Windward side of Oahu. Off Mokapu Blvd and Kaneohe Bay Drive. Clearly marked. RM: p-7, G/6. NMC: Honolulu, 14 mi SW.

GENERAL INFORMATION: MCAS Kaneohe Bay, home for the 1st Marine Amphibious Brigade which is composed of the 3rd Marine Regiment (Rein), Brigade Service Support Group-1 and Marine Aircraft Group-24. The Air Station is composed of Hqs & HgsSqdn and Station Operations and Maintenance Squadron.

TEMPORARY MILITARY LODGING: Bldg 3038, inside the H-3 gate, 0700-2200 hrs dly, Comm: 808-257-2808/3513, ATVN: 457-2808. All ranks welcome.

LOGISTICAL SUPPORT: Complete spt fac avail. Key phone numbers follow.

MCX-257-2111	Auto Ctr-257-2039	Chap-257-3552/3506
Chld C-257-2608	Commiss-257-2418	EM C-257-2873/2657
Fam Svc-257-3168	Legal-257-2168	Med-257-2697
Off C-257-2081	Pkg S-257-1537/8/9	PMO-257-2367
7-Day Store-257-3378	Snack B-257-2665	SNCO C-257-2592
Theat-257-3668		

HEALTH & WELFARE: Clinic/Emerg, 257-3133. Inpatient, Tripler Army Med Ctr, 488-5811. Chapels serving all faiths, 257-3506.

Kaneohe MCAS, Cont'd

RECREATION: Athletic Fields, 257-3550, Beaches, 257-3520, Bowling (K-Bay Lanes), 257-3375, Golf, 257-2914, Gym, 257-3550/3622, Handball/Racquetball, 257-3550, Library, 257-3583, Marina, 257-2219, Scuba Locker, 257-2548, Skeet Range, 257-2868, Station Pool, 257-2922, Tennis, 257-3520, Ticket Ofc, 257-3108. Beach cottages, 13 units, Bldg 3030, 257-3513, 4 camper spaces w/o hookups, 257-2808. Reservations for cottages - 60 days in advance.

SPACE-A: See Hickam AFB and Barbers Point NAS listings.

ATTRACTIONS: Kaneohe Bay, beaches, outdoor sports, historical sites, and wildlife refuges.

KILUAEA MILITARY CAMP (HI17R6)

Hawaii Volcanoes National Park, HI 96718-5000 — **Comm:** 808-543-2626 — **ATVN:** None

LOCATION: On island of Hawaii, 216 air mi SE of Honolulu, 32 mi fr Hilo IAP. Scheduled bus trnsp to Camp, resv req Hilo to KMC. RM: p-7, H/11. NMC: Hilo, 32 mi NW.

GENERAL INFORMATION: Armed Forces Recreation Center, Kilauea Military Camp (KMC).

TEMPORARY MILITARY LODGING: Biltg Ofc, 0800-1630 hrs, M-F, Comm: 808-543-2626. All ranks. 40 cabins, 5 apts. Resv req. DV/VIP, none.

LOGISTICAL SUPPORT: Limited spt fac avail. Key phone numbers follow.

Caf-967-7418	Gas S-967-8132	Gen S-967-7474
Pkg S-967-7474	Res-543-2626	

HEALTH & WELFARE: Dispensary/967-7467. Inpatient, Tripler Army Med Ctr, 488-5811. Chapel, Catholic and Protestant services each Sunday.

RECREATION: Bowling, Bicycle Rental, Golf, Rec Room, Koa Bar, Lectures, Mini Health Center, Multi-Purpose Course (Basketball, Volleyball), Tennis (2 courts on Camp), Rental Equip for Hiking, Tennis, Golf, and other sports, and Tours - check with registration desk. Call 543-2626.

SPACE-A: See Hickam AFB listing.

ATTRACTIONS: Tours of the big island, Hawaii, Hilo, Black Sands Beach, Kona, and Volcanoes National Park.

PEARL HARBOR NAVAL BASE (HI19R6)

Pearl Harbor Naval Base, HI 96860-5020 — **Comm:** 808-471-7411 — **ATVN:** 430-0111

LOCATION: Off H-1 adjacent to Honolulu IAP. Clearly marked. RM: p-7, H/4. NMC: Honolulu, 5 mi E.

GENERAL INFORMATION: Hq Pacific Fleet, Hq Submarine Force Pacific, Naval Logistical Command Pacific, Naval Shipyard, Support Units and homeport for more than 40 ships.

TEMPORARY MILITARY LODGING: Biltg Ofc: USS Arizona Hall, Bldg 1623, 24 hrs dly, Comm: 808-471-8053/7112, ATVN: 474-5210. All ranks. DV/VIP, EX-5210.

LOGISTICAL SUPPORT: Complete spt fac avail. Key phone numbers follow.

NX-423-3330	Caf-423-3243	Chap-471-3971

Pearl Harbor NB, Cont'd

Chld C-422-7133	Commiss-471-0263	Conv S-423-3217/3218
CPO C-474-3147	EM C-471-0841	Fam Svc-474-2220/4222
Gas S-423-3229/3226	Legal-474-5249	Loc-471-7411
Med-471-9541	Off C-471-8455	Pkg S-471-9048
Pol-474-1237	Snack B-474-6119/432-3246	Theat-471-0726

HEALTH & WELFARE: Clinic/Emerg, EX-9541. Inpatient, Tripler Army Med Ctr, 488-5811. Chapels serving all faiths, EX-3971.

RECREATION: Athletic Fields, 471-9139, Arena, 474-6156, Bowling, 474-2162, Dependents Activites Center, 471-0392, Golf, 472-8495 & 471-0142, Gym, 474-6156 & 471-9183, Fishing, 471-9150, Marina, 471-9680, Spec Svc Ofc, 474-6156 & 471-3169, Swimming, 471-9723/9181, 472-8250, 471-9307 & 474-7122, Tennis/Racquetball 471-0510/9183, Tickets, 474-6156/1190 & 422-9444, Video Games, 423-3391.

SPACE-A: See Hickam AFB and Barbers Point NAS listings.

ATTRACTIONS: USS Arizona Memorial and Visitor Center, and USS Bowfin Submarine Exhibit.

SCHOFIELD BARRACKS (HI13R6)

Schofield Barracks, HI 96857-5000

Comm: 808-655-4930
ATVN: 455-4930

LOCATION: Off H-2 or HI-99 in the center of the island of Oahu. Clearly marked. RM: p-7, F/3. NMC: Honolulu, 20 mi SE.

GENERAL INFORMATION: Infantry Division and support units.

TEMPORARY MILITARY LODGING: Biltg Ofc, PO Box 995, Bldg 692, 24 hrs dly, Comm: 808-624-9877. All ranks.

LOGISTICAL SUPPORT: Complete spt fac avail. Key phone numbers follow.

PX-622-1773	Caf-624-2753	Chap-655-0301/0739
Chld C-655-4297	Commiss-655-0946	Conv S-624-3158/2191
EM C-655-9273	Fam Svc-655-4663	Gas S-624-2496/9543
Legal-655-9262	Loc-438-1904/1803	Med-655-4747/9268
NCO C-624-5360	Off C-655-9795	Pkg S-624-2692
Pol-655-5116	Snack B-624-9316/4230	Theat-655-0603

HEALTH & WELFARE: Clinic/Emer, EX-4747, Appoint, 655-4027. Inpatient, Tripler Army Med Ctr, 433-5811. Chapels serving all faiths, 655-0301/0739.

RECREATION: Arts & Crafts, EX-9042, Archery, EX-4804, Athletic Fields, EX-0900, Bowling, EX-0757, Golf, EX-9833, Gym, EX-0900, Hobby Shops, EX-0898/9042/9368, MWR Act, EX-4804, Sports Rental, EX-0143, Rec Ctr, EX-9091, Library, EX-0145, Swimming, EX-9698, Youth Ctr, EX-4641, Scuba, EX-0143, Tennis, EX-0852, Tickets, EX-9972. Waianae Rec Ctr, operated by Fort Shafter, 35 mi NW of Honolulu, year round, 808-696-2494, 3 camper spaces w/o hookups, 33 cottages.

SPACE-A: See Hickam AFB and Barbers Point NAS listings.

ATTRACTIONS: Sugar cane and pineapple fields. Tropic Lightning Museum.

TRIPLER ARMY MEDICAL CENTER (HI03R6)

Tripler Army Medical Center, HI 96859-5000

Comm: 808-433-6661
ATVN: 315-433-6661

LOCATION: Take H-1 W from Honolulu to Tripler exit. Turn right on Jarrett White Rd to Tripler AMC. RM: p-7, J/8. NMC: Honolulu, 3 mi SE.

Tripler AMC, Cont'd

GENERAL INFORMATION: Tripler Army Medical Center.

TEMPORARY MILITARY LODGING: Biltg Ofc, Bldg 228B, Jarrett White Rd, 0600-1900 hrs dty dys, Comm: 808-433-6963, ATVN: 315-433-6963. All ranks. DV/VIP, 433-6963.

LOGISTICAL SUPPORT: Complete spt fac avail. Key phone numbers follow.

PX-833-1267	**Chap-433-5727**	**Commiss-438-1367/2903**
Gas S-833-1250	**Legal-433-5311**	**Loc-433-6661**
Med-433-6620	**NCO-433-6661**	**Off C-833-1268**
PMO-433-6671	**Snack B-833-1259**	

The Aliamanu Military Reservation, nearby housing area has Shoppette-833-6997, Snack B-833-0977, Gas S-833-6997.

HEALTH & WELFARE: Army Hospital, Ambulance, 433-5700, Emerg, 433-6620/6629, Appoint, 433-5811. Chapels serving all faiths, 433-5727.

RECREATION: Bowling, 433-6420, Swimming, 433-5257, Gym, 443-6443, Tennis, 433-6443, Racquetball/Handball, 433-6443, Golf Driving, Spc Svc, 433-5772, Library, 433-6968.

SPACE-A: See Hickam AFB listing.

ATTRACTIONS: USS Arizona National Memorial, Pearl Harbor. Great view of Honolulu from here.

WAHIAWA NAVAL COMMUNICATIONS STATION (HI16R6)

Comm: 808-653-5385
ATVN: 453-5385

Wahiawa, HI 96786-5000

LOCATION: Take H-2 to Wahiawa, Station is 3 mi N. Clearly marked. RM: p-7, F/3. NMC: Honolulu, 20 mi SE.

GENERAL INFORMATION: Naval Communications for Pacific area.

TEMPORARY MILITARY LODGING: No temporary military lodging available. This Station is TLA (Temporary Living Allowance) approved. BEQ, ATVN: 453-5442.

LOGISTICAL SUPPORT: Limited spt fac avail. Key phone numbers follow.

BX-653-5364	**Chap-653-5577**	**CPO C-653-5320**
EM C-653-5470	**Gas S-653-5451**	**Loc-653-5385**
Med-653-5350/5340	**Off C-653-5486**	**PAO-653-5571**
Pol-653-5385	**SDO-653-5385**	**Tickets-653-5535**

HEALTH & WELFARE: Clinic/Emerg, EX-5350, Appoint, EX-5340. Inpatient, Tripler Army Med Ctr, 433-5811. Chapels serving all faiths, EX-5577.

RECREATION: Bowling, EX-5576, Gym, EX-5542, Arts/Crafts, EX-5574, Racquet Sports, EX-5542, Sports Equip, EX-5542, Spec Svc, EX-5535, Swimming, EX-5306, Tickets, EX-5535.

SPACE-A: See Barbers Point NAS and Hickam AFB listings.

ATTRACTIONS: Short drive to Honolulu, and beach areas.

WAIANAE ARMY RECREATION CENTER (HI05R6)

Comm: 808-696-2494
ATVN: None

Waianae, HI 96792-5000

LOCATION: On the west coast of Oahu. Take H-1 W to HI-93 (Farrington Hwy), W to Waianae. RM: p-7, G/2. NMC: Honolulu, 35 mi SE.

GENERAL INFORMATION: Recreation center on the beach operated by the Army for all uniformed personnel and their families both active and retired.

TEMPORARY MILITARY LODGING: Biltg Ofc, 85010 Army Street, 0900-1600 hrs, Mon - Fri (only), Comm: 808-696-2494. 33 cottages. All ranks. DV/VIP, none.

LOGISTICAL SUPPORT: Limited spt fac avail. Key phone numbers follow.

PX-696-7142	**Beach Club-696-4778**	**Pol-696-2811**
Pkg S-696-7142	**Rec Ctr-696-2494**	

HEALTH & WELFARE: All care, Tripler Army Med Ctr, 433-5811. No chapel.

RECREATION: All beach front sports/recreation, some equipment available, 3 camper spaces w/o hookups.

SPACE-A: See Hickam AFB listing.

ATTRACTIONS: Great beaches on this part of the island.

WHEELER AIR FORCE BASE (HI14R6)

Comm: 808-655-1414
ATVN: 455-7411

Wheeler Air Force Base, HI 96854-5000

LOCATION: Off H-2 or HI-99 in the center of the island of Oahu. Adjacent to Schofield Barracks. RM: p-7, G/3. NMC: Honolulu, 20 mi SE.

GENERAL INFORMATION: Pacific Air Forces Base. 326th Air Division 15 Air Base Squadron, Tactical Air Support Squadron, Air National Guard Aircraft Control and Warning Squadron and support units.

TEMPORARY MILITARY LODGING: Limited to BOQ/BEQ. Call EX-0531.

LOGISTICAL SUPPORT: Most spt fac avail. Key phone numbers follow.

BX-624-3151	**Chap-655-1385**	**Chld C-655-1911**
Fam Svc-655-1211	**Gas S-624-3451**	**Legal-655-1997**
Loc-655-1414/7411	**Med-655-1139**	**NCO C-655-1229**
Off C-655-1817	**Pkg S-655-1795**	**Pol-655-1161**
Snack B-624-3251	**Theat-655-1825**	

HEALTH & WELFARE: Dispensary/Emerg, EX-1693. Inpatient, Tripler Army Med Ctr, 433-5811. Chapels serving all faiths, EX-1384.

RECREATION: Athletic Fields, EX-1690, Bowling, EX-1690, Golf, 622-1776, Hobby Shops, EX-1861, Gym, EX-1690, Spec Svc, EX-1663, Tennis, EX-1853, Racquetball/Handball, EX-1690, Youth Act, EX-1853, Library, EX-1867.

SPACE-A: Army exec aircraft supporting Schofield Barracks, local HI flights, Base Ops, 655-1414. See Hickam AFB listing for CONUS and overseas.

ATTRACTIONS: Pineapple and sugar cane fields, local color, quaint places.

MIDWAY ISLAND

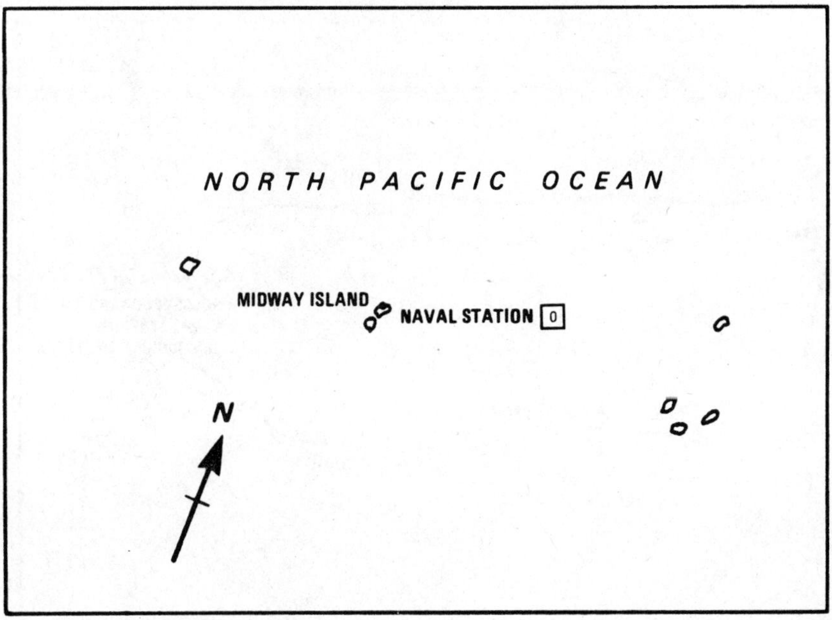

MIDWAY ISLAND NAVAL AIR FACILITY (MW01R8)

FPO San Francisco 96114-1200

Comm: Call Small Is Oper, ask for Midway Is, EX-415/400/814
ATVN: 430-0111

LOCATION: On Midway Island in the Pacific Ocean, 1,300 air miles NW of Honolulu, HI. NMC: Honolulu, HI, 1,300 mi SE.

GENERAL INFORMATION: Naval Defensive Sea Area. Requires entry approval to visit which is obtained thru the Comdt, 14th Naval District, Honolulu, HI, 45 dys in advance.

TEMPORARY MILITARY LODGING: Officer, Bldg B, 24 hr desk, Comm: EX-648. No lodging is available except for military personnel unless prior arrangements have been made through Comdt, 14th Naval District, Honolulu, HI.

LOGISTICAL SUPPORT: Very limited spt fac avail: BX, Caf, Snack B, all Clubs, Pkg S, Theat. SDO-EX-541/400, ATVN: 430-0111.

HEALTH & WELFARE: Limited. Dispensary. Inpatient via MEDEVAC.

SPACE-A: MAC Term, EX-436, CONUS and overseas.

ATTRACTIONS: Water sports, beaches.

PUERTO RICO

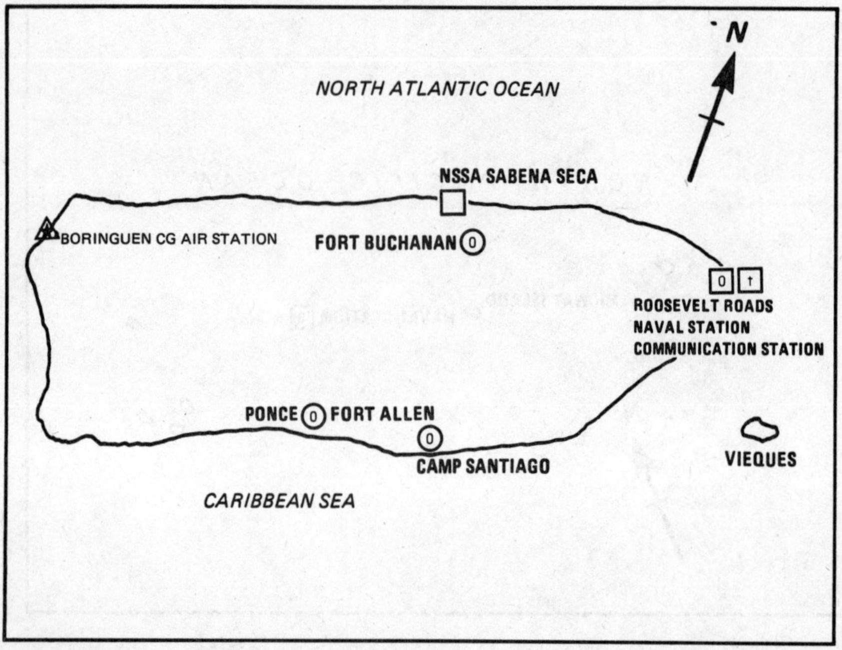

BORINGUEN COAST GUARD AIR STATION (PR03R1)

Aguadilla, PR 00604-5000

Comm: 809-890-5201
FTS: 890-5201/7109

LOCATION: At the old Ramey AFB, N of Aguadilla. Take PR-2 W from San Juan or N from Mayaguez to PR-110 N to CGAS. RM: p-119, B/18. NMC: San Juan, 65 mi E.

GENERAL INFORMATION: Coast Guard Air Station for the Puerto Rico area.

TEMPORARY MILITARY LODGING: Housing Compound, dty hrs, Comm: 809-890-3137 and 809-890-2581, FTS: 890-3137/2581. All ranks. DV/VIP, EX-5201.

LOGISTICAL SUPPORT: Limited spt fac avail. Key phone numbers follow.

BX-890-3127	**Caf-890-7109**	**Cons C-890-2581**
Loc-890-5201/7109	**Med-890-5201**	**Pkg S-890-3127**
Pol-890-5201	**Snack/Vending-890-7109**	

HEALTH & WELFARE: Emrg 890-5201. Inpatient, Roosevelt Roads Naval Hosp, 809-865-4133. No chapel.

RECREATION: Limited.

SPACE-A: CG Hangar, dty hrs, Comm: 809-890-5201/7109, FTS: 890-5201/7109, flts to PR.

ATTRACTIONS: Local beaches and restaurants, excellent surfing dur winter

Boringuen CGAS, Cont'd

mos, 9-hole golf course avail to mil, excellent snorkel and scuba diving.

FORT BUCHANAN (PRO1R1)

Fort Buchanan, PR 00934-5026

Comm: 809-783-2424
ATVN: 740-2424

LOCATION: From San Juan IAP toward San Juan, exit Cagvas/Bayamon to PR-18, exit Bayamon right to PR-22, 1.5 mi to Ft on left. RM: p-119, B/23. NMC: San Juan, 6 mi SW.

GENERAL INFORMATION: The major US Army Post in Puerto Rico providing support to NG and Reserve units.

TEMPORARY MILITARY LODGING: Biltg Ofc, Bldg 152, 0700-1600 hrs dly, Comm: 809-783-2424, EX-7289/3256, ATVN: 704-7289/3256. All ranks. DV/VIP, EX-3205/7281/8291.

LOGISTICAL SUPPORT: Complete spt fac avail. Key phone numbers follow.

PX-792-7047	Caf-792-7049	Chap 783 2424 EX-8251/2136
Chld C-783-2424 EX-7169	Commiss-783-2424 EX-2250	Fam Svc-783-2424 EX-8285/2218
Conv S-792-7405	EM C-792-8431	Med-783-2424 EX-8888
Gas S-792-4297	Legal-783-2424 EX-8164/2145	Pkg S-792-8046
Loc-783-2424 EX-8296	Off C-792-6234	Theat-783-2424 EX-7198
NCO C-792-8431	Snack B-792-8116	
Pol-783-2424 EX-7777		

HEALTH & WELFARE: Clinic/Emerg, EX-8888, Appoint, EX-8236. Inpatient, Roosevelt Roads Naval Hosp, 809-865-4133. Chapels serving all faiths, EX-8251.

RECREATION: Auto Hobby, EX-8172, Ceramics Shop, EX-7172, Golf, EX-7252, Health Club, EX-7167, Library, EX-3208, Little Theater, EX-7133, Photo Hobby, EX-8160, Rec Svc, EX-2101, Equip Issue, EX-7134, Swimming, EX-8182, Woodworking Shop, EX-8160, Youth Ctr, EX-7187.

SPACE-A: See Roosevelt Roads Naval Station listing.

ATTRACTIONS: Near capital of PR, El Morro Castle, Plaza Las Americas Shopping Center, El Condado, Old San Juan shopping area, Bacardi Rum distillers and beaches.

ROOSEVELT ROADS NAVAL STATION (PRO2R1)

FPO Miami 34051-5000

Comm: 809-865-2000
ATVN: 831-2000/4311

LOCATION: From San Juan IAP left (east) onto PR-3 for 45 mi, sign on right indicating exit to Naval Station. RM: p-119, C/26. NMC: San Juan, 50 mi NW.

GENERAL INFORMATION: Hq Naval Forces Caribbean, Fleet Air Caribbean, Marine Barracks, Naval Air Station, and support units.

TEMPORARY MILITARY LODGING: Biltg Ofc, Bldg 1708, FDR Drive, 24 hrs, Comm: 809-865-4145/4147, ATVN: 831-4145. All ranks.

LOGISTICAL SUPPORT: Complete spt fac avail. Key phone numbers follow.

NEX-865-5202/5243	Chap-865-4326/28	Chld C-865-4699/4509
Commiss-865-4287/5605	Conv S-865-5544	CPO C-865-5279
Deli-865-5285	EM C-865-4142	Fam Svc-865-4377/5377
Gas S-865-4609/5449	Legal-865-4315/4320	Loc-865-2000

Roosevelt Roads NS, Cont'd

Med-865-4133/7133	Off C-865-7200	Pkg S-865-7140
Pol-865-4123/4195	SDO-865-6383	Theat-865-4380

HEALTH & WELFARE: Naval Hospital, Ambulance, EX-4144, Info, EX-4133/7133, 120 bed. Chapels serving all faiths, EX-4326.

RECREATION: Auto Hobby, EX-5611, Bowling, EX-4524, Crafts Ctr, EX-4421, Comm Ctr, EX-5231/5384, Fitness Center, EX-4033, Flying Club, EX-7525, Golf, EX-5584, Library, EX-4353, Marina, EX-4476, Stables, EX-5344, Swimming, EX-4033, Yacht Club, EX-5283, Youth Ctr, EX-5231/5384.

SPACE-A: Pax Term and Lounge, Air Ops fac, dty hrs, Comm: 809-865-5208 and 809-865-4263, ATVN: 831-5208/4263, flts to CONUS and overseas.

ATTRACTIONS: El Yunque (L-UN-KEE) rain forest with waterfalls, hiking trails, and restaurants.

SABANA SECA NAVAL SECURITY GROUP ACTIVITY (PR04R1)

Comm: 809-795-2255
ATVN: 831-7202

FPO Miami 34053-1000

LOCATION: From San Juan IAP, take Rt 3W to Bayamon exit to PR-22W to PR-886N (Las Arenas exit) to Base. RM: p-119, A/23. NMC: San Juan, 14 mi E.

GENERAL INFORMATION: Naval Security Group Activity for Caribbean.

TEMPORARY MILITARY LODGING: Biltg Ofc, dty hrs, Comm: 809-795-2255, EX-345 (Off & Enl), very limited facilities. All ranks. DV/VIP, EX-210.

LOGISTICAL SUPPORT: Limited spt fac avail. Key phone numbers follow.

BX-795-2255 EX-213	Chap-795-2255 EX-300	CPO C-795-2255
Loc-795-2255	Med-795-8755/2255 EX-291	Off C-795-2255 EX-277
Pol-783-2424 EX-7777		

HEALTH & WELFARE: Clinic/Emer, 795-4165/795-2255 EX-290/291/292. Inpatient, Roosevelt Roads Naval Hosp, 809-865-4133. Chapels serving all faiths, 795-2255, EX-300.

RECREATION: Rec Ctr, EX-270 & EX-371, Swimming, Weight Room, Basketball, EX-371, Bowling, EX-372, Auto Hobby, EX-279, Ceramics, EX-278, Handball Court, Tennis, Sports Fields, Rec Svc, 795-2255 EX-270.

SPACE-A: See Roosevelt Roads Naval Station and Boringuen CGAS listings.

ATTRACTIONS: Great beaches, El Yunque Rain Forest, El Morro Fort and Casinos.

WAKE ISLAND

WAKE ISLAND AIR FORCE BASE (WK01R8)

Comm: Small islands oper in Honolulu (430-0111) can dial Wake exts
ATVN: 430-0111, ask for Wake Island

APO San Francisco 96501-5000

LOCATION: A US Trust Territory in the Pacific, 2100 mi NW of Hawaii. Access by military aircraft only from Honolulu.

Wake Island AFB, Cont'd

GENERAL INFORMATION: Host is Det 4, 15 ABW/PACAF. If not on official orders, a letter may be required from Commander of Det 4 authorizing visit.

TEMPORARY MILITARY LODGING: Very limited; usually filled by those on duty.

LOGISTICAL SUPPORT: Limited spt fac avail. Key phone numbers follow.

PX-EX-395 | Barber-EX-345 | Laundry-EX-341
PO-EX-259

HEALTH & WELFARE: Bldg 443, 0800-1700 M-S, EX-455.

RECREATION: Swimming, fishing, boating.

SPACE-A: PAX info, bldg 1502, 0800-1700 M-S, EX-210.

ATTRACTIONS: Tropical climate, nice beaches.

FOREIGN COUNTRIES

ANTIGUA

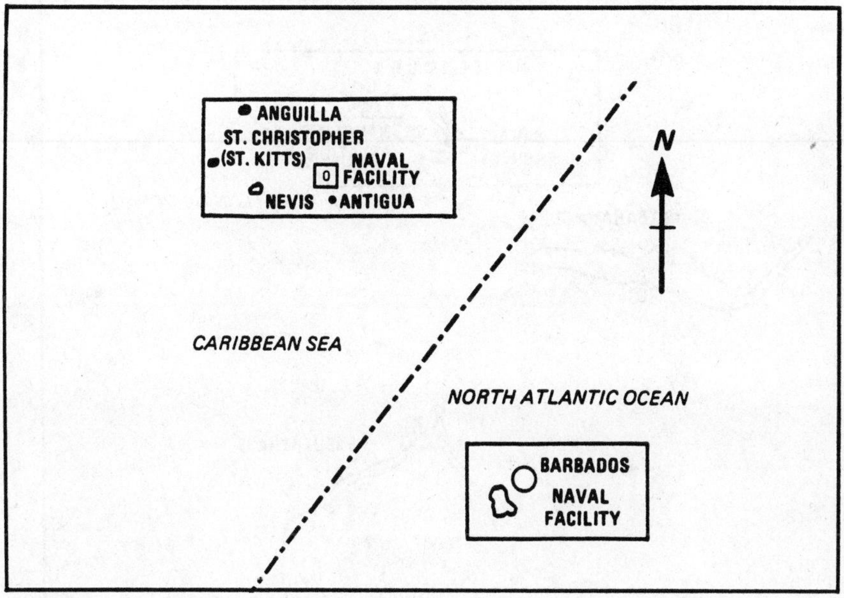

ANTIGUA NAVAL FACILITY (AN01R1)

FPO Miami 34054-5000

Comm: 305-494-23171
ATVN: 854-1110 (ask for Antigua)

LOCATION: One mi from Coolidge Airport, on NE side of island of Antigua. NMC: St. John's, 9 mi W.

GENERAL INFORMATION: AD on leave are restricted to $20/day purchases ex-

Antigua Naval Fac, Cont'd

cept at snack bar and clubs. Retirees may use only Club and Med fac due to SOFA.

TEMPORARY MILITARY LODGING: Adm Bldg, EX-451, 0800-1800 M-Th. Other hrs EX-398. Resv acpt. DV/VIP: Adm Off, EX-386, 0-6 plus. Good availability most of the year. Gov Civ Emp no biltg. No pets.

LOGISTICAL SUPPORT: Limited spt fac avail. Key fac follow.

BX, Theat, Pkg S, Snack B, E5-6 Club, All Hands Club, O Club-EX-356, Spc Svc.

HEALTH & WELFARE: Med Emerg care and Chaplain available.

RECREATION: Swimming, Boating, Deep Sea Fishing, Golf, Tennis, Hiking.

SPACE-A: All PAX info through Patrick AFB, FL, Comm: (305) 494-5631; ATVN: 854-1110 (ask for Antigua). PAX Lounge, Bldg 5631.

ATTRACTIONS: Great beaches. Forts and museums. Casino Gambling.

BERMUDA

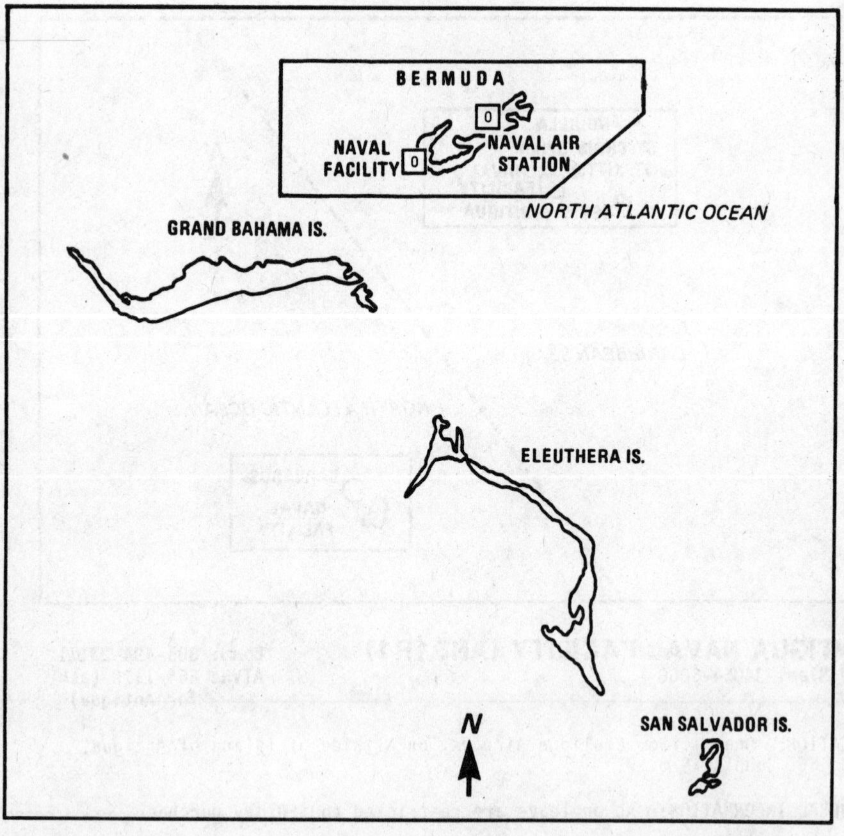

BERMUDA NAVAL AIR STATION (BM01R1)

FPO New York 09560-5000

Comm: 809-293-5131
ATVN: 938-5131

LOCATION: Bermuda is 600 mi E of Cape Hatteras, NC, in the Atlantic Ocean. NAS is on the N end of St David's Island near St George City. NMC: Hamilton, 6 mi SW.

GENERAL INFORMATION: Naval Weather Ser, ASW Squadrons, Nav Reg Med Ctr, NASA Tracking Sta, Canadian Forces, Naval Oceanographic Comm Det, NIS, USN Per Sup Act Det, Tac Supp Ctr.

TEMPORARY MILITARY LODGING: Bldg 550, McCloskey Road, EX-4126/4235. No Civ Emp biltg. Retirees, Space-A. DV/VIP: Cmdr, Bldg 101, EX-2130. O-6 plus. Good availability Nov-Jan. No children or pets allowed.

LOGISTICAL SUPPORT: Most spt fac avail. Key phone numbers follow.

BX-293-3106	**Caf-293-3100**	**Commiss-293-4209**
Conv S-293-3211	**CPO C-293-5292**	**EMC-293-6137**
Gas S-293-3215	**Legal-293-4121**	**Loc-293-2116**
Off C-293-5202	**Pkg S-293-5298**	**Pol-293-7231**
Theat-293-7109		

HEALTH & WELFARE: Naval Branch Med Annex, 293-1117. Chaplain serving all faiths, 293-2111.

RECREATION: Auto Hobby Shop, Bowling, Ceramics Shop, Marina (charter boats), 293-2235, Moped Rental, 293-7237, 9 Golf courses, Handball, Racquetball, Tennis.

SPACE-A: Scheduled and nonscheduled MAC flights, Bldg 1079. PAX Term (0800-1600), 293-3139. Recording, 293-5190. ATVN: 938-5190

ATTRACTIONS: Great beaches, deep sea fishing, boating.

Please, see Special Order Coupon Pages 259 & 261

CANADA

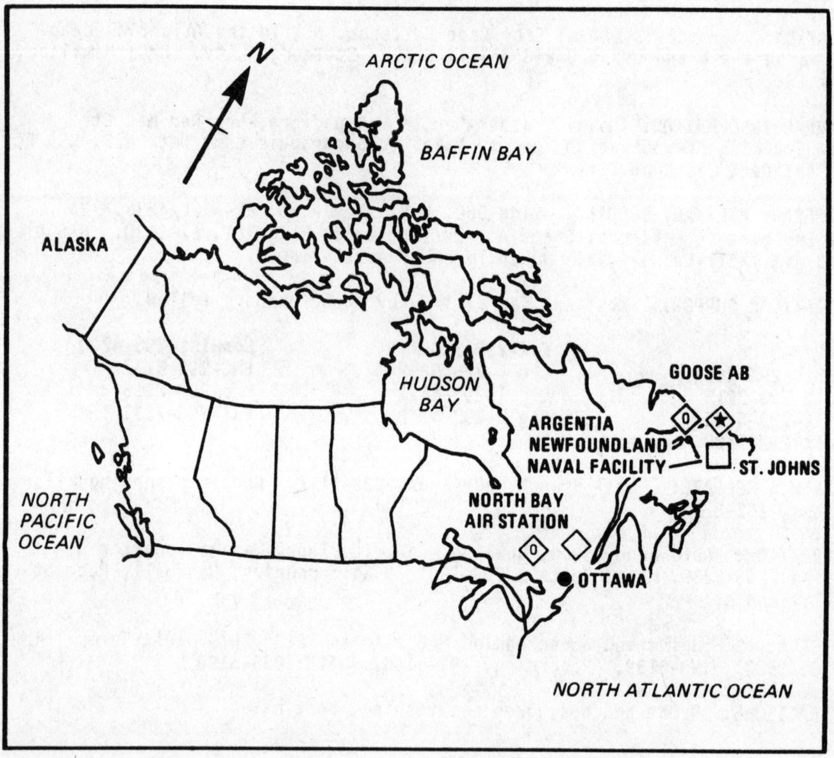

ARGENTIA NAVAL FACILITY (CNO2R1)

FPO New York 09597-5000

Comm: 709-227-5872/5873
ATVN: 622-1690 EX 30

LOCATION: At terminus of North Sydney (Nova Scotia) to Argentia (Newfoundland) Ferry Service. There is access to Trans Canadian Highway (TCH)-1, via CN-101, 25 mi NE. The Naval fac are in Argentia and are clearly marked. RM: p-117, G/24. NMC: St John's, 100 mi NE.

GENERAL INFORMATION: Commands currently located at Argentia are jointly US-Canadian manned Naval Facility.

TEMPORARY MILITARY LODGING: Biltg coordinated at Strain Memorial Hall, Bldg 848M, 24 hrs dly, phone same as above. All ranks. Resv acpt. Some DV/VIP avail, EX-5862.

LOGISTICAL SUPPORT: A small command with minimum spt fac avail. Key phone numbers follow.

NX-227-5841	**CDO Emerg-227-5777**	**Chap-227-5402**
Chld C-227-5372	**Commiss-227-5543**	**Customs-227-2691**
Din Hall-227-5331	**Off C-227-75331**	

HEALTH & WELFARE: Dispensary located on first floor of Strain Mem Hall, Bldg 848. Med Emrg: 227-5532; Chaplain: 227-5402.

Argentia Naval Fac, Cont'd

RECREATION: Silver Harbor CMO Club, Windjammer EMO Club, Argentia Teen Club (age 13-19), Archery Club, Square Dancing, Bowling-227-5482, Movies-227-7542, Gym, Hobby Shops, Skeet Range, and Charter Boat Fishing. North East Arm Rec Camp, 8 mi frm Base, Comm: 709-227-2017; ATVN: 622-1690, 8 1-rm cabins.

SPACE-A: CONUS PAX Info: (709) 576-4981; OCONUS PAX Info: (709) 576-4981; PAX Info: (Other Svcs, Navy, Resv, NG) 622-1690 EX-35. PAX Svc Off: 622-1690 EX-35 or (709) 227-5173.

ATTRACTIONS: Great hunting and fishing area.

CUBA

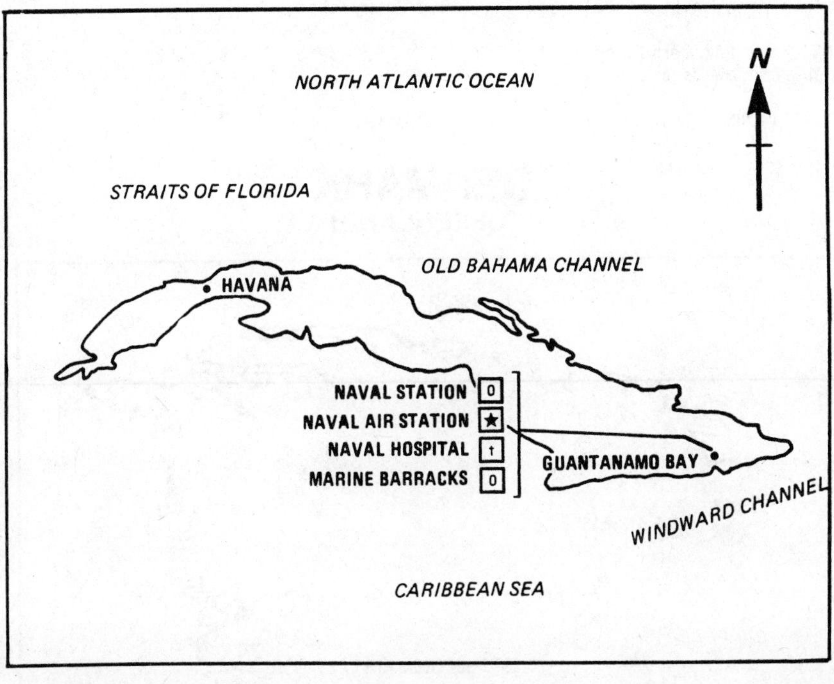

GUANTANAMO BAY NAVAL STATION (CU01R1)

Comm: 011-53-99-xxxx
ATVN: 723-3690/564-4063

FBPO Norfolk 23593-5000

LOCATION: In the SW corner of the Republic of Cuba. Accessible only by air. NMC: Miami, FL, 525 air mi NW. NOTE: All personnel not assigned must have permission of the commander to visit.

GENERAL INFORMATION: Navy Fleet Training Group, 2nd Bn, 8th Marines at Camp Bulkley, ship repair fac, 100 bed hospital.

TEMPORARY MILITARY LODGING: Biltg ofc, ATTN: UOPH Mgr, Bldg 1670, Deer Point Rd, Comm: 011-53-99-2402/3826/2238, ATVN: 564-4063. 24 hr dly. Navy Lodge. All ranks. DV/VIP, EX-2402/4063.

Guantanamo Bay NS, Cont'd

LOGISTICAL SUPPORT: Complete spt fac avail. Key phone numbers follow.

MX-EX-2682	NX-EX-4461	Caf-EX-4815
Chap-EX-4550	Chld C-EX-2205	Commiss-EX-4438
Conv S-EX-2464	EM C-EX-2304	Fam Svc-EX-4141
Gas S-EX-4670	Legal-EX-4600	Loc-EX-4453
Med-EX-4424	NCO C-EX-2501	Off C-EX-2132
Pkg S-EX-4525	Pol-EX-4105	Snack B-EX-2327
Theat-EX-2173		

HEALTH & WELFARE: Naval Hospital, Comm: 011-53-99-7201, ATVN: 723-3960, EX-7201. 100 beds. Chaplains serving all faiths, EX-4550.

RECREATION: Spc svc, EX-5449, boat and fishing rental, swimming, golf, picnic area, bike rental, bowling, auto hobby shop.

SPACE-A: PAX term, Comm: 64-204/5, ATVN: 723-3960, EX-8850/8523. Flts to CONUS and overseas.

ATTRACTIONS: Beaches, fishing, warm climate.

DENMARK
(GREENLAND)

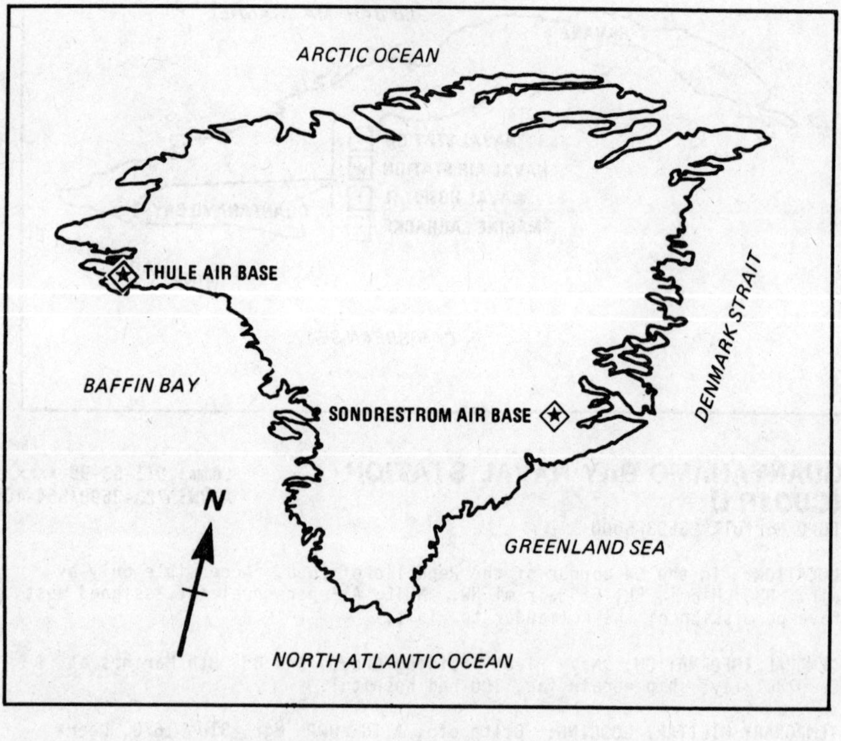

SONDRESTROM AIR BASE (DNO1R7)

APO New York 09121-5000

Comm: None
ATVN: 834-1211 (EX-513/246)

Sondrestrom AB, Cont'd

LOCATION: On the SW coast of the island of Greenland, 33 mi above the Arctic Circle. 7 mi inland on long fjord. NMC: Holsteinbourg, 100 mi NW.

GENERAL INFORMATION: Remote SAC Base. Families not allowed. Security clearance normally reqd. 4684 ABG, TAC radar sites, 109 TAG, Royal Danish AF, 2004 Comm Sqd, 5049 Broadcasting Sqd. Supports the DEW Line.

TEMPORARY MILITARY LODGING: Very limited availability, mainly filled by TDY personnel. Comm hotel nearby.

LOGISTICAL SUPPORT: Limited spt fac avail. Key phone numbers follow.

BX-EX-283	**Med-EX-604**	**NCO C-EX-661**
Off C-EX-497	**Pol-EX-320**	

HEALTH & WELFARE: All care, Hospital, EX-604. Protestant Chaplain, EX-508. Catholic priest visits from Thule AB.

RECREATION: Bowling, EX-644, Gym, EX-684, Library, EX-351, Rec Ctr, EX-243. For photo, ceramics, woodworking and lapidary, EX-592. Also boating, fishing and hunting at Lake Ferguson.

SPACE-A: Base Ops, 24 hrs, EX-513/246, flts to CONUS and overseas. Flts avail from McGuire AFB, NY.

ATTRACTIONS: Hunting, fishing, hiking.

THULE AIR BASE (DNO2R7)

APO New York 09023-5000

Comm: None
ATVN: 834-1211, EX-2155

LOCATION: NW Coast of Greenland. 800 mi from the North Pole. 700 mi N of the Arctic Circle. 2500 mi from McGuire AFB, NJ. Closer to Seattle than New York City by 5 mi.

GENERAL INFORMATION: Host unit is the 1012th Air Base Group. 1983 Comm Squadron-Air Traffic Control and Communications. Det 3, AFSCF is AF Systems Command providing support for Dept of Defense and other agencies.

TEMPORARY MILITARY LODGING: No TML, June-Sept. EX-3276.

LOGISTICAL SUPPORT: Complete spt fac avail. Key phone numbers follow.

BX-2732	**Din Halls-2614**	**Barber-3127**
APO-2615	**Laundry-2249**	**Tailor-2706**

Consolidated Open Messes: Blue Nose Club-2509, Top of the World Club-2418

HEALTH & WELFARE: Hospital operated under contract by DAC personnel. Provides care and treatment to all assigned military, American and Danish personnel, EX-2696. Chaplain serving all faiths, EX-2211.

RECREATION: Rec Ctr (pool tables, TV lounge and recording ctr), EX-3171, Library, EX-2539, Theater, EX-2427, Gymnasium, EX-2519, Bowling, EX-2435, Hobby Shops-2228, Golf, Skeet, and Aero Club. Excellent sports facilities.

SPACE-A: PAX Info: EX-2155. Scheduled flights to McGuire AFB, Goose AB, Sondrestrom AB, & St John's Airport, CN. Only AD and ret personnel can travel to Thule. No dependents. Prior approval by EXO, 12th MWG is required for travel to Thule by Space-A PAX.

ATTRACTIONS: Excellent.

PANAMA

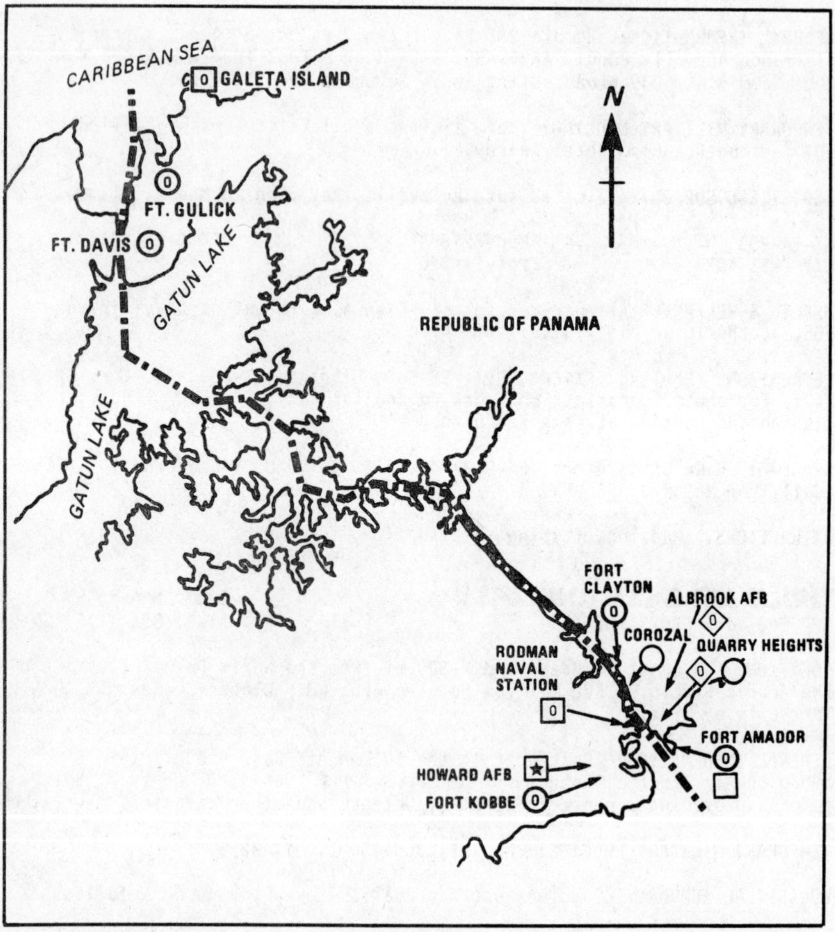

FORT CLAYTON (PNO2R3)

APO Miami 34004-5000

Comm: 507-87-6112
ATVN: 285-6112

LOCATION: Near Pacific entrance to the Panama Canal. Take Galliard Hwy toward the Miraflores locks, clearly marked. RM: p-119, I/24. NMC: Panama City, 8 mi SW.

GENERAL INFORMATION: HQ US Army Forces in Panama, HQ 193rd Inf Brig, 1-187th Infantry Batt, 154th Signal Batt, 324th Supp Gp, 193rd Supp Gp, 142nd Med Batt, 534th MP Comp and 79th Army Band.

TEMPORARY MILITARY LODGING: Biltg Off, Bldg 518, Hospital Rd. 24 hrs dly. Resv req, Ext 87-4454/3251. TML EXTREMELY LIMITED. Best: Oct-May. DV/VIP, Bldg 77, Ext 82-3558. No pets.

Ft Clayton, Cont'd

LOGISTICAL SUPPORT: Complete spt fac avail. Key phone numbers follow.

PX-285-4133	**Caf-287-4718**	**Chap-282-3610**
Chld C-287-5657	**Commiss-285-5405**	**Conv S-282-3131**
EM C-287-4716	**Fam Svc-287-3119**	**Gas S-282-3131**
Legal-287-5202	**Loc-287-4716**	**Med-282-5111**
NCO C-287-4716	**Off C-282-3534**	**Pkg S-282-3637**
Pol-287-4401	**Snack B-285-5420**	**Theat-282-4287**

HEALTH & WELFARE: Dispensary-87-7223, Med Emrg-282-5111. Inpatient, Gorgas Army Comm Hosp, 252-3969. Chaplain-282-3610

RECREATION: Arts & Crafts, 287-5957, Boat Shop, 287-6453, Bowling, 285-6366, Ceramics, 287-4360, Gymnasium, 287-3862, Library, 287-3853, Swimming, 287-6660, SCUBA Rentals, 287-5858, Youth Center, 287-6451.

SPACE-A: Space-A at Howard AFB, MAC PAX Term, Bldg 228, 0500-2100 dly, flight info, Comm: 284-3608/4306; ATVN: 285-3608/4306.

ATTRACTIONS: The Panama Canal, Panama City, 16th Century Cathedrals, Indian Ruins, Pacific and Atlantic beaches.

HOWARD AIR FORCE BASE (PN01R3)

APO Miami 34001-5000

Comm: 507-284-6110
ATVN: 284-1110

LOCATION: Adjacent to Thatcher Hwy on Pacific side of Panama. Ten mi from Albrook AFS across the Thatcher Ferry Bridge. RM: p-119, I/24. NMC: Panama City, 10 mi W.

GENERAL INFORMATION: HQ USAF Southern Air Division, TAC.

TEMPORARY MILITARY LODGING: Biltg Off, Bldg 708, Comm: 312-84-4914/5306; ATVN: 285-4914/5306. DV/VIP, Bldg 119, Resv Req, 312-84-4640/4663.

LOGISTICAL SUPPORT: Complete spt fac avail. Key phone numbers follow.

BX-84-3944	**Caf-84-3944**	**Chap-84-3220**
Commiss-84-3071	**EM C-84-6123**	**Fam Svcs-84-5492**
Gas-84-3983	**Loc-84-6110**	**NCO C-84-7189/2111**
Off C-84-4105	**Taxi-84-5285**	

HEALTH & WELFARE: Med Clinic-84-3092; Dental Clinic-84-4869; Gorgas Hospital, operated by Canal Zone Government, provides all care for military personnel and dependents. Admission is arranged through the base clinic, except in emergencies. Chaplain-84-3220.

RECREATION: Recreation Area at Gamboa, Canal Zone, near buoy 46. Comm: 84-7198. Activities include sunbathing, fishing and boating. On base: Arts & Crafts, 84-3295, Library, 84-3182, Stables, 84-5170, Auto Hobby Sh, 84-2370 Gym, Bowling, Tennis, Kobbe Beach Complex.

SPACE-A: MAC PAX Term Bldg 228, 0500-2100 dly, Comm-84-4306/3608; ATVN: 313-284-4306/3608. Resv-84-4857. PAX Svc-84-5497.

ATTRACTIONS: Panama City, 10 mi W, The Panama Canal, 16th Century Cathedrals, Indian ruins, Atlantic and Pacific beaches.

QUARRY HEIGHTS POST (PN03R3)

APO Miami 34003-5000

Comm: 507-82-3602
ATVN: 285-6110

Quarry Heights Post, Cont'd

LOCATION: On the Pacific side of Panama. Clearly marked. RM: p-119, I/24. NMC: Panama City, adjacent to city.

GENERAL INFORMATION: USCINCO, USCOUTHCOM.

TEMPORARY MILITARY LODGING: Biltg Off, Bldg 119, 0800-1700 hrs dly; EX-3439/3602.

LOGISTICAL SUPPORT: Limited spt fac avail. Key phone numbers follow.

PX-282-3281, Bar Sh-282-3517, O Club, Pkg S, Pol-287-4401.
Complete spt fac at other bases in Panama. See other listings.

HEALTH & WELFARE: Med Emrg, Fort Clayton, 2-5111. Inpatient, Gorgas Army Comm Hosp, 252-3969. Chaplain, Fort Clayton, 2-3610.

RECREATION: Craft Shop, 82-3073, Library, 82-4224, Theater, 82-3560.

SPACE-A: Scheduled and unscheduled flights available at Howard AFB, Comm: 284-4306/3608; ATVN: 285-4306/3608.

ATTRACTIONS: The Panama Canal, beaches, 16th century cathedrals and ancient Indian ruins.

FORT KNOX (KY01R2) cont'd

Theat-942-0822/0275/7999/6832

HEALTH & WELFARE: Ireland Army Community Hospital, clinics and depts, Emerg, EX-9000, Appoint, EX-9148. Chapels serving all faiths, EX-4524.

RECREATION: Arts & Crafts, EX-4725, Auto Hobby, EX-5410/5338, Bowling, EX-4740/6840, Golf, EX-4218/942-0984, Gym, EX-3316/3641, Sports Rental, EX-2314, Outdoor Rec, EX-7754, Field House, EX-2214, Library, EX-1232/4723, Swimming, EX-5416/7230, Rec Ctr, EX-8254, Tennis, EX-2214/4033. Camp Carlson Travel Camp, on Post, year round, Comm: 502-624-4836, ATVN: 464-4836, 25 camper spaces w/hookups, 10 camper spaces w/o hookups, 4 cabins, 4 youth cabins (18 bunks ea.), 2-32' mobile homes, complete facilities and all activities.

SPACE-A: Godman Army Airfield, flts via exec aircraft, EX-1181.

ATTRACTIONS: Mammoth Cave, 50 mi S, Louisville, 30 mi N, Hodgenville, birthplace of Abraham Lincoln, 30 mi S, and Bardstown, site of "My Old Kentucky Home," 30 mi E. Patton Museum on Fort Knox is open at 0900, holidays and weekends at 1000.

- NOTES -

APPENDIX A

STATE ABBREVIATIONS

AK-Alaska
AL-Alabama
AR-Arkansas
AZ-Arizona
CA-California
CO-Colorado
CT-Connecticut
DC-District of Columbia
DE-Delaware
FL-Florida
GA-Georgia
HI-Hawaii
IA-Iowa
ID-Idaho
IL-Illinois
IN-Indiana
KS-Kansas
KY-Kentucky
LA-Louisiana
MA-Massachusetts
MD-Maryland
ME-Maine
MI-Michigan
MN-Minnesota
MO-Missouri
MS-Mississippi
MT-Montana
NE-Nebraska
NC-North Carolina
ND-North Dakota
NH-New Hampshire
NJ-New Jersey
NM-New Mexico
NV-Nevada
NY-New York
OH-Ohio
OK-Oklahoma
OR-Oregon
PA-Pennsylvania
RI-Rhode Island
SC-South Carolina
SD-South Dakota
TN-Tennessee
TX-Texas
UT-Utah
VA-Virginia
VT-Vermont
WA-Washington
WI-Wisconsin
WV-West Virginia
WY-Wyoming

POSSESSIONS ABBREVIATIONS

GU-Guam
MW-Midway Island
PR-Puerto Rico
WK-Wake Island

COUNTRY ABBREVIATIONS

AN-Antigua
BM-Bermuda
CN-Canada
CU-Cuba
DN-Denmark
PN-Panama

FREQUENTLY CALLED NUMBERS

NAME	NUMBER	COMMENTS

APPENDIX B
General Abbreviations

This Appendix contains general abbreviations used in this book. Commonly understood abbreviations, i.e., M-F, Monday to Friday, have not been included in order to save space. Address abbreviations are also not included in most cases in order to save space and because they have proven to be accurate mail addresses when used as stated in this book.

A
AAF-Army Air Field
AAFES-Army Air Force Exchange System
AB-Air Base
A/C-Air Conditioning
Acpt-accepted
Acf-Aircraft
Act-activity
AD-Active Duty
Add-additional
Adj-adjacent
Adm-Administration
Adv-advance
AF-Air Force
AFB-Air Force Base
AFN-Armed Forces Network
AFRC-Air Force Reserve Center
AFRC-Armed Forces Recreation Center
AFS-Air Force Station
Alcoh-alcohol
AMC-Army Medical Center
AMEM-American Embassy
AMEXCO-American Express Company
ANG-Air National Guard
APO-Army Post Office
Appoint-appointment
Approx-approximately
Apt(s)-apartment(s)
AR-Army
ARC-American Red Cross
AS-Air Station
ATC-Air Traffic Control
ATCO-Air Traffic Coordinating Organization
ATVN-Automatic Voice Network (Autovon)
Aux-auxiliary
Avail-available
Ave-avenue
Avl-available
Avn-aviation

B
BAQ-Basic Allowane for Quarters
Barks-barracks
Bdrm-bedroom
BEQ-Bachelor Enlisted Quarters
Bev-beverage
Biltg-billeting
Bldg-building
Blks-blocks
Blvd-boulevard
BOQ-Bachelor Officers Quarters
Btwn-between
B/W-black and white
BW-Bomb Wing
BX-Base Exchange

C
Caf-cafeteria
Cap-capacity
Cat-categories
CB-communal bath
CBPO-Civilian Base Personnel Office
CG-Coast Guard
CG-Commanding General
CGAS-Coast Guard Air Station
CGLS-Coast Guard Loran Station
CGX-Coast Guard Exchange
Chap-Chaplain
Chg-change
Chld-children
Chld C-Child Care Center
Civ-civilian
Class VI-liquor package store
Cmd-command
Cmdr-commander
Cmmd-commanding
COM-Commissioned Officers Mess
Comm-commercial
Comm-commercial telephone system
Comm-community
Commiss-commissary
Comp-complete
Comp-complimentary
Comss-commissary
Consol-consolidated
CONUS-Continental United States
Conv-convenience
Conv S-convenience store
Cp-camp
CPO (C)-Chief Petty Officer (Club)
CQ-Charge of Quarters
CSM-Command Sergeant Major
Ctr-center

D
DAV-Disabled American Veteran
Dblrm-double room
Dep-dependent
Dept(s)-department(s)
Det-determined
Diff-difficult
Dir-direction
Dly-daily
DOD-Department of Defense
Drv-driving

Dty-duty
DV-Distinguished Visitor

F
Ea-each
Eff-efficiency
El-electricity
Elig-eligible
EM-Enlisted Personnel
EM C-Enlisted Men's Club
Emer(g)-emergency
EML-Environmental & Morale Leave
Emp-employee
Eq(uip)-equipment
Evap-evaporation
EX-extension
Exch-exchange (money)
Excel-excellent
EXO-Executive Officer
Ext-telephone extension

F
F-female
Fac-facilities
Fam-family
Fam Svc-Family Services
FCFS-First Comes First Served
Fm-from
FMR-fair market rent
FPO-Fleet Post Office
Flt(s)-flight(s)
Freq-frequency
Ft-fort
FTS-Federal Telephone System
Furn-furniture/furnished

G
Gas S-gas station
Gen-general
GH-Guest House
GO-General Officer
Gov-government
grp-group
Grtr-greater
GS-General Schedule
GTR-Government Travel Request

H
Hndcp-handicapped
HB-hall bath
Helo-helicopter
HH-Hostess House
Hol(s)-holiday(s)
Hq-Headquarters
Hrs-hours

I
I-Interstate
IAP-International Airport
Ice V-ice vending
Ind-indicated
Info-information
Inst-installation
Inter-interior
Int-intersection
Intl-International
ITT-Information Tour and Travel

J
None

K
Kit-kitchen

L
L-left
Lic-license
Lim-limit
Lndry-laundry
Loc-location (locator)
LR-living room
Ltd-limited
Lv-leave

M
M-male
MAC-Military Airlift Command
Mach-machine
MAP-Military Assistance Program
MARS-Military Affiliated Radio Station
MATCU-Military Air Traffic Coordinating Unit
Max-maximum
MC-Marine Corps
MCAS-Marine Corps Air Station
MCDEC-Marine Corps Development & Education Command
MCLB-Marine Corps Logistical Base
MCRD-Marine Corps Recruit Depot
Med-medical
Mem-member(s)
MH-mobile home
Mi-miles
Mil-military
MILGP-Military Group
Min-minimum
Mo-monthly
Mod-modern
Msn-Mission
MTA's-Military Transportation Authorization
MTS-Military Telephone System
Muni-Municipal (airport)
MW-Microwave
MX-Marine Exchange

N
NA-Navy
N/A-not applicable
NAB-Naval Amphibious Base
MAEC-Naval Air Engineering Center

NAF-Naval Air Facility
NAS-Naval Air Station
Natl-national
NB-Naval Base
NCBC-Naval Construction Battalion Centers
NCO (C)-Noncommissioned Officer (Club)
NCU-Naval Communication Unit
NETC-Naval Education & Training Center
NMC-nearest major city
NMI-nearest military installation
NOS-Naval Ordnance Station
Nr-near
NS-Naval Station
NSA-Naval Support Activity
NSGA-Naval Security Group Activity
NSWC-Naval Surface Weapon Center
NSY-Naval Shipyard
NTC-Naval Training Center
Nur-nursery
NWC-Naval Weapons Center
NES-Naval Weapons Station
NX-Navy Exchange

O
O Club-Officers' Club
Occ-occupancy
OCONUS-outside the Continental United States
OD-Officer of the Day
Ofc-Officer
Ofce-office
Off-official
Off C-Officers' Club
OIC-Officer in Charge
Oper-operation
Ops-Operations
Org-organization

P
PAX-passengers
PB-private bath
PCS-Permanent Change of Station
Pers-personnel
Pkg-package
Pkg S-Package Store
PO-Post Office
POC-Proof of Citizenship
Pol-police
POV-privately owned vehicle
Prefab-prefabricated
Pres-President
Pub wrks-public works
PX-Post Exchange

Q
Qtrs-quarters

R
R-right
RAAF-Royal Autralian Air Force
RAF-Royal Air Force
Rec-recreation
Rec Svc-Recreation Services
Refr-refrigerator
Regis-registration
Remod-remodeled
Renov-renovated
Req-required
Resv-reservation
Resvs-Reservist
Retd-retired
Riv-river
Rm-room
RM-Rand McNally Road Atlas
Rntl-rental
RV-recreational vehicle

S
S & S-Stars and Stripes
SATO-Scheduled Airline Ticket Office
SB-Semi-private bath
Sch-school
Schd-schedule
SDO-Staff Duty Officer
Sec-security
Sew-Sewer
Sgl rm-single room
Sl-sleeps
Snk B-Snack Bar
SNCO-Senior Noncommissioned Officer
SOFA-Status of Forces Agreement
SP-Security Police
Space A-Military Air Space Available
Spc-special
Spc Svc-Special Services
Spon-sponsor(s)
Spt fac-support facility
Sta-station
Stat-statement
Stes-suites
Struc-structure
Sugg-suggested
Svc-service

T
T&R-travel and recreation
TAD-Temporary Attached Duty
TAQ-Temporary Airmen's Quarters
TDY-temporary duty
Tec-technical
Tel-telephone
TEQ-Temporary Enlisted Quarters
Term-terminal
Theat-theater
TLF-Temporary Lodging Facility
TLF-Transient Lodging Facility
TLQ-Temporary Living Quarters
TML-Temporary Military Lodging
Tng-training
TOQ-Temporary Officers Quarters
Trans-transient
TUAF-Turkish Air Force
TV-television

Tvl-travel
TVOQ-Transient Visiting Officers Quarters

U

Un-unit
Una-unaccompanied
Unif-Uniformed
Unk-unknown
UPH-Unaccompanied Personnel Housing
Uten-utensils
US-United States
USA-United States Army
USAF-United States Air Force
USDAO-United States Defense Attache Office
USMAAG-United States Military Assistance Advisory Group
USMC-United States Marine Corps
USMCA-United States Military Community Activity
USN-United States Navy
USO-United service Organizations
USPHS-United States Public Health Service

V

Vac-vacancy
Vaca-vacation
VAT-Value Added Tax
Veh-vehicle
Vend-vending
VEQ-Visiting Enlisted Quarters
VIP-Very Important Person
VOQ-Visiting Officer Quarters

W

Wa-water
W/D-washer/dryer
Wid-widow(er)
Wk-week
Wknd-weekend
Wkly-weekly
w/o-without

X

None

Y

Yr-yearly

Z

None

STATE & FOREIGN COUNTRY MAPS

Legend

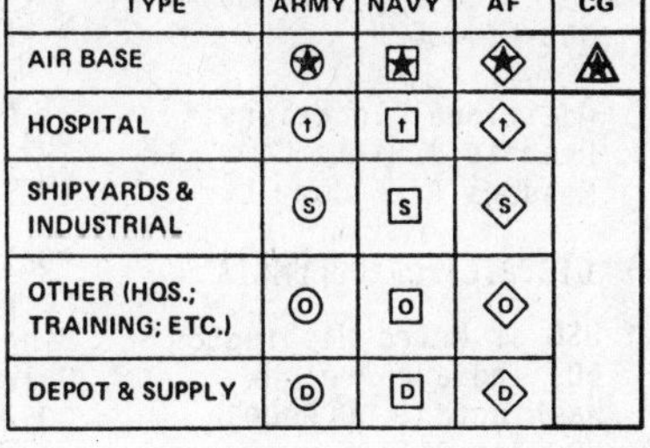

APPENDIX C

UNITED SERVICE ORGANIZATIONS

IN THE UNITED STATES

ALASKA

USO Fairbanks
507 Gaffney
Fairbanks, AK 99707
907-456-4522

Mailing address:
P O Box 1228
Fairbanks, AK 99707

CALIFORNIA

***LOS ANGELES AREA USO**
1641 N Ivar Avenue
Los Angeles, CA 90028
213-462-6904/5
(24 hr Emerg Svc LA)

USO-LAX
Jimmy Doolittle Center
%American Airlines
Term 4, 400 World Way
Los Angeles Intl Arpt
Los Angeles, CA 90045
213-642-0188

USO Los Angeles
Greyhnd Bus Term/Outreach
(Desk) (Weekends Only)
Main Terminal
208 E Sixth Street
Los Angeles, CA 90014
213-624-0066/462-0747

USO-MEPS Outreach
5051 Rodeo Road
Los Angeles, CA 90016
213-292-3219

Mailing address for LAX,
Bus & MEPS USOs is:
1641 N Ivar Avenue
Los Angeles, CA 90028

Bob Hope USO-Hollywood
1641 N Ivar Avenue
Los Angeles, CA 90028
213-462-0747/467-3876

Ventura County USO
418 West 4th Street
P O Box 773
Oxnard, CA 93032
805-483-3710

San Diego USO
433 Harbor Drive
San Diego, CA 92101
619-235-6503

***BAY AREA USO (SF)**
Bldg 7, Rm 161
Naval Sta, Treasure Island
San Francisco, CA 94130
415-391-1657

Downtown S F USO Center
1107 Mission Street
San Francisco, CA 94103
415-552-2436

S F Intl Arpt USO Center
Intl Term, 4th Floor
San Francisco, CA 94128
415-761-4611

Oakland IAP USO Center
1 Airport Dr, Box 20
Oakland, CA 94614
415-562-3448

Travis AFB USO Center
Passenger Terminal
P.O. Box 1663
Travis AFB, CA 94535
707-438-3316

COLORADO

Pikes Peak Region USO
207 N Nevada Avenue
P O Box 1694
Colorado Springs, CO 80911
303-471-9790, Ext 222

USO Information Desk
Colorado Spgs Muni Aprt
Colorado Spgs, CO 80916
303-574-9626

Additional locations:
Deerfield Hills Comm Ctr
Meadows Park Comm Ctr

DISTRICT OF COLUMBIA

USO of Metro Washington
601 Indiana Ave, NW
Washington, DC 20004
202-783-8117

USO Information Desk
Andrews AFB Ops Term
Andrews AFB, MD 20331
301-981-2525

USO Arpt Ctr & Info Desk
National Arpt, North Term
Washington, DC 20001
703-920-6990/2705

BWI Airport USO
(desk & lounge)
Baltimore/Wash Intl Arpt
Baltimore, MD 21240
301-859-3188

Additional locations:
Bellevue Housing Comm,
Fort Myer, Kiosk on the
Mall, Southern Manor

FLORIDA

Homestead USO
Keys Rec Ctr, 2nd Floor
P O Box 365
Homestead, FL 33030
305-258-0055

***Greater JAX Area USO**
P O Box 108, NAS
Jacksonville, FL 32212
904-778-2821

USO Center-JAX Beach
P O Box 51000
234 6th Ave North
Jacksonville Bch, FL 32240
904-246-3481
(Serving Mayport NS)

USO Center-St Augustine
Lightner Museum
8 Granada Street
P O Box 4
St Augustine, FL 32085
904-829-6481, Ext 204
(Weekends only)

***Orlando USO Center**
Recruit Reception Center
Bldg 102, Nav Tng Ctr
Orlando, FL 32814
305-647-2972

Orlando USO
Bldg 2008, Nav Tng Ctr
Orlando, FL 32814
305-647-0407

Emil B Miller USO Arpt Ctr
Orlando IAP, Landside "B"
6000 McCoy Road
Orlando, FL 32809
305-855-7192

Mailing address for Orlando USOs:
P O Box 19472
Orlando, FL 32814

USO Center of Tampa
318 Madison Street
Tampa, FL 33602
813-229-9401

GEORGIA

USO Arpt Ctr Atlanta
Hartsfield Intl Arpt
North Term, Intl Concourse
Atlanta, Ga 30320
404-761-8061

USO MEPS Center
80 Tenth St
Atlanta, GA 30309
404-761-8061

Mailing address for both:
P O Box 20963
Atlanta, GA 30320

USO Camden County
Naval Submarine Base
Kings Bay, GA 31547
912-673-4013

Augusta USO
609 Broad St
Augusta, GA 30902
404-722-8371

Warner Robins USO
101 Market Street
Warner Robins, GA 31093
912-922-5885

Mailing address is:
P O Box 688
Warner Robins, GA 31099

HAWAII

USO Aloha Airport Center
Honolulu Intl Arpt
Terminal Box 48
Honolulu, HI 96819
808-836-3351

Hickam USO Lounge
Hickam MAC Term
Hickam AFB, HI 96853
808-449-2887

ILLINOIS

Chicago USO
225 North Wabash
3rd Fl, Oxford House
Chicago, IL 60601
312-781-0730

USO Airport Center
O'Hare Intl Arpt
P O Box 66434
Chicago, IL 60666
312-686-7396

Great Lakes USO
Bldg 927
Recruit Tng Cmmd
Great Lakes, IL 60088
312-688-2274

INDIANA

USO Airport Center
Indianapolis Intl Arpt
Rm F-117, P O Box 51657
Indianapolis, IN 46251
317-241-6070

KENTUCKY

Patton USO Airport Center
13 Lee Terminal
Standiford Field
Louisville, KY 40209
502-361-1888

USO Info Desk
Greyhound Term, 1st Fl
720 W Muhammad Ali Blvd
Louisville, KY 40203
502-583-1945
(Weekends Only)

MARYLAND

BWI Airport USO
(Desk and Lounge)
Baltimore/Wash Intl Arpt
Baltimore, MD 21240
301-859-3188

MASSACHUSETTS

Boston USO
46 Joy Street
Boston, MA 02114
617-720-4949

USO Logan Airport Ctr
Logan Intl Arpt, C Term
Ground Level
Boston, MA 02128
617-567-9330

MISSOURI

J S McDONNELL USO Arpt Ctr
P O Box 10367
Lambert Field
St Louis, MO 63145
314-429-1234

NEW YORK

***USO of Metro N Y**
140 West St, Suite 2730
New York, NY 10007
212-962-1440

USO Gen Douglas MacArthur Memorial Center
1540 Broadway
(45th St Entrance)
New York, NY 10036
212-719-2364

USO JFK Airport Center
Pan Am Worldport Term
J F Kennedy Airport
Jamaica, NY 11430
718-656-4616

Watertown USO
YWCA Building
50 Public Square
Watertown, NY 13601
315-788-4400

Mailing address:
Mrs Walter Parry
46 Public Square
Watertown, NY 13601

NORTH CAROLINA

Joe Barr USO of NC
333 Ray Avenue
Fayetteville, NC 28302
919-483-3179

Mailing address:
P O Box 427
Fayetteville, NC 28302

Jacksonville USO
9 Tallman Street
Jacksonville, NC 28540
919-455-3411

OHIO

USO MEPS Center
1240 East 9th Street
Cleveland, OH 44199
216-621-4120

USO Airport Center
Hopkins Intl Airport
Rocky River Drive
Brookpark, OH 44135
216-433-7313

USO Airport Center
Port Columbus Intl Arpt
Room 102
614-231-7300

Mailing Address:
4600 E 17th Avenue
Columbus, OH 43219

Columbus MEPS USO
3333 Indianola Ave
Columbus, OH 43214
614-267-3131

PENNSYLVANIA

Naval Base USO
Bldg 500, 2nd Deck
Naval Base Philadelphia
Philadelphia, PA 19112
215-336-0908

USO Airport Center
Terminal D
Philadelphia Intl Arpt
Philadelphia, PA 19153
215-365-8889

SOUTH CAROLINA

Charleston USO
Charleston Intl Arpt
P O Box 120
Charleston, SC 29411
803-767-3963

TENNESSEE

Memphis/Millington USO
7918 Church Street
Millington, TN 38053
901-872-7722

TEXAS

USO of South Texas, Inc.
Bldg 3, Naval Air Sta
Corpus Christi, TX 78419
512-939-2391

USO Airport Center
Houston Intercont Arpt
Term C, North Concourse
713-443-2451

Mailing address:
P O Box 2986
Houston, TX 77252

San Antonio USO
410 South Alamo
San Antonio, TX 78205
512-227-9373

VIRGINIA

Virginia Peninsula USO
Coliseum Mall (entrance A)
1800 W Mercury Blvd
Hampton, VA 23666-3203
804-838-4182, 804-827-1063

Mailing address:
P O Box 7250
Hampton, VA 23666

WASHINGTON

***USO Arpt Ctr & Adm Off**
Seattle/Tacoma Intl Arpt
Military Lounge
Terminal, 2nd Floor
Seattle, WA 98158
Admin Off: 206-246-1908
Ctr: 206-433-5438

McChord Air Force Base
MAC Terminal
McChord AFB, WA 98438
206-984-2400

WISCONSIN

USO War Memorial Center
750 N Lincoln Mem Dr
Milwaukee, WI 53202
414-271-3133
(weekends & holidays only)

PUERTO RICO

Puerto Rico USO
U.S. NS Roosevelt Roads
P O Box 3594
Ceiba, PR 00635
809-865-7350
809-865-7275
809-865-2000
Ext-4491 (mil)

Mailing address:
Box 3594, USNS
FPO Miami 34051

USO Puerto Rico
P O Box 34295
U S Army Garrison
Fort Buchanan
San Juan, PR 00934
809-781-7590
809-783-2424, Ext 8217

VIRGIN ISLANDS

St. Croix USO
114 Market Street
P O Box 616
Frederiksted
St. Croix, VI 00840
809-772-0523

St. Thomas USO
75 Veterans Drive
P O Box 1857
Charlotte Amalie
St. Thomas, VI 00801
809-774-1012

GUAM

USO Guam
Hoover Park
Piti, Guam
Marianas Islands
Civilian Tel: (671) 332-2021/2022

Mailing address:
Box 21 COMNAVMAR
FPO San Francisco 96630
ATVN: 322-1100
Ext 332-5300/6447

Guam USO MAC Term Lounge
Andersen AFB, Guam
Civilian Tel: (671) 366-4246 Ext 7246

Mailing address same as USO Guam

APPENDIX D

ARMED SERVICES YMCA BRANCHES

Armed Services YMCA
609 "F" Street
Anchorage, AK 99501
(907) 277-8522

Armed Services YMCA
1776 Depot Square
Ayer, MA 01432
(617) 772-2681

Armed Services YMCA
PO Box 3675, Amador Rd
Balboa, Republic of Panama
(011) 507-52-2839

Armed Services YMCA
32 City Square
Boston, MA 02129
(617) 242-2660

Armed Services YMCA
155 First Street
Bremerton, WA 98310
(206) 377-3741

Armed Services YMCA
1300 Madison Street
Clarksville, TN 37040
(615) 647-2376

Armed Services YMCA
PO Box 5096
Bolivar Avenue
Cristobal,
Republic of Panama
(011) 507-41-6868

Armed Services YMCA
315 E. Franklin Avenue
El Paso, TX 79901
(915) 532-4957

Armed Services YMCA
Fort Wood I-44 Spur
PO Box 350
Fort Leonard Wood, MO 65473
(314) 329-4513

Armed Services YMCA
PO Box 216, Ruckman Road
Fort Monroe, VA 23651
(804) 723-1852

Armed Services YMCA
250 South Hotel St
Honolulu, HI 96813
(808) 524-5600

Armed services YMCA
PO Box 288, 111 E 16th St
Junction City, KS 66441
(913) 238-2972

Armed Services YMCA
PO Box 115
Killeen, TX 76541
(817) 634-5445

Armed Services YMCA
201 South 4th St
Lawton, OK 73501
(405) 355-5520

Armed Services YMCA
PO Drawer 906
Fort Polk, LA 71459-0906
(318) 537-7823

Armed Services YMCA
50 Washington Square
Newport, RI 02840
(401) 846-3120

Armed Services YMCA
500 East Ocean View Ave
Norfolk, VA 23503
(804) 480-3744

Armed Services YMCA
Whidbey Island
PO Box 794,
1182 W Pioneer Way
Oak Harbor, WA 98277
(206) 675-2771

Armed Services YMCA
604 4th Street
Oceanside, CA 92054
(619) 722-7671

Armed Services YMCA
509 King Street
Portsmouth, VA 23704
(804) 397-1675

Armed Services YMCA
312-320 East Wabash Ave
Rantoul, IL 61866
(217) 892-2137

Armed Services YMCA
500 West Broadway
San Diego, CA 92101
(619) 232-1133

Armed Services YMCA
124 John Sims Parkway
Valparaiso, FL 32580
(904) 678-1408

Affiliated Programs

YMCA
1088 Quail Dr
Charleston, SC 29412
(803) 795-0820

Pikes Peak Region YMCA
207 N. Nevada Ave
Colorado Springs, CO 80901
(303) 471-9790

YMCA-Southeastern Conn
Dolphin Center-Rt 12
Groton, CT 06340
(203) 449-4702

YMCA-Monterey Peninsula
600 Camino El Estero
Monterey, CA 93940
(408) 373 4166

YMCA-Philadelphia Area
Armed Services Branch
Flag Hall, Bldg 537
Naval Base Philadelphia, PA 19112
(215) 897-5858

Watertown YMCA
6 Public Square
Watertown, NY 13601
(315) 782-3100

Two new affiliates will come on-line soon:

YMCA of Greater Lake City
Waukegan, IL

Savannah YMCA
Savannah, GA

APPENDIX E

LOCATION OF UNIFORMED SERVICES HOSPITALS & VA HOSPITALS

Western United States

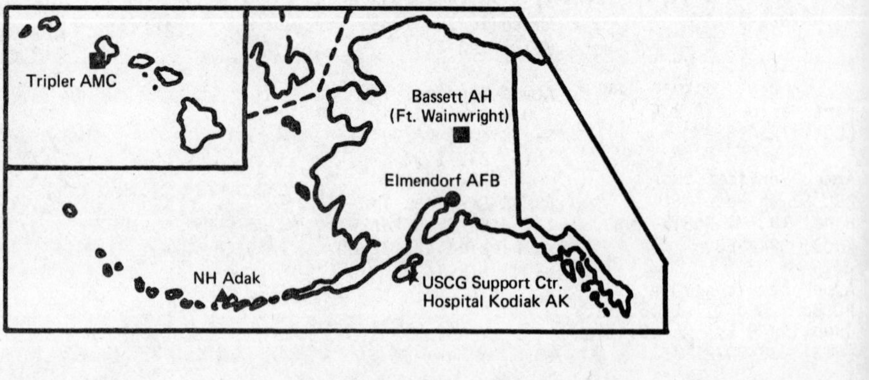

LOCATION OF UNIFORMED SERVICES HOSPITALS & VA HOSPITALS

Western United States

LOCATION OF UNIFORMED SERVICES HOSPITALS & VA HOSPITALS

Eastern United States

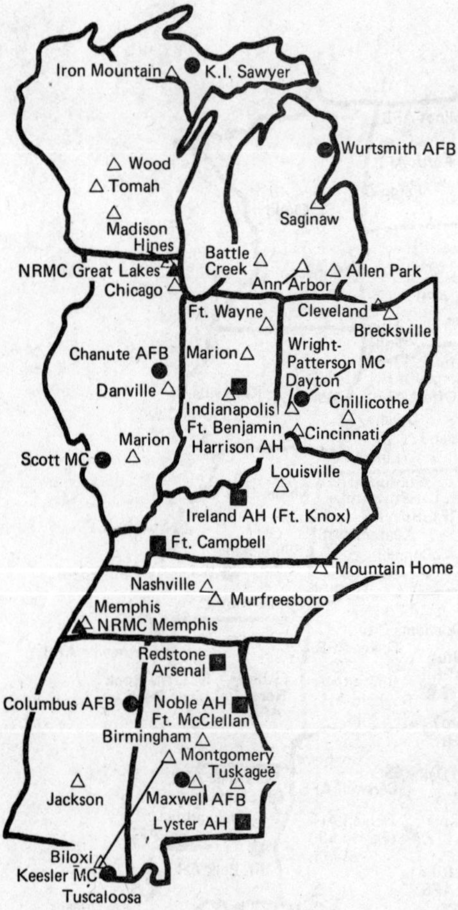

NOTE: There is a Veterans Canteen (Exchange) Service at each VA hospital, which can be used by Uniformed Services ID Card holders. Also, Navy Exchanges at all Navy Regional Medical Centers.

LEGEND

- ★ U.S. Coast Guard
- △ VA Hospital
- ▲ U.S. Navy
- ■ U.S. Army
- ● U.S. Air Force
- □ Uniformed Services Medical Treatment Facility

LOCATION OF UNIFORMED SERVICES HOSPITALS & VA HOSPITALS

Eastern United States

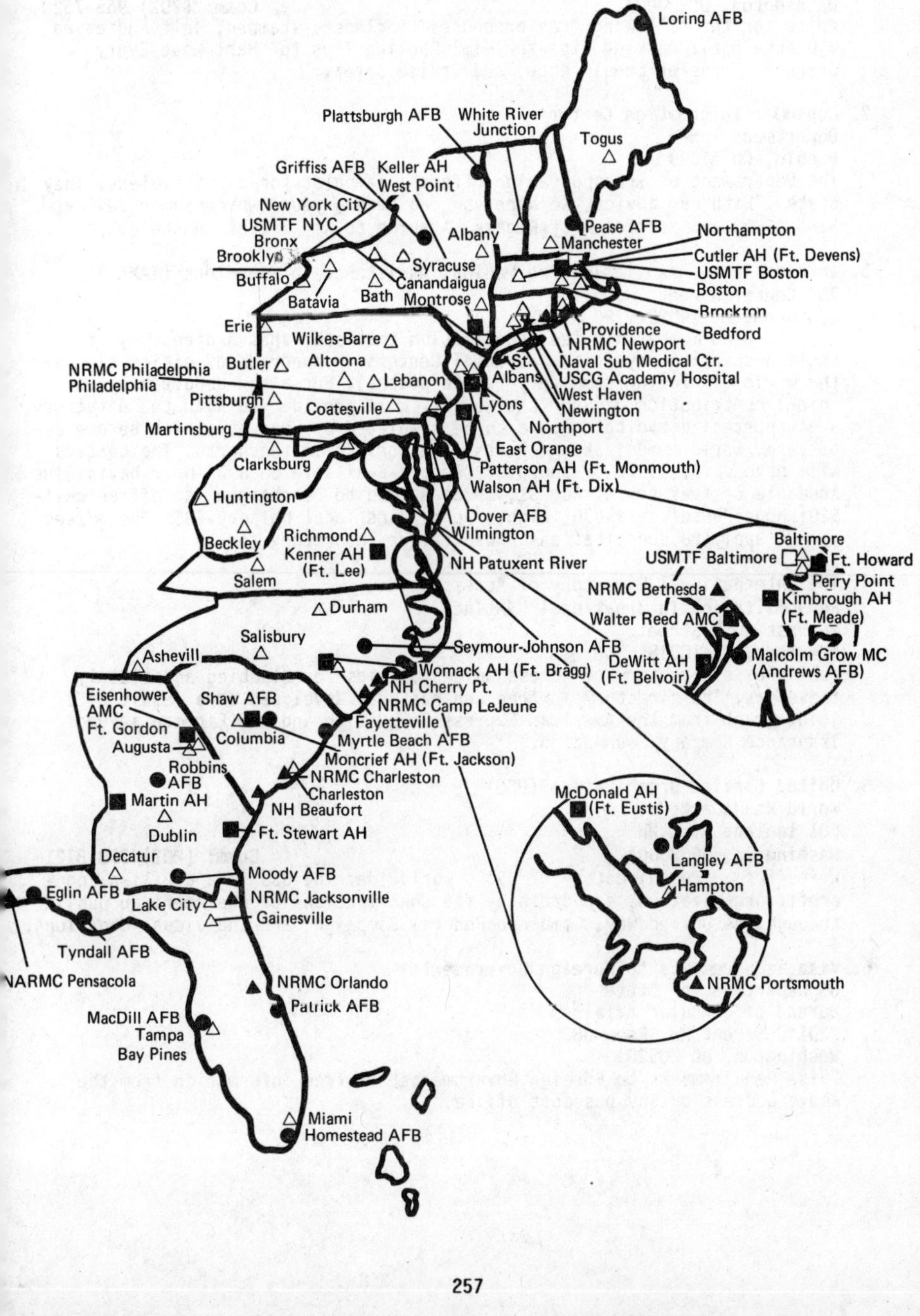

APPENDIX F

SPECIAL TRAVEL AIDS

1. **American Society of Travel Agents**
   Public Relations Department
   4400 MacArthur Blvd, NW
   Washington, DC 20007 **Comm: (202) 965-7520**
   Write for the following free brochures (include a stamped, self-addressed, #10 envelope): Packing Tips for Him; Packing Tips for Her; What Every Overseas Traveler Should Know; and Cruise Safety.

2. **Consumer Information Center**
   Department 165-P
   Pueblo, CO 81009
   The Department of Transportation offers a pamphlet for air travelers. They state, "With our advice, we hope you can become your own consumer advocate." Send $1.00 and ask for "Fly Rights, A Guide to Air Travel in the U.S."

3. **International Association for Medical Assistance to Travelers (IAMAT)**
   736 Center Street
   Lewiston, NY 14092-1796
   This is a no membership fee organization that publishes a directory of English-speaking physicians at IAMAT Centers in hundreds of cities all over the world (including some very remote places). For a tax deductible (we think) contribution of your choice they will provide you with the directory, a passport-size medical record to be completed by your physician before departure, world immunization charts, and world climate charts. The centers will have daily lists of approved doctors available on a 24-hour basis. The schedule of fees (as of Aug 31, 1983) agreed to be charged is: office call-$20; house/hotel call-$30; night, Sunday and local holiday-$35. These fees do not apply to consultations, hospital or laboratory fees.

4. **The International Directory of Access Guides**
   Rehabilitation International USA Inc.
   20 West 40th Street
   New York, NY 10018
   Publishes a directory of Access Guides as aids for disabled and elderly travelers. The directory is free and has been developed as a result of a joint grant from the American Express Foundation and the Fireman's Fund Insurance Company Foundation.

5. **United Service Organization (USO)**
   World Headquarters
   601 Indiana Ave, NW
   Washington, DC 20004 **Comm: (202) 783-8121**
   Write for a free directory of USO's worldwide. The USO is a civilian, non-profit organization, supported by the contributions of the American public through the United Way, Combined Federal Campaign, and individual donations.

6. **Visa Requirements to Foreign Governments**
   US Department of State
   Bureau of Consular Affairs
   2201 C Street NW, Room 683
   Washington, DC 20520
   "Visa Requirements to Foreign Governments" is free information from the above address or any passport office.

CENTRAL ORDER COUPON

MILITARY Living™

MILITARY LIVING PUBLICATIONS
P.O. Box 2347, Falls Church, VA 22042
Telephone: (703) 237-0203.

Publications	Qty.	Publications	Qty/Yrs.
The One Everyone is Talking About **Military Space-A Air Opportunities Around the World*** **$15.45 ea.**		**Assignment Washington A Guide to Washington Area Military Installations** **$7.45 ea.**	
You Can Have Fun With This Book! **Military RV, Camping & Rec Areas Around the World*** **$9.45 ea.**		*The World-Wide Travel Newsletter* **Military Living's R&R Report** 5 yrs. - $42.00 2 yrs. - $20.00 3 yrs. - $27.00 1 yr. - $12.00 (6 issues)	
Our All Time Best Seller! **Temporary Military Lodging Around the World*** **$11.45 ea.**		*Local Washington Area Magazine* **Military Living** 3 yrs. - $16.00 1 yr. - $7.00 (12 issues) 2 yrs. - $12.00	
U.S. Forces Travel Guide U.S.A. & Caribbean Areas* **$9.45 ea.**			
Subtotal: $		Subtotal: $	

*If you are an R&R subscriber, you may deduct $1.00 per book. (No discount on R&R Report itself)

For 1st Class Mail, add $1.00 per book

Mail Order Prices are for the U.S., APO & FPO Addresses. Please consult Publisher for International Mail Price. Sorry, no billing. GREAT FUND RAISERS! Please write for wholesale rates.

Total: $________

VA addresees add 4½% sales tax $________ (Books only)

Total Amount $________ Enclosed

We're as close as your telephone. . .by using our Telephone Ordering Service. We honor Visa, American Express, Choice or Mastercard. Call us at (703) 237-0203 and order today! Sorry, no collect calls. Or. . .fill out the info in the mail-order coupon below.

Name: ______________________

Street: ______________________

City/State: ______________________

Phone: ______________________
(with area code)

Signature: ______________________

- ☐ Active Duty
- ☐ Retired
- ☐ Widower
- ☐ 100% Disabled Veteran
- ☐ Guard
- ☐ Reservist
- ☐ Other ________

Rank: ____________ or Rank of Sponsor: ____________

Branch of Service: ______________________

Please mail check or money order to: Military Living, P.O. Box 2347, Falls Church, VA 22042.